Y0-ABG-488

ANIMAL COMMUNITIES
IN TEMPERATE AMERICA

This is a volume in the Arno Press collection

HISTORY OF ECOLOGY

Advisory Editor
Frank N. Egerton III

Editorial Board
John F. Lussenhop
Robert P. McIntosh

*See last pages of this volume for a
complete list of titles.*

ANIMAL COMMUNITIES IN TEMPERATE AMERICA

AS ILLUSTRATED IN THE CHICAGO REGION

VICTOR E. SHELFORD

ARNO PRESS

A New York Times Company

New York / 1977

QL
173
.S53
1977

Editorial Supervision: LUCILLE MAIORCA

———◆———

Reprint Edition 1977 by Arno Press Inc.

HISTORY OF ECOLOGY
ISBN for complete set: 0-405-10369-7
See last pages of this volume for titles.

Manufactured in the United States of America

———◆———

Library of Congress Cataloging in Publication Data

Shelford, Victor Ernest, 1877-
    Animal communities in temperate America.

    (History of ecology)
    Reprint of the 1937 ed. published by the University of
Chicago Press, which was issued as no. 5 of the Geographic
Society of Chicago bulletin.
    Bibliography: p.
    1. Animal ecology--Illinois--Chicago metropolitan area.
I. Title. II. Series. III. Series: Geographic
Society of Chicago. Bulletin ; no. 5.
QL173.S53   1977   591.5'09773'11   77-74252
ISBN 0-405-10421-9

ANIMAL COMMUNITIES IN
TEMPERATE AMERICA

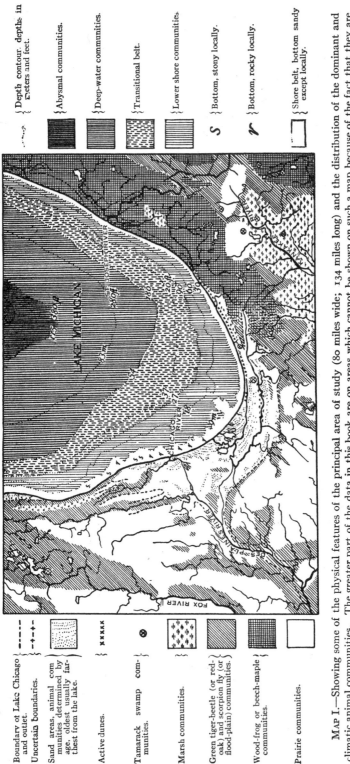

Boundary of Lake Chicago and outlet. -------
Uncertain boundaries. -+-+-

Sand areas, animal communities determined by age, oldest usually farthest from the lake.

Active dunes. x x x x x

Tamarack swamp communities. ⊗

Marsh communities.

Green tiger-beetle (or red-oak) and scorpion fly (or flood-plain) communities.

Wood-frog or beech-maple communities.

Prairie communities.

Depth contour, depths in meters and feet. ......

Abysmal communities.

Deep-water communities.

Transitional belt.

Lower shore communities.

Bottom, stony locally. S

Bottom, rocky locally. r

Shore belt, bottom sandy except locally.

LAKE MICHIGAN

ST. JOSEPH RIVER

CALUMET RIVER

DES PLAINES RIVER

FOX RIVER

MAP I.—Showing some of the physical features of the principal area of study (80 miles wide; 134 miles long) and the distribution of the dominant and climatic animal communities. The greater part of the data in this book are on areas which cannot be shown on such a map because of the fact that they are small and local, lying within the dominant or climatic communities. The distribution of most of the communities shown here is based upon a study of literature. In the case of Lake Michigan they are not accurate because of the lack of adequate investigation; on the land the distribution of forest communities is based upon (a) works on travel and industry of a half century and more ago which are of necessity inaccurate (for sources of information see chap. iii for Lake Michigan and references numbered 180 to 185 in the Bibliography), and (b) some observation. The distribution of the sand areas of Lake Michigan is fairly accurate but many types of animal communities are included; these belong to stages of forest development, to ponds of various sizes and ages, to marshes, and to open sand.

THE GEOGRAPHIC SOCIETY OF CHICAGO
BULLETIN No. 5

ANIMAL COMMUNITIES IN TEMPERATE AMERICA

AS ILLUSTRATED IN THE CHICAGO REGION

A STUDY IN ANIMAL ECOLOGY

BY

VICTOR E. SHELFORD

The University of Illinois

PUBLISHED FOR THE GEOGRAPHIC SOCIETY OF CHICAGO

BY

THE UNIVERSITY OF CHICAGO PRESS
CHICAGO ILLINOIS

COMPOSED AND PRINTED BY THE UNIVERSITY OF CHICAGO PRESS
CHICAGO, ILLINOIS, U.S.A.

PREFACE

Courses in field zoölogy usually lack the convenient background of organization which one finds in the doctrine of evolution when presenting the animal series from a structural standpoint. The need of some logical and philosophical background for the organization of natural history instruction into something more unified than haphazard discussions of such animals as were encountered in chance localities, was keenly felt at the beginning of the author's experience as a teacher of field zoölogy. Evolutionary background was tried, but failed and was rejected; genetics and faunistics proved inadequate. Behavior as presented and studied by zoölogists was incomplete. Plant ecological methods were, when unadapted, applicable only in part, while much of physiology dealt with organs and internal processes.

The organization of the data here presented is the result of many attempts and failures which at times made the task seem hopeless. The literature relating to this subject has been written almost exclusively from points of view which are very different from the one here presented. It is scattered, and the bibliography has never been brought together. Accordingly its incorporation here has called for the expenditure of much time, and often for reinterpretation, which is always fraught with danger of error. The time consumed in working over the literature has been great, but, for the reason stated, the amount covered has been relatively small, and the literature in foreign languages has not received its share of attention. Furthermore, since the bulletin is not written primarily for investigators, much of the literature not in English has been omitted from the Bibliography but some of it will be found in the papers cited. To present such a subject as we have before us without constant reference to the writings of such naturalists as Buffon, White, Darwin, Wallace, Bates, Belt, Hudson, Romanes, Audubon, Brehm, Fabre, Claude Bernard, Huber, Giard, Forel, Schmarda, Janet, Haase, Möbius, Dahl, and others (35a) seems at first thought quite unjustified, but a complete study of the works of such men would be almost a life's work in itself. The writer does not claim to have a detailed knowledge of all the articles written by these men. He knows them only in part. Their facts, in so far as they are known to him and relate to the questions at hand, tend to support the main contentions. But the successful organization of such a subject depends more upon the investigation of the particular species and localities covered than upon work done from different points of view in remote localities. This bulletin is not intended as a textbook. Several years of work would be necessary to give it the completeness and form which a textbook should have, and the physiology which should be included in such a textbook is almost entirely omitted.

The organization here presented has in the main grown out of three lines of thought: (*a*) the physiology of organisms as opposed to the physiology of organs (51)[1]; (*b*) the phenomena of behavior and physiology, as illustrated by the studies of Loeb (72), much of the data of which can be related to natural environments; and (*c*) the organized comparable data of plant ecology, as set forth by Cowles (58) and Warming (12). The results of these five years of labor will not be pleasing to many zoölogists because the principles of evolution, heredity, etc., have not been correlated. Their omission, however, has not been due to any prejudice against their introduction, but rather to the fact that they can only occasionally be related to this line of organization. It was thought also that the complexity of the problems and concepts here treated made separation a necessity to clearness.

The number of problems thrown open by the investigation is infinite. Naturalistic observation and survey work could be carried much farther along the lines here blocked out. The chief lesson which the author has drawn from his labors is that experimental study, conducted with due reference to the relations of the animals to natural environments, with conditions carefully controlled, and a single factor varied at a time, is one of the stepping-stones to future progress. We are confronted with centuries of animal and human geography, with only inference or speculation as to controlling factors for a background, and the experimental study of factors in the case of man and other land animals only at its beginnings. Though man is a land inhabitant, all the best work along these and many other lines has been done upon aquatic animals. The writer's course in the future will probably be determined by the needs of the science, and will be turned from the purely naturalistic method of study to a method made up of naturalistic observations and controlled experiments.

In undertaking a new line of work, one must have first, inspiration, next, method and motive, and finally, in the case of ecological work, the assistance of a large number of persons in various departments of knowledge. For such assistance I wish to express my indebtedness to

[1] Numbers in parentheses, scattered through this work, refer to references in the Bibliography at the end (pp. 325–36).

the following: to Professor C. M. Child, of the University of Chicago, for my first serious inspiration in natural history, and for my opportunity to develop ecology; he has also rendered important assistance in connection with the preparation of this work, by giving information regarding animals about Chicago; his assistance with the worms and other lower invertebrates has been of particular importance; to Dr. H. C. Cowles, of the University of Chicago, for constant assistance with the plants and all matters relating to plant ecology. Various graduate students and assistants at the University of Chicago have also aided materially in the preservation of notes, specimens, and records. Mr. Beniah H. Dimmot, Dr. W. C. Allee, Mr. G. D. Allen, Mr. S. S. Visher, and Mr. M. M. Wells should be mentioned especially. Mabel Brown Shelford collected the data on the former occurrence of animals now extinct, and on other historical matters.

The following have furnished identifications and important advice in connection with the various groups in which they are specialists:

Dr. N. A. Harvey, Ypsilanti, Mich., Sponges.
Dr. R. C. Osburn, Columbia University, Polyzoa.
Dr. J. P. Moore, University of Pennsylvania, Leeches.
Mr. F. C. Baker, Chicago Academy of Sciences, Mollusca.
Dr. C. D. Marsh, U.S. Department of Agriculture, Copepods.
Mr. Richard W. Sharpe, Brooklyn Institute, Ostracoda.
Dr. Chauncey Juday, University of Wisconsin, Cladocera.
Dr. E. A. Ortmann, Carnegie Museum, Crayfishes.
Miss A. L. Weckel, Oak Park, Ill., Amphipods.
Miss Harriet Richardson, U.S. National Museum, Isopods.
Mr. O. F. Cook, U.S. Department of Agriculture, Myriopods.
Dr. R. H. Wolcott, University of Nebraska, Water Mites.
Mr. Nathan Banks, U.S. Department of Agriculture, Spiders.
Mr. C. A. Hart, University of Illinois. All groups of insects.
Dr. J. G. Needham, Cornell University, Aquatic insects.
Dr. Cornelius Betten, Lake Forest University, Caddis-flies.
Mr. W. J. Gerhard, Field Museum, Hemiptera and general entomology.
Mr. A. B. Wolcott, Field Museum, Beetles.
Prof. H. F. Wickham, University of Iowa, Beetles.
Dr. Joseph Hancock, Chicago, Orthoptera.
Mr. W. S. Blatchley, Indianapolis, Orthoptera.
Dr. A. D. MacGillivray, University of Illinois, Sawflies and insect larvae.
Dr. S. E. Meek and Mr. S. F. Hildebrand, Field Museum, Vertebrates.
Mr. Alexander Kwiat, Chicago, Lepidoptera.
Miss Clara Cunningham, South Bend, Tamarack Swamps.
Dr. Frank Smith, University of Illinois, Annelids.

Mr. S. S. Visher and Mr. Ralph Chaney contributed most of the habitat data on birds. Dr. R. M. Strong verified those included here which were also compared with Butler's account (108). T. C. Stephens supplied the photographs of nests.

Dr. P. G. Heinemann, University of Chicago, Bacteria.

Dr. Susan P. Nichols, Oberlin College, Algae.

Mrs. Elva Class and Mr. M. M. Wells of the University of Chicago, and Dr. W. C. Allee, of the University of Illinois, Gas analysis.

Mariner and Hoskins, Commercial Chemists, Analysis of Water.

The original records upon which the work is largely based could not all be presented. Those placed at the end of the chapters are believed to be representative, in that they include some characteristic animals, some which are numerous but occur elsewhere also, and some of wide distribution. The records in the text are also largely original, except in the case of mammals, the habitat locations of which are based upon literature. Mr. W. H. Osgood of the Field Museum has assisted in the editing of the data on mammals. Original records in this group are especially indicated. Data on the nesting habits of birds have likewise depended upon compilation, though the locality records are those of the persons mentioned. Mr. W. S. Stahl, assistant United States attorney, edited the paragraphs on the legal restrictions upon field study and collection of animals.

The matter of scientific names is one presenting unusual difficulties because of the scattered and incomplete character of catalogues. The work of identification having occupied several years, changes in nomenclature may have led to some confusion and duplication of records under different names. The matter of correcting spelling is unusually difficult because of numerous works which it is necessary to consult for verification in dealing with representatives of nearly all groups from Protozoa to mammals. The specialists on the different groups have been very kind in answering any question, but the final responsibility rests with the author. In the main the nomenclature in the following works has been followed (numbers refer to Bibliography at the end of this work): mammals, 21; birds, 108; reptiles, 157, 157a; *Amphibia*, 139 and 152; fishes, 79; flies, Aldrich's ('oo) Catalogue (N.A.); beetles, 156 and Samuel Henshaw's ('85) checklist; *Hemiptera (Heteroptera)*, Bank's ('11) Catalogue; aquatic insects, 95 and 96; ants, 54; insects not included in the special lists, 177; *Hymenoptera* not in 177, E. T. Cresson's ('87) Synopsis; spiders, 159; *Phalangidae* and land mites, 172 and 184; water-mites, 149; myriopods, 183; mollusks, F. C. Baker's ('06) Catalogue for Illinois; leeches, 91a; crayfishes, 101, 101a; amphipods, 102; isopods,

182; copepods, 146, 146a; ostracods, 147; other *Entomostraca*, Herrick and Turner's ('95) synopsis for Minnesota.

In the case of several names not included in any of these works there are contradictory spellings, authors, etc., and we have used some name which we believe will be understood.

In bringing together the illustrations, material assistance has been rendered as follows:

Dr. S. W. Williston, loan of Figs. 30, 31, 32, 126, 132, 174, 186, 187, 188, 210, 267, 269, 270, 271, 272, 273, 274, 275, 282, 283, 284, 285, 286, from his *Manual of North American Diptera*.

Dr. F. R. Lillie and the *Biological Bulletin*, loan of Figs. 66, 67, 68, 69, 83, 84, 85, 101, 251, 252, 253, previously published by the author in the *Biological Bulletin*.

Professor S. A. Forbes, the Illinois State Laboratory, and the State Entomologist's Office, loan of Figs. 35, 36, 44, 45, 46, and 72, which appeared in Vol. III of the *Natural History Survey of Illinois*, and for electrotypes of Figs. 261, 262, 264, 265, 288, 289, 290, 291, 292, 296, 297, 301, 302, 303, 304, 305, 306, which appeared originally in the Annual Reports and Bulletins of the State Entomologist and other state and national publications.

Professor S. E. Meek and the Field Museum, loan of Fig. 37.

Professor J. M. Coulter and the *Botanical Gazette*, loan of Fig. 115.

Professor F. L. Washburn and the Minnesota State Entomologist's Office for electrotypes of Figs. 136, 137, 156, 189, 194, 211, 212, 213, 229, 256, 263, 266, 268, 276, 277, 278, 293, 298, 299, 300.

Professor Vernon L. Kellogg, privilege of using Figs. 188, 270, 271, 274 from *North American Insects*, which appear also in Williston's *Manual of North American Diptera*.

Professor J. H. Emerton, privilege of using Figs. 207, 208, 224, 225 from *Common Spiders*.

Figures after Lugger appeared originally in *Bulletins 55, 66*, and *69* and the *Fourth Annual Report* of the Minnesota Agricultural Experiment Station.

Figures after Marlatt, Riley, and Chittenden appeared originally in publications of the U.S. Department of Agriculture; after Gorham, Smith, Jennings, and Reighard, in publications of the U.S. Fish Commission.

The author is also indebted to Dr. J. P. Goode, Dr. Otis W. Caldwell, Dr. H. C. Cowles, Professor R. D. Salisbury, Professor C. M. Child, and Mr. M. M. Wells for assistance in editing the manuscript and reading proof. Mr. W. J. Gerhard rendered special assistance in the reading of the proof of the scientific names.

It is evident from the number of persons who have assisted in the working over of material and the accumulation of the data on which this work is based, that the survey aspect of ecology is a subject for co-operative investigation. Because of the complexity of the problems, it has been deemed advisable to publish this work even in its present preliminary and necessarily incomplete form, in order to make the material accessible as soon as possible to teachers, investigators, and others who are interested.

DEPARTMENT OF ZOÖLOGY
UNIVERSITY OF CHICAGO
September 9, 1912

PREFACE TO THE SECOND IMPRESSION

The second impression of this book is unchanged except for the correction of typographical and clerical errors. An annotated Bibliographical Appendix has been added which will enable the reader to go on with the subject. The community nomenclature has changed materially in the twenty-five years which have elapsed since the book was written. These changes are, however, not serious and it has been possible to provide for the correction of them in a table. These corrections were previously published in the journal *Ecology*. The taxonomic nomenclature has been left in the original form. In many cases the species mentioned have been separated as two or more, and the natural areas on which the book is based have, to a considerable extent, been destroyed so that it would not be possible to discover the relation of species now recognized. The termites afford an example of this kind since several species now take the place of what was considered one in 1912. Their arrangement, however, makes even a better case of distribution correlated with succession than was indicated by the supposed single species.

The author is indebted to Dr. W. C. Allee and Dr. A. E. Emerson of the University of Chicago for suggestions regarding the Bibliographical Appendix and to the Illinois Natural History Survey for the loan of a halftone block. Dr. Allee also provided notes on the stations described on pages 52–56. Station 30a has an automobile road through its center and is partially filled and Stations 37, 44, and 49 are no longer available for study, while Station 42 is now occupied by buildings.

DEPARTMENT OF ZOÖLOGY
UNIVERSITY OF ILLINOIS
March 4, 1937

TABLE OF CONTENTS

INTRODUCTION

Just at the beginning of the present century, there seems to have been a revival of interest in plants and animals in relation to their environments, and various workers have turned from the study of anatomy and classification in the laboratory to the study of organisms in nature. In this, the botanists have preceded the zoölogists, in success if not in time. In 1901 Dr. H. C. Cowles published a bulletin on the *Plant Societies of the Chicago Area.* This was one of the first attempts of an American biologist to treat all the plants of a given area in a strictly ecological manner. This study of all the organisms of an area, from the point of view of their relations to each other and to their environment, is still a new or at least a renewed idea. Zoölogists have devoted most of their attention to the study of animals from the standpoint of a single individual and of single species. Practically all of the more general study has been comparative. We have comparative anatomy, comparative embryology, comparative physiology, and comparative psychology. These are comparisons of the structure or physiology of one species, or group of species, with that of another species or group of species.

Our point of view is very different. We shall deal with many species from the standpoint of their dependence upon each other and their relations to their environments. We shall attempt to present what has been learned upon this subject during several years of investigation and field teaching. In the spring of 1903, the writer made his first field excursion in the Chicago area, and from that time has been engaged in further study of the subject.

The study of organisms in relation to environment is entitled *ecology.* The definition of ecology, like that of any growing science, is a thing to be modified as the science itself is modified, crystallized, and limited. At present, *ecology is that branch of general physiology which deals with the organism as a whole, with its general life processes, as distinguished from the more special physiology of organs* (51), *and which also considers the organism with particular reference to its usual environment.*

Undertaking such a study from the point of view of many organisms involves matters of both ecological and taxonomic classification. Classification of animals is difficult because animals are so exceedingly numerous. There are probably from 10,000 to 20,000 species of animals which the naturalist may encounter in the area which we are treating, while

1

in the same area the botanist would probably find only about 2,000 conspicuous plant species. Representatives of all animal species must be submitted to specialists for identification, that is, the specialist gives the correct scientific name to the animal. Scientific names are definitely arranged as below, if man is taken as an example.

Phylum	-	-	-	-		*Chordata* or *Vertebrata*
Class	-	-	-	-	-	*Mammalia*
Order	-	-	-	-	-	*Primates*
Family	-	-	-	-	-	*Hominidae*
Genus	-	-	-	-	-	- *Homo*
Species	-	-	-	-	-	- *sapiens*

The young of many insects and of some other animals cannot be placed in the proper species because animal life histories are very imperfectly known. Such animals are merely placed in the proper genus or family. The common names of animals rarely apply to single species but to whole genera, families, or even orders. "Caddis-worm" is a name applied to a whole order of insect larvae and as these are very imperfectly known the term caddis-worm is applied to many species, and, applied in this way, appears in many places in the text.

Because of the large number of animals and the difficulty in naming them, it is quite impossible to deal with the data in the specific way that might be possible with plants. Furthermore, while the data for plant distribution are not well known, those for animal distribution are much less well known. Therefore in most cases it is necessary to speak in general terms. It is impossible and undesirable to discuss each community of animals in detail. The facts are not known, and even if they were known, their volume would be such as to exclude the great majority of them from the limits of this treatise. In most cases it is best to make a statement of the leading facts, and a few statements about the specific situations to give an idea of the kinds of animals that are characteristic or common there. It should be noted also that the most characteristic animals are often not generally known and are in some cases rare.

The scientific names of characteristic and common animals are included, not so much for geographers at present, as to form a basis for further work and comparison by zoölogists and zoögeographers. Where given in the form of tables they present the actual scientific background for the facts here stated. Much greater detail would be needed for a full zoölogical treatment. Scientific names are usually used where the common names apply to many species. The names of authors of species are added in the text and description of figures only where they do *not* appear in either the lists and tables or in the descriptions of figures. No attempt has been made to include the same animals in the text, tables, and illustrations, as the only aim has been to make each part as useful as possible.

While the amount of work that might have been done along the lines here represented is infinite, this work represents only a general survey. The data are incomplete, but we believe them to be adequate for the purpose of illustrating the principles involved. Considerable experimental work has been conducted with reference to animal communities,[1] but it has served only as a background, and in comparing them we have relied upon comparison of (*a*) habitats and (*b*) species. The latter is fraught with many dangers, for it assumes, in the absence of evidence to the contrary, that the physiological character of a species is the same in the different situations in which it is taken. Observation has shown this to be true for most species within rather uncertain limits. There are, however, many well-known exceptions to this, some of which are cited in the text. Such use of species is certainly to be avoided in the study of the extensive or geographic distribution of animals, and it remains to be seen how far it may be employed locally. Certainly ecology cannot reach its best development if it relies upon such a method. Whatever further investigation may prove on this point, it is hoped at least that we may be able to suggest problems which may be attacked from new points of view. Should this object be accomplished, the work will have served its purpose.

[1] The term community, as used here, refers to all the animals living in the same surroundings.

CHAPTER I

MAN AND ANIMALS

I. Introduction

1. Culture and Nature

In this discussion we are concerned with nature and our relations to nature.

Nature is an enormous aggregation of things—objects—each having certain metes and bounds, certain qualities and powers, beyond which it cannot go. Now, knowledge of nature, sanity toward nature, consists exactly not only in ever increasing the extent of our inventory of these objects, but of recognizing, without addition or subtraction, that is, accurately and justly, the forms, the qualities, and the forces of these objects—what they are and what they are not; what they can do and what they cannot do.

Is there anything worse than mild folly in the belief in the "sea serpent"? That depends. If the belief involves the notion "monster," then yes, decidedly, for the belief is of the self-same kind that has prevented men from being sane, that has filled them with dread, in all ages. It is a question, not of nature, but of state of mind. The person whose mental attitude is such that he easily and unwittingly puts into the sea from his own consciousness a creature that does not exist in the sea, and holds it to be as real as those that do exist there, is also in a state of mind to attribute to all sorts of innocent creatures and persons qualities and powers they do not have and hold these powers to be as real as the ones they actually do possess.—Ritter (1).[1]

We have all heard of the octopus or devil-fish, with its long arms covered with powerful suckers, which is always waiting to seize the unsuspecting, choke and bite him, always grasping with another arm when the grip of one of them is loosened—suitable symbol of the trust.

A person wading in the water among rocks where there are devil fishes is about as likely to be attacked and bitten by one of the animals as he is to be injured by the explosion of a watermelon, when walking through a melon patch. Both things are possible.

The octopus secretes a great quantity of black fluid and makes use of this by squirting it into the water to envelop itself in "pitch darkness" against the approach of enemies. But the fluid is not poisonous, nor the leastwise injurious to anybody or to any creature, so far as we know.

[1] Numbers in parentheses, scattered through this work, refer to references in the Bibliography at the end (pp. 325–36).

In short, the animal is not a "horrid thing," as it is painted in story and in many a dimly lighted imagination. There is nothing devilish about it.

And here is the moral of the "devil fish": If there is a corner of your mind that wants to attribute to the octopus malevolent qualities and powers that it does not possess, and is content to overlook or deny to it qualities and powers of interest and beauty that it does possess, mark my word, the same corner of your mind will tend to treat such at least of your fellow-men as you do not know well, in the same way. This unfortunate corner of your mind will, like all other corners, be true to itself—to its own qualities. It is the old impossibility of blowing hot and blowing cold at the same time.—Ritter (1).

We may accept this as one of our relations to nature and general culture, and sanity toward nature as one of the benefits to be derived from study of science and nature.

2. SCIENCE AND NATURE

All biological problems are problems of nature. Evolution became a problem only when a large knowledge regarding the number and diversity of animal species had been acquired. This has been the problem around which most zoölogical facts have been accumulated. Indeed, most zoölogists have little interest in problems not throwing light on evolution. The development of zoölogy has therefore been one-sided. Had geology clung as closely to the origin of the earth as zoölogy to evolution, it would not be the unified science which we see it today. The lack of unity in zoölogy has been caused in part by the neglect of the aspects which we are to take up here. In this connection, Thompson (2) has said of Brehm, one of the older students of natural history: "He [Brehm] had unusual power as an observer of the habits of animals. His particular excellence is his power of observing and picturing animal life *as it is lived in nature*, without taking account of which, biology is a mockery, and any theory of evolution a one-sided dogma." It follows also that sanity in science is dependent upon a knowledge of nature. Our first steps in the task before us must accordingly be a consideration of wild nature as it really is. This can perhaps best be accomplished by comparing the reality with some of our conceptions of it.

II. THE STRUGGLE IN NATURE

The first step toward an understanding of our relation to nature, or rather the animals and animal communities of natural conditions, is to acquire a knowledge of the conditions of animals in a state of nature. There is much literature on this subject, but our conception of the struggle for existence and the survival of the fittest is too often entirely forgotten when we are considering our relation to animals. Nature is cruel and heartless, and to die to become food of another organism is the fate of the vast majority of animals. Mr. Roosevelt has said:

Watching the game, one was struck by the intensity and evanescence of their emotions. Civilized man now usually passes his life under conditions which eliminate the intensity of terror felt by his ancestors when death by violence was their normal end, and threatened them during every hour of the day and night. It is only in nightmares that the average dweller in civilized countries now undergoes the hideous horror which was the regular and frequent portion of his ages-vanished forefathers, and which is still an everyday incident in the lives of most wild creatures. But the dread is short-lived, and its horror vanishes with instantaneous rapidity. In these wilds the game dreaded the lion and the other flesh-eating beasts rather than man. We saw innumerable kills of all the buck and of zebra, the neck usually being dislocated, it being evident that none of the lion's victims, not even the truculent wildebeeste or huge eland, had been able to make any fight against him. The game is ever on the alert against this greatest of foes, and every herd, almost every individual, is in imminent and deadly peril every few days or nights, and of course suffers in addition from countless false alarms. But no sooner is the danger over than the animals resume their feeding, or love-making, or their fighting among themselves. Two bucks will do battle the minute the herd has stopped running from the foe that has seized one of its number, and a buck resumes his love-making with ardor, in the brief interval between the first and second alarm from hunter or lion. Zebras will make much noise when one of their number has been killed; but their fright has vanished when once they begin their barking calls.

Death by violence, death by cold, death by starvation—these are the normal endings of the stately and beautiful creatures of the wilderness. The sentimentalists who prattle about the peaceful life of nature do not realize its utter mercilessness; although all they would have to do would be to look at the birds in the winter woods, or even at the insects on a cold morning or cold evening. Life is hard and cruel for all the lower creatures, and for man also in what the sentimentalists call a "state of nature." The savage of today shows us what the fancied age of gold of our ancestors was really like; it was an age when hunger, cold, violence, and iron cruelty were the ordinary accompaniments of life. If Matthew Arnold, when he expressed the wish to know the thoughts of earth's "vigorous, primitive" tribes of the past, had really desired an answer to his question, he would have done well to visit the homes of the existing representatives of his "vigorous, primitive" ancestors, and to watch them feasting on blood and guts; while as for the "pellucid and pure" feelings of his imaginary primitive maiden, they were those of any meek, cow-like creature who accepted marriage by purchase or of convenience, as a matter of course.—From *African Game Trails*, by Theodore Roosevelt; Copyright, 1910, by Charles Scribner's Sons (3).

III. Man's Relation to Nature

Mr. Roosevelt's statement is quite different from much of the poetry about nature, still it is a true picture. We live in a man-made nature from which the conspicuous animals and their deadly struggles have been eliminated (4, 5). Of the admirers of the beauties of nature I fancy that many, perhaps the majority, think of it as a series of lawn-like pastures, well-trimmed hedges, such as finds its ideal expression in some of the older countries like England.

> The trees, round, woolly, ready to be clipped;
> And if you seek for any wilderness
> You find, at best, a park, a Nature tamed
> And grown domestic like a barnyard fowl.
> —E. B. Browning, "Aurora Leigh."

The close observer of nature, even in such man-made conditions as in Bedfordshire or in the Chicago parks, sees all the struggle which Mr. Roosevelt has depicted for the birds and mammals of primeval conditions. To kill is nature's first law.

1. MAN'S CONDUCT TOWARD ANIMALS

There is much sentimental nonsense about nature, about animals and cruelty to animals, as well as much actual cruelty and wanton destruction of useful animals. With some people birds obscure all else in the animal world. The destruction of squirrels, which are equally if not more interesting than birds, is sometimes advocated because of their alleged destruction of birds' eggs. The friend of the squirrel would plead equally hard for the destruction of certain hawks and owls as enemies of the squirrel. Certainly all lovers of the insect world might advocate the destruction of birds to protect their particular zoölogical pets.

That birds save the harvests of every season is believed by many. The student of mammals is equally sure that certain mammals are the balance wheel, while the herpetologist is convinced of the importance of snakes, and the entomologist's economic world turns about predatory and parasitic insects and spiders. The fact is that each view, even thus extremely stated, contains its elements of truth. The whole truth is hardly knowable. Each animal is dependent upon many others. The dependences are so numerous that we find it necessary to isolate particular animals and construct them into a society of real but limited relations for purposes of discussion (see p. 170). Still there are a few things that we can be reasonably sure of. The first is that we cannot interfere with any animals or the habitats of any animals without interfering with many others. The second is that all animals are of some economic importance. The third, that few animals can be said to be either wholly beneficial or wholly noxious, excepting those reared or preserved for their direct utility, and those directly and perniciously attacking the necessities of man's existence.

Considering the first, we note that civilized man's operations interfere with animals and animal habitats. His first work is to destroy all large, dangerous animals. He clears and cultivates the land, bringing death and destruction to many more, and gradually substitutes domestic animals for wild game (5a). Vegetarians often argue for the exclusive use of vegetable food on the ground that animals should not be killed, but to secure more plants for this purpose they of necessity would clear more land to grow more corn and thus destroy myriads of animals by methods more cruel than those of the butcher and huntsman. Our relations to animals are *not simple, but very complex* and our conduct often inconsistent. We cease wearing aigrettes because the collecting of them often leaves young birds to die, and kill every mouse and mole that happens to come our way, though their young must die as do those of the birds. Some of us wear leather shoes while arguing for a vegetarian diet because animals should not be cruelly slaughtered.

Turning to the second and third ideas stated above, we note that few animals which feed upon a variety of foods, both plant and animal, can be said to be of any great usefulness, except when the plants eaten are useless to man. In other words, the good done the farmer by an animal which eats many insects, including noxious ones, may be offset by a destruction of grain. Birds eat a variety of food. Those feeding upon useful plants are not rated as of great economic importance. The bobolink, for example, eats grain and weed seeds in the spring when insects are scarce; soft-bodied insects in June and July when seeds are not available. In August the insects mature and are hard shelled. The birds now reject them for the grain seeds. This bird, furthermore, eats that which is available and most easily secured during the different seasons. This is also true of many, probably the vast majority of animals. The food of fishes is to a considerable extent determined by the kind of food available where they are living (6). Ruthven (7) has found this true of garter-snakes; the same is true of men.

Many animals, birds (8), mammals, reptiles (9), toads (10), and insects destroy quantities of noxious insects, but along with them many insects that are enemies and parasites of the noxious ones are also destroyed. The parasites, especially, are often more beneficial to man's interests than the animals which devour them, and which take good and bad without the slightest discrimination from our economic point of view. Because of their destruction of parasitic insects Severin (8) argues that birds should not be protected. Certain mammals and reptiles often show a decided superiority over certain birds in this respect, in that they are strictly predatory and are not directly noxious at any time of year as are some birds which feed upon grain.

Many animals feed extensively upon insect pests when they are numerous and accordingly threaten a crop. This is true of spiders, insects (11), amphibians, reptiles, mammals, and birds (8), especially those that are largely predatory. This fact is the only sure guaranty of the economic value of many birds, and is perhaps overworked by the fanciers of the group. This value belongs equally to certain insects, so that if birds were not devouring such insects along with pests, these hexapods would probably be able to put the pests down. The other vertebrates also would probably be able to put down the pest without the aid of the birds; Forbes has said that a balance would finally be reached if all the vertebrates were exterminated (see 26).

In the preceding pages we mention "sanity toward nature." Sanity toward nature is based upon a full knowledge of available facts. Partial knowledge, if fully depended upon, is as dangerous as falsehood, for it leads to false interpretations. We must *know* nature, not a part, but the whole, if we wish to treat the simplest everyday problem of our relations to animals intelligently and justly.

Why protect birds? Is the present attempt justified? In the answer to these questions all sentimentalism should be laid aside. It is sometimes urged that birds have a greater aesthetic value than other animals. This it seems is unjustified unless the songs of some constitute the justification. Persons with only a small acquaintance with insects, mollusks, fishes, amphibians, reptiles, and mammals find as much beauty in these groups as the bird fancier does in his. All groups should be preserved for their aesthetic value as the appreciation of it depends entirely upon temperament,[1] training, and especially a knowledge of the

---

[1] A few persons known to the writer are repelled by birds because of their claws, scaly legs, and other reptilian characters. Many admirers of nature and animals are not attracted by birds because as a rule they must be seen from a distance. Inquiry at close range necessitates either shooting or capturing the bird and neither is a particularly aesthetic operation. In the case of capture, only a short period of necessary neglect usually renders the surroundings and often also the bird not only not aesthetic

group in question. From the economic viewpoint there is not a complete agreement as to bird protection. France does not co-operate with England in bird protection because her leaders in economic thought (Severin and others) have ably opposed it on economic grounds; still France is more progressive than England in agricultural matters. Other things being equal there are but two more reasons for special measures for the preservation of birds than for the preservation of reptiles, amphibians, or insects. First, birds are subject to destruction by reckless gunners. Second, they are less dependent upon natural conditions on the ground and are better able to survive after land has been put under cultivation than some other groups. Many other animals whose diets are varied have been exterminated or will be so by agriculture, leaving the birds as the most easy point for protective effort. The protection of birds should not be urged at the expense of the extermination of other animals because of their alleged occasional attacks upon birds. The great danger of acting on partial truth regarding animal interdependences makes societies for the protection of birds alone scientifically and educationally unjustified. The protection of all groups should be urged, in particular through the preservation of the natural features upon which they depend. It is well to protect fishes from seiners and birds from gunners but this often only delays their fate. We must also consider where they will breed a few years hence.

When one comes to love an animal or a group of animals, he is in no position to draw scientific conclusions regarding it. For this reason bird enthusiasts are not always to be trusted. It was the persistent efforts of such "benefactors" which gave us that detestable avian rat, the English sparrow, the feeding or sheltering of which is now a misdemeanor in some of our states.

Should we slaughter animals? As members of a system of nature in which to kill is the first law, we must answer in the affirmative. Man is the master of all destroyers. Where are the bison, the beaver, the elk, the thousand and one denizens of the primeval forest and prairie? We scarcely walk over a path or lawn without bringing "death" and "suffering" to animals of some sort. The crime of their destruction can be no

but malodorous and repulsive. Thus to those who wish to examine objects closely other animals have a greater aesthetic value. Claims for a greater aesthetic value for birds must be based upon impressionistic appreciation of them in connection with landscape. There is no reason to desire or assume that this interest will decrease with time, but it is reasonable to suppose that with further dissemination of scientific ideas and methods among the people a comparable amount of more serious interest will develop in connection with other groups and perhaps with birds as well.

crime at all, in so far as the destruction is absolutely unavoidable. The wanton and useless destruction of animals not condemned as noxious by years of investigation, though probably not forbidden by the example of the animal world, is forbidden by the best sensibilities of every civilized man and woman. When the value of an animal to us is in question, the animal should have the benefit of the doubt, and we should hesitate long before introducing animals of supposed value. Certainly, also, every animal condemned by careful investigators should be destroyed whenever opportunity is presented. Mistaken and sentimental ideas cause the killing of many useful animals and the protection of many noxious ones. The farmer kills snakes and skunks whenever he has the opportunity, though they are among the most useful animals. Shrews are master destroyers of mice. Still many people mistake shrews for meadow mice and destroy them. Likewise the housewife kills the house centipede, the enemy of household pests, as a dangerous and repulsive creature even in the absence of any knowledge of the questionable charge that it bites young infants. Mistakes are not confined wholly to uninitiated individuals. Misjudgment by the officials of the Brooklyn Institute of Arts and Sciences, possibly influenced by the sentiment of Longfellow's mistaken poem on the "Birds of Killingworth" brought about one of the first official introductions of the English sparrow. Thus we see that the complexity of the problem demands careful study and conservative action.

2. MAN-MADE COMMUNITIES

Animal communities are divisible into primeval or primary communities, and man-made, or secondary communities (12, 13). As has been noted when civilized man enters a new territory, he first destroys all large game which threatens himself and his domestic animals. He then destroys the natural vegetation and other animals by clearing the timber, burning all woody débris, and plowing and putting out plants which are entirely new to the region. Under primeval conditions, plants are arranged irregularly, as roughly indicated by the letters in Diagram 1; after being put to agricultural purposes, they are arranged as in Diagram 2. The plants are all of one kind and are arranged in rows. A grove of the original vegetation is sometimes left. The rate at which these changes take place is directly related to the rate at which man occupies and cultivates the new territory. As compared with natural changes, this process is rapid and is accompanied by an equally rapid decline of primeval or primary communities.

heddeb hei cd efg cbe mi
e mefg nm bc de fg fgbn
ghi be co dp eqfr gohifb
bdcviwhxgyfzembndoc e ih
efgxny uinh fgbhjnk nsfg
ghia dftghtyb hfj tkibhc
sdftunmgkiuoht hyfgtrdcg
dfgythufbnjks vdg fhtgry
hfgt fhgty sdswaq nfhjdl
ghtyuwiokp fbndhutbs gtu
vdfxzabjfmua fgh yfs j i
edfgrthfinbghb fgvnzxvcb
erffghtjk vbxzzasxscdfge
thigjszxlkm, j hytfsdtrfb

edbeddgjcdbgdcgdcbedcdgebc
feceiejfadfeedefadfcecdede
cbaaaaaaaaaaaaaaacb
edaaaaaaaaaaaaaaed
fgaaaaaaaaaaaaaafg
dcaaaaaaaaaaaaaadc
ebaaaaaaaaaaaaaaeb
dgaaaaaaaaaaaaaadg
fdaaaaaaaaaaaaaafd
dcaaaaaaaaaaaaaadc
feaaaaaaaaaaaaaafe
egaaaaaaaaaaaaaaeg
fci bedfg bcg bdg ded jef gdj fc
cgj cde f d e d f d f e bf cg

DIAGRAM 1.—Showing the arrangement of plants and animals on a plot of
ground under primeval conditions. The
letters are fortuitously chosen to represent the fortuitous arrangement of
plants and accordingly the animals associated with them. Thus m, n, x, and
z may be taken to represent oak, maple,
basswood, and cherry, respectively, and
the animals associated with each. The
other letters may be taken to represent
herbs and shrubs and the animals associated with them.

DIAGRAM 2.—Showing the arrangement under agricultural conditions.
Here the plants which are put out in
rows are represented by a's arranged in
rows. There are certain animals associated with such plants and the a's represent these also. Land is not usually
cultivated close to the fences and thus
each field is surrounded by a border of
original shrubs, herbs, and sprouts from
the original trees. These and the animals associated with them are still fortuitously arranged.

3. THE DECLINE OF PRIMEVAL COMMUNITIES AT THE HEAD OF
LAKE MICHIGAN

BY MABEL BROWN SHELFORD

When the white man first appeared near Chicago no secondary
community existed, as the aborigines lived almost entirely by hunting
and fishing. They cultivated the land only a little, and are accordingly
to be ranked with the larger animals as a part of the original communities.
The Indians of this region were chiefly Potawatomi, although there
were a few Chippewas and Ottawas (14, 15). Early in 1833 (15) about
5,000 assembled in Chicago to treat for the sale of their entire remaining
possessions in Illinois and Wisconsin. A treaty was finally ratified and
in 1835–36 (14, 15) they left the region forever. They settled in Iowa
for a time, but the advancing tide of civilization drove them
farther and farther west. In 1890 (16) the larger part of the Potawatomi, about 950, occupied land in Kansas and Oklahoma. The region
about Chicago was particularly adapted to the life of the Indians, and
it was probably an important region for them, as well as their successors.
The innumerable water courses and ponds afforded an abundance of

muskrats, mussels, fish, etc., and the larger game of the land was particularly abundant and diversified, because of the numerous habitats represented. Unfortunately, a fragmentary record is all we have of the decline of the primeval communities and the development of the present ones. These records apply mainly to the large animals of Cook County. The time of the disappearance from Southern Michigan, Northern Indiana, and Lake County, Illinois, was probably much later and, with the exception of the bison, bear, and elk, the more numerous kinds of game nearly all still occur in the thinly settled portions of Illinois (5a). The earliest explorers of this region, Marquette, LaSalle, and others, speak repeatedly of the great abundance of large game (17, p. 34). LaSalle, in the autumn of 1679, sailed along the western shore of Lake Michigan until the end of the lake was reached. Landing, he found deer, bear, and wild turkeys in great abundance. Grapevines loaded with clusters of ripe grapes hung from the tall forest trees and provided a rich feast for the bears. Continuing toward the headwaters of the Kankakee River, one stray buffalo was found sticking in a marsh. It was the beginning of winter and the remainder of the herd had probably migrated South, but on entering the headwaters of the Illinois River, in the autumn of the following year, LaSalle says that he found the great prairies "alive with buffalo" (18).

The Indians claimed that bison were very plentiful on the prairies until the Storm Spirit, becoming angry at the Indians, sent a great snowfall and very cold weather, which drove the buffaloes away and they never returned (19). The time of the great storm seems to have been between 1770 and 1780. There is good evidence, however, that they were found in considerable numbers in this part of the state as late as 1800 (20). Soon after this they entirely disappeared. As late as 1838 traces of them were still to be found in buffalo paths, well-beaten trails, leading generally from prairies in the interior of the state to margins of large rivers. These paths were very narrow, showing that the animals went in single file (20).

In 1800 and for many years afterward, bears, deer, and elk, especially deer, were very plentiful. For some time deer continued to increase with the population because of the protection found in the neighborhood of man from the beasts of prey, and the gradual thinning-out of the animals which preyed upon them (21). Elk had almost entirely disappeared in 1837, although a few were seen occasionally (22, 20, 20a, 23). John Reynolds, an early settler of Chicago, tells of being one of a hunting party that wounded an elk (20a). In 1837 bears were seldom seen (20,

20a). Panthers and wildcats were found occasionally in the forests (20a). Beavers and otters, once numerous, had almost gone (20). Among the rodents, the varying hare disappeared about 1834.

In 1838 timber wolves and coyotes were still numerous (20a). The deer was the most common prey of the timber wolf, but these failing, they attacked sheep, pigs, calves, poultry, and even young colts (20). For some time the increase of wolves kept pace with the increase of live stock. Reptiles were most common in the heavily timbered country. As this was cleared, they disappeared, while the prairie reptiles were destroyed largely by prairie fires.

The coyote disappeared about 1844, while the timber wolf did not entirely disappear until about ten years later (22). The red fox, quite common at one time along Lake Michigan, also disappeared from this locality, about this time, although still found occasionally throughout the state. The gray fox, once quite common, was no longer to be seen after 1854 (22). The black bear and badger had entirely disappeared at the same date, although the latter was still common farther south (22). The fisher, formerly seen frequently in the heavy timber along Lake Michigan, was no longer to be found. The mink, skunk, otter, and weasel were still common (22).

The pocket gopher and the badger, once very abundant, were very rare in 1854 (22). The Canada lynx and wildcat were still abundant, but of the panthers a single individual was known to have been seen in Cook County previous to 1854 (22). The decline and disappearance of the carnivores was followed by the greatest abundance of the deer. According to Wood (21) the deer began to disappear from Central Illinois about 1865 and had totally disappeared in 1870. Their disappearance from Cook County probably antedated this. The opossum, at one time not uncommon in this vicinity, was now rare except in Southern Illinois. The only trace left of beavers was the remains of their dams in several streams (22).

4. RECOGNIZABLE SECONDARY COMMUNITIES

We may recognize the following communities in the order of their degree of difference from the primeval ones:

a) *Communities of roadside, fence-row, and abandoned field vegetation.* —These are composed chiefly of animals which commonly inhabit weeds and thickets along the edges of woods. Since these are most nearly like the thicket or forest-margin communities treated in chap. xiii, they are not discussed here.

b) Communities of parks and pastures.—The ground and subterranean animals of both pastures and lawns are (near Chicago) chiefly such prairie animals as can live under the conditions of close grazing or close mowing. This type of community is probably better developed in the pastures than in the parks and lawns. The thirteen-lined squirrel, the May beetle grub, and the earthworms are among the common species. On the lawns a few grass-feeding species have a hazardous existence. On the pasture land prairie animals are more abundant, and an occasional prairie bird nests in a clump of weeds which the cattle have not eaten.

Shrubs, when present, are inhabited by the forest-margin species. The trees present are inhabited by such forest animals as are able to live without the characteristic ground conditions of a forest and under the more severe atmospheric conditions. There are various facts pointing to a difference in the animals attacking trees differently located with respect to other trees; for example, trees standing alone in open pastures probably have a very different fauna from trees of the same species growing in the woods. This has not been fully investigated, however. The trees of the parks and lawns are often somewhat different from those of pastures, because of the introduction of many trees not native to the region. The animal communities of trees frequently include species introduced from Europe.

c) Communities of lands devoted to cultivated annuals.—The communities of farm lands are made up of animals from the prairies, the forest margin, and marsh vegetation, together with introduced species, such as the cabbage butterfly, the wheat aphis, the Hessian fly, etc.

d) Communities of orchards.—The communities of fruit-growing lands are made up of the animals from the wild haw, wild crab, wild plum, and other forest trees, the greater number of which are commonest on floodplains. There are also a number of introduced species.

e) Communities of buildings.—The communities of barns, factories, and dwellings include the common bedbug (introduced), the silver fish, the cockroaches (introduced), various buffalo bugs of which several are introduced; one (*Dermestes lardarius* Linn.) is dangerous to stored materials and has been known to eat holes in lead pipe; while various spiders, centipedes, and camel crickets occur. The house mouse, the Norway rat, and the English sparrow have all been introduced. About 75 household species are to be expected in and about Chicago.

f) Communities of polluted waters.—In connection with the building of cities, we always find the introduction of sewage and industrial wastes

(24) into streams, ponds, and lakes. The effect of the industrial wastes differs with their character. Sewage practically destroys all the life of a stream or lake near the point of entrance, through the introduction of many poisonous substances, through the increase of carbon dioxide and ammonia and through the lowering of oxygen content. Nichols (25) states that the oxygen above the entrance of the Paris sewer into the Seine was 9.23 c.c. per liter, and immediately below 1.05 c.c. per liter, a reduction of almost 90 per cent. The typical swift-water fauna of Thorn Creek at Thornton was reduced to practically nil by the opening of the Chicago Heights sewage system. The common isopod (*Asellus communis*) was the only animal able to withstand the conditions. At a distance from a point of entrance of sewage the amount of plankton is increased by its introduction because of the nitrogen and other food for plants which it contains. Forbes (see 5a) reports that the amount of plankton near Havana in the Illinois River has doubled since the opening of the Drainage Canal.

5. EQUILIBRIUM IN THE SECONDARY COMMUNITIES

Equilibration means a restoration of balance in the numbers of contending organisms of the community. For instance, as has already been noted, the deer reached their maximum number with the corresponding destruction of the carnivores by man. This indicated that the primeval balance between the carnivores and the herbivores had been disturbed. An entirely new balance has now been established through the complete destruction of both the large herbivores and carnivores, by man. Most of our knowledge of equilibration in communities has resulted from the study of the secondary communities of parks and agricultural lands. Concerning these Forbes (26, p. 15) has said:

There is a general consent that primeval nature, as in the uninhabited forest or the untilled plain, presents a settled harmony of interaction among organic groups which is in strong contrast with the many serious maladjustments of plants and animals found in countries occupied by man. [All our serious outbreaks of insect pests are instances of these maladjustments.]

To man, as to nature at large, the question of adjustment is of vast impor-tance, since the eminently destructive species are the widely oscillating ones. Those insects which are well adjusted to their environments, organic and inor-ganic, are either harmless or inflict but moderate injury (our ordinary crickets and grasshoppers are examples); while those that are imperfectly adjusted whose numbers are, therefore, subject to wide fluctuations, like the Colorado grasshopper, the chinch bug, and the army worm, are the enemies which we have reason to dread. Man should then especially address his efforts, first to prevent any unnecessary disturbance of the settled order of the life of his region which will convert relatively stationary species into widely oscillating ones; second, to destroy or render stationary all the oscillating species injurious to him; or, failing in this, to restrict their oscillations within the narrowest limits possible. For example, remembering that every species oscillates to some extent and is held to relatively constant numbers by the joint action of several restraining forces, we see that the removal or weakening of any check or barrier is sufficient to widen and intensify this dangerous oscillation, and may even convert a perfectly harmless species into a frightful pest.

Forbes mentions that cottony scale, a common pest in our parks, was rare in natural conditions. The close setting of trees has favored its increase. Close setting is nearly always a factor which has to be considered.

How do pests arise? The recent rise of the wheat aphis may be taken as an example. The spring of 1907 was very warm in the southern part of the wheat belt, and the grain aphis, which is said to reproduce freely at temperatures from 100° F. to below freezing, was accordingly able to reproduce without interruption from its parasites and enemies, which do not become active at such low temperatures as occurred. When the weather grew warmer and the enemies appeared the aphids were so numerous that the work of the enemies was hardly appreciable. But since they too, like the aphids, are rapid reproducers, with such favorable conditions they were able to increase rapidly. With their great increase the aphids decreased and soon their numbers were far too great for the available aphid food. The enemies therefore decreased because of the absence of sufficient food, and this portion of the society was accordingly restored to an approximate equilibrium. It is to be understood that such an oscillation in the society is far-reaching in its effects. It has been noted that such oscillations affect the whole community. The birds and mammals find certain kinds of food abundant and accordingly eat things different from what they do under different conditions. Such fluctuations in the animal communities are constantly going on.

The whole process may be summarized as follows:

1. Weather conditions unfavorable to enemies and favorable to plant pest.
2. Increase in pest.
3. Increase in enemies.
4. Decrease in pest.
5. Decrease in enemies.
6. Balance.

6. DISTRIBUTION OF SECONDARY COMMUNITIES ABOUT CITIES AND VILLAGES

The secondary communities of the regions about Chicago are those typical of the forest-border area; some of them are found throughout the temperate world. The communities in and about cities are not particularly different from those discussed in general terms in the preceding pages. This is a topic for special study and we can give but the briefest outline here.

When a city is in the village stage the communities of barns and dwellings are crowded together and the area of cultivated land and park is proportionally larger than in the country. As a village grows into a city, usually a central area of business houses, factories, and cheap tenements, dominated by the communities of dwellings, succeeds, practically all others being excluded. This type usually radiates from this center for a short distance along the principal lines of railroad and river transportation. Except for these narrow radiations, the central business section is surrounded by a belt of residences, which are of the park-lawn type, usually with the garden or cultivated type very much reduced or entirely eliminated. This type extends outward along all lines of passenger transportation. Toward the outskirts of this, and often quite irregularly arranged, are vacant lots and squares allowed to grow up to weeds and shrubs, and which are usually occupied by forest-margin animals. Outside of and adjoining these is the area of market gardening on the lower and better soils. Other types of agricultural land are usually poorly cultivated in the vicinity of cities.

A succession of conditions dominated by one or another of the secondary communities may be seen as the pioneer farm passes into the city stage. The pioneer-farm type is succeeded by the village type, with its park-lawn and dwelling combination. The village gives way to the business center, dominated by the "dwelling" animals. As these processes take place, a succession of the various grades of human society is noticeable. In dwellings probably the first resident pest is the clothes moth. This is probably succeeded by the silver fish and an occasional cockroach before the succession of the various grades of society has begun. Cracks appear in the woodwork as the building becomes "run down," and the introduction of a lower grade of society begins. The bedbug next makes its appearance and marks the beginning of a rapid lowering of standards on the part of occupants. The house mouse makes its appearance and is followed later by rats and vermin which mark the final stages in the degeneration into a cheap tenement.

IV. THE ECONOMIC IMPORTANCE OF ANIMALS

Why study bugs? Why waste your time upon that which can bring in no money? Why study insects, worms, birds, or snakes? These are questions which are often asked of the zoölogist, especially such as go into the field to study and collect animals and accordingly meet the public. They are questions which the zoölogist seldom can answer to the satisfaction of the inquirer, who not infrequently thinks the observer, if alone, is somewhat insane. Indeed, the conduct of one Chicago entomologist led to a police inquiry into his sanity. His offense was that of collecting insects under an electric light. The questions above we shall not attempt to answer here, except by asking, "Why study anything?"

We have already noted the complexity of the problems of our relation to nature. We have noted the disturbed balance, the ravages of species introduced by accident and by official act. We have noted that knowledge is necessary as a basis for "sanity toward nature." We have still to call attention to some of the economic values of animals.

It follows from the nature of the animal community and the close interdependence of the various species that *every* species is of *some* importance in the chain of food, space, and other relations, and every species is therefore of *some economic* importance. A few are of *great* economic importance. In addition to this we have certain definite practical uses and well-known matters of importance attached to each of the animal groups. Taking the various groups in their taxonomic order, we note the following:

The protozoa are one of the important sources of food of larger forms. Also about a half-dozen human diseases are known (27) to be due to them, and the list is continually growing. The shells of extinct species are an important part of chalk.

The uses of sponges are familiar. Aside from their importance as food of other forms, the coelenterates furnish us with corals of all sorts. Among the echinoderms the starfish is an important enemy of the oyster and mussel beds (28). The flatworms are important as parasites, many species having been recorded in the body of man (29). The round worms are of considerable importance in the same way, and some are serious enemies of grain. The earthworms are of much value to the soil (30). The crustaceans are the most important aquatic invertebrates, the *Entomostraca* being, from the standpoint of food supply, to the waters what rooted plants are on the land, one of the things to which nearly all food interaction can be traced. Some are used as food (lobsters, shrimps, crabs). Some are quite extensively used as fertilizers (horseshoe crabs). The mollusks, aside from importance to other animals, give us our pearls, pearl buttons, shell work of all kinds (31), fertilizers, important food, ink, cuttlebone, etc. (32).

The insects are of such importance that nearly every state and civilized country maintains an expensive staff of trained men whose business it is to advise the public in regard to their treatment and to investigate the relations of insects to industry. Their ravages or fear of the same are the basis of some of the speculation which enriches some and pauperizes other speculators in the necessities of human life. Aside from this we have the numerous products from insects—tincture of cantharides, honey, wax, lac (33), carmine (34), and cochineal. Many are used as human food in the tropics (locusts, water-bugs, flies, larvae of the palm weevil, etc.). Some few, such as the scorpions, are poisonous. Many diseases are known to be carried from person to person by insects and arachnids (cholera, yellow fever, malaria, sleeping sickness, typhoid, typhus, bubonic plague, mountain fever, perhaps leprosy) as well as a great host of larger parasites.

We turn now to the vertebrates, which are familiar and their uses quite well known. From this group we get our leather, furs, animal oils (snake oil, fish oil, turtle oil, lard, whale oil, skunk oil, woodchuck oil, neatsfoot oil), all of which have recognition in the markets and some of which have peculiar properties which adapt them to particular purposes (32). Glue, gelatin, bone meal, fertilizers, bone black, etc., are extensively used in industries; meats, dairy products, furs, leather, etc., are necessities.

We must not, however, fail to call attention to animals as the basis for nearly all experimental study of life processes, of heredity, of behavior and psychology, of diseases and their cure and prevention. The public should disabuse itself of the idea that biological investigators are wasting their time on bugs, for lower animals are the only material upon which the problems of our race can be solved, and until we are prepared to submit ourselves to be used in the solution of our own problems, biologists will be compelled to use lower animals as material.

CHAPTER II

THE ANIMAL ORGANISM AND ITS ENVIRONMENTAL RELATIONS

I. Nature of Living Substance

The bodies of living plants and animals are made up of living matter known as protoplasm (35, chap. ii). Protoplasm is a chemical substance or a mixture of chemical substances. It is very difficult to distinguish living and non-living matter by definition. However, we experience little difficulty in separating living from non-living things. This is because living things usually possess certain definite forms and ability to reproduce and move (especially animals). They also possess irritability. This is the property by virtue of which the force applied to living substance is not in proportion to the force resulting (35, p. 124). One strikes a horse with a whip; the energy which the horse exerts in running is not proportional to the force of the blow, but is far greater.

In considering the environmental relations of animals, we shall separate our discussion into that concerned with form and that concerned with movement (motor activity) and other functional manifestations. The term function is understood to cover all action on the part of the various parts of the organisms, motor activity included (35a, chaps. vii, viii, and ix).

II. The Relation of Form or Structure to Function

The term animal calls forth a mental picture of activity and movement. The animals with which we are most familiar are those of large size, such as fishes, birds, and mammals. They and the groups to which they belong represent only a very small part of the animal kingdom, but we may consider one of these familiar animals as an example of animals in general. The black bass will serve our purpose.

Such a fish is a complicated, highly organized animal (36, p. 183), possessing many organs, such as fins, gills, teeth, a stomach, an intestine, a liver, a heart, and a brain and spinal cord harnessed to the rest of the body by a series of small nerves which control all the organs. The fins, which are the external organs of locomotion, are sufficient in number to control the body and force it forward. The muscles which move the fins must receive nourishment in order to do their work. The nourishment is carried in blood-vessels, and the fluid which bears the nourishment is propelled by the heart, which is an organ possessing definite form and a certain type of activity. In the case of a complex animal like the black bass, we might elaborate upon the relations of form and structure to activity and function almost indefinitely. It is obvious that the two features are related in the bass. When we consider animals which possess less elaborate structure, the relations become less obvious upon mere inspection because organs are less clearly differentiated, but they are still more easily demonstrable through methods employed by the biologist.

In both the lower and the higher organisms, structure may be controlled by activity. If one cuts off the posterior end or tail of a flat-worm, a new tail is formed. Professor Child (37) found that if the animals were permitted to crawl on the bottom of the containing vessels while the new part was growing, the tail was pointed. If they were not allowed to crawl, the tail was rounded. There are many other pieces of experimental work which show that structure may be modified by function. In but few cases, however, has the modified structure been found to be inherited.

At present the relations between function and structure have not been investigated in many cases, but Child has made their relation quite clear by comparing the organism to a river. "The relation between structure and function in the organism is similar in character to the relation between the river as an energetic process and its banks and channel. From the moment that the river began to flow it began to produce structural configurations in its environment, the products of its activity accumulated in certain places and modified its flow." It deposits and removes, and thus continually "moulds its banks and bottom, forming here a bar, there an island, here a bay, there a point of land, but still flowing on, though its course, its speed, its depth, the character of the substances which it carries in suspension and in solution all are altered by the structural conditions which it has built up by its own past activity" (37a). Thus we see that function and structure are mutually interactive and mutually interrelated, and, for the sake of clearness only, we shall separate the two rather sharply in our discussion.

III. The Basis for the Organization of Ecology

We have already noted that ecology deals with animal life as lived in nature, or, in other words, with the relations of animals to their environments. The question of what aspects of these relations are most important and best suited as a basis for the organization of ecology at once confronts us. The selection of the basis for organization is the most important step before us, because if we may judge from the history of previous attempts, success or failure depends upon this selection. It appears from the preceding pages that we must choose between emphasizing structure and form on the one hand, and function and activity on the other.

I. FORM AND STRUCTURE IN RELATION TO ENVIRONMENT

Each article of furniture in the room where I am sitting, each garment which I am wearing, and the watch in my pocket were made for a purpose, and are adapted to the purpose for which they were made. This is so generally true of everything with which we have to do in our daily lives that we come to think of the phenomena of nature in the same terms, often without stopping to consider whether or not it can be true of nature.

The reading into nature of the idea of purpose and of adaptation has been a common thing since the earliest records of science (38, pp. 52–56). Two centuries ago the idea that animals were created to fit their particular place in nature, just as a watch is made for a purpose, was the idea held by scientists; indeed, such is often the idea of non-scientific people today. Later, Lamarck conceived the idea that the animal was not necessarily adapted to a given place, but became adapted to such a place by trying to live in that place, or, while not able to do a certain thing, became structurally able to do that thing by trying to do it, just as the flatworm's tail becomes pointed, and the blacksmith's arm becomes strong through use. Lamarck (38, p. 169; 39, chap. vi) believed that the changes brought about by the uses which the organism made of its parts were inherited, but science has found chiefly evidence that such changes in structure are not inherited, and this idea of the origin of adaptation has been quite generally rejected.

Following Lamarck came Darwin, who conceived the idea that all the individuals of a species which came into existence were not equally adapted to the mode of life that was necessary for them and those best adapted survived. Their characters, being born with the individual, were inheritable and the adaptation of species to which the individuals belong became perfected through the destruction of the unadapted. The destruction of the poorly adapted and the survival of the best adapted is called "natural selection" or the "survival of the fittest."

Following Darwin, a large number of investigators set to work to apply his theory to the phenomena of nature in detail. The ideas of "protective resemblance," "mimicry," and "warning coloration" were developed (40). The idea of protective resemblance is as follows: A certain insect is green and lives on green leaves. The natural-selection observer at once theorizes to the effect that the animal is green because, at a time when not all the individuals of that species were green, the birds secured all those not green and left the green ones because they were difficult to see; now therefore only green ones occur. In the case of mimicry, one species of insect (or other animal) resembles another. The theorist finds or thinks that one of them is distasteful to birds and other animals. He further discovers or concludes that the species not having a bad odor or taste is not eaten by enemies because it resembles the distasteful species. The species having the bad odor or taste is the model. The species not having the bad odor or taste is the mimic. The mimic arose and attained its perfection because those individuals of the mimic species which resembled the model species survived.

In the case of warning coloration, the animal supposed to be distasteful has bright colors. The birds, learning that certain bright colors are associated with bad tastes, avoid such strikingly colored forms. Accordingly, the most brilliantly colored distasteful forms survive.

More detailed study in recent years has tended to show such speculations to be of questionable value. Such ideas must remain matters of speculation at present, because of the difficulty of applying experimental methods to their study. Based on a theory with few facts to support it, and not withstanding critical analysis, the ideas of structural adaptation, including any of the ideas just mentioned, are not a good basis for the organization of a science of ecology.

The revival of an old idea that animal species arose in places and by methods unknown, and by chance found places to which they were adapted, now constitutes the central idea of the most recent theory of the origin of adaptation and is to be favored as a working hypothesis, because it may be tested experimentally (41).

Another reason for the inadvisability of attempting to organize ecology on the basis of structure lies in the fact that structural changes resulting from stimulation by the environment are rarely of advantage or disadvantage to the animal, and further that the structure of motile animals is not readily modified by the environment. A considerable number of animals are larger or smaller, lighter or darker, according to conditions surrounding them during development (42), but few biologists see any advantage or disadvantage to the animal in these changes.

2. FUNCTION AND ACTIVITIES IN RELATION TO ENVIRONMENT

We have just noted that from the point of view of structural adaptation, structure cannot be separated from function. It is equally true that from the point of view of physiology, function and behavior cannot be separated from structure.

Turning again to the black bass, which we have already used to illustrate some points, we note that for the simple act of swimming, the digestive tract, gills, heart, blood-vessels, brain, and muscles are necessary. They must all be present to furnish the animal with the necessary energy and impulses to make motions for swimming. It is obvious that there is a division of labor between the different organs, and if any of them are impaired or injured, or interfered with, the work cannot go on in its proper manner, or perhaps not at all. The organism is a complex of correlated parts and processes. If we interfere with any of these, e.g., the circulation of the blood of the fish, which might be done in many ways, the whole system of interdependent processes is interfered with. The fish is a highly organized animal, but the same general laws of relations of processes, such as respiration, circulation of food materials, digestion, etc., apply to animals in which there are no special or definite organs to take care of each separate process. The interdependence of processes in the organism is sometimes called *physiological proportionality* (37*a*), i.e., the work accomplished by any one set of organs or processes is proportional to that of another set or all the sets of processes in the organism. When the processes are going on in perfect accord and in proper proportionality, the organism is said to be in *physiological equilibrium*. The conception of the organism stated above may best be used in considering the relations of animal activities and functions to the environment.

It is generally held that the various animal forms are made up of different kinds of protoplasm and that the eggs of no two species are alike as to protoplasmic character. They may differ only slightly in appearance, as for example the eggs of a frog and of a salamander, but even if the eggs of these two animals are laid in the same pool at the same time, and the conditions are essentially the same surrounding the two masses, one mass of eggs develops into frogs and the other into salamanders (43, p. 8). The only logical conclusion is that the composition or protoplasm of the eggs is different.

It must be noted at the outset, therefore, that different organisms are made up of different kinds of protoplasm; furthermore, combinations of the same living substances into different special organs would of necessity give different organisms different properties.

Different chemical substances often behave differently under a given condition of temperature, pressure, or light, etc. Likewise, if a cockroach and a house fly are liberated in the center of a room, the fly goes to a window and the cockroach into a shadow; furthermore, a cold night will kill the house fly, while to dispose of the cockroach the proverbial two wooden blocks are necessary. Both differences in physiological character (behavior) are due to differences in the organisms. Different organisms often behave differently in the same intensity of the same physical factor, for example, the same temperature or light, just as the different chemical substances do. Different chemical substances often undergo different changes with variations in temperature, pressure, or light. Each has its characteristic reactions. Still whole groups may behave quite similarly. Changes in conditions affect organisms. We have all noted the effect of a cool day upon the activity of animals such as the insects. Different organisms usually behave differently in some respects, while whole communities may behave quite similarly in other respects.

a) The organism as unaffected by the environment.—When all of the external conditions continue approximately the same, the activities of the organism are called *spontaneous* (35, p. 347). As has been stated, the organism is naturally active. Accordingly, movements may possibly take place as a manifestation of the released energy inside the animal, or of disturbances and changes in the organism which are not directly initiated by the environment. Probably animals often move without any external stimulation (44, chap. xvi). One who has observed the wonderful Japanese dancing mice knows that their constant movement may not be the result of the external conditions, but of the energy which is expended within the organism.

Jennings (44) stated that these spontaneous movements must be recognized in the study of behavior, and that many errors have arisen from their neglect. If we see an animal moving, we should not assume that it is moving because of some external condition acting on it at the time. It may be due to previous stimulation or it may be the result of internal conditions. Growth, maturity, reproduction, and death are accompanied by changes in behavior, structure, etc. All may take place without great change in environment.

b) The organism as affected by the environment.—Many organisms are not sensitive to slight changes in the external environment. Having developed in, and never having been separated from, fluctuating conditions, they do not respond to all environmental fluctuations. The terms approximately constant and spontaneous used above are then both relative. Any change in the external conditions sufficient to alter the internal processes of the organism is called a *stimulus*. The visible movement of the organism or other *phenomena* resulting from stimulation is called the *reaction*. The reaction may be: (*a*) cessation of movement, (*b*) initiation of movement, or (*c*) change in kind or direction of movement.

Fluctuations in the environmental conditions in nature usually involve more than one factor. Experiments are necessary to determine which factor is affecting the activities of the animal. The effect of the various factors taken singly upon a few animals has been determined. These factors are pressure, including currents and contact with other bodies, shock, vibrations and sound, temperature, water, chemicals, light, etc. For example, if we lower the temperature surrounding an insect sufficiently, it will become apparently stiff and lifeless (35, p. 396). If the temperature is raised again, the animal becomes active. The activity is increased as the temperature is raised until a degree of heat nearly high enough to kill it is reached, when the animal becomes inactive again. If the temperature is raised only a little more, the animal dies. In general changes in any factor produce either excitation or depression, or in other words, an acceleration or retardation of the activities. In connection with the acceleration or retardation of activity, animals frequently turn toward or away from the source of light or sound, or in the direction of a current of air or water. Or they congregate at a point where the temperature or the light or the chemical conditions interfere least with their internal processes.

Such turnings or congregations are called tropisms or taxes (45). If the animals turn toward or go toward the source of stimulation they are said to be *positive*. If they turn away or go away or congregate at a distance they are called *negative*. The names applied to the reactions are given below. There are various theories as to the exact manner in which these turnings and congregations are brought about, but, as a rule, animals congregate where their internal processes are least interfered with, and random movements nearly always play some part in the process. There are two sets of terms applied to such responses as described above; they are given in parallel columns below (p. 29). Taxis means arrangement. Tropism means turning.

Reactions to light are called   phototaxis   or heliotropism
    "   temperature are called   thermotaxis   " thermotropism
    "   moisture are called   hydrotaxis   " hydrotropism
    "   gravitation are called   geotaxis   " geotropism
    "   chemicals are called   chemotaxis   " chemotropism
    "   contact are called   thigmotaxis   " stereotropism
    "   pressure are called   barotaxis   " barotropism
    "   electric currents   galvanotaxis " galvanotropism
    "   current in medium   rheotaxis   " rheotropism

If we place a number of common pond snails in a dish which is dark at one end and grades to sunlight at the other, we find that most of the snails are found after a time in faint light. The explanation of this phenomenon is that the snails are stimulated by intense light and by very weak light, i.e., either of these conditions of illumination interferes with some of the internal processes of the animal, and the random movements which result bring the animal into various conditions, one of which (faint light) relieves the disturbance. The animal then ceases to move at random, because its internal processes are no longer interfered with by the stimulus. The snail's activity is lessened, or it turns back from regions of either too strong or too weak light; accordingly, most of the snails are found in faint light. The internal processes have been adjusted or *regulated*. The snails are said to be *negatively phototactic* to strong light and *positively phototactic* to weak light.

The animal lives in an environment which is constantly changing. Its spontaneous movements are constantly bringing it into different conditions. It tends to regulate its internal processes by selecting the point in the environment in which its internal processes are not disturbed. The writer has observed snails in ponds. They move into their optimum light, i.e., the light which does not disturb them. On dark days they are found in the light. On sunny days they are found in the shade of the vegetation. They shift their position according to conditions and their distribution at any given time is a better index of conditions than the distribution of plants in the same pond.

c) Modifiability of behavior and different physiological states.—We all know that our actions may be modified by *experience*. There are but few people who have not been greatly frightened by some accident accompanied by a characteristic noise. For days afterward, one starts at the slightest unexpected noise. His response has been modified. It is a well-known fact in animal training that an animal may be "spoiled." A horse may be ruined for some purposes by an accident which has caused it to run away, for it thereby acquires the habit of running away. In the lower animals we find the same condition; their behavior may be modified, but the modifications are less permanent than in man and other mammals.

Changes within the organism cause approximately fixed environmental conditions to act as stimuli. Changes are going on all the time within the organism. Such changes may result from growth, maturation of sexual products, or other causes. The organism may be in physiological equilibrium (see p. 26) in a given set of conditions before the development of the eggs and spermatozoa begins, but these processes are accompanied by other great physiological changes, which frequently put the animal out of adjustment to its surroundings.

The queen ant is in physiological equilibrium in the darkness of the nest until she becomes sexually mature. She then becomes positively phototactic, goes toward the light, flies from the nest with the males, and, being negatively geotactic, stays away from the ground. When fertilized, she at once becomes negatively phototactic, positively geotactic, and positively thigmotactic. Accordingly she places her body in contact with the ground, and burrows into it and starts a new colony. The ant is then in a different physiological state after becoming sexually mature, and in a third state after fertilization.

3. ENVIRONMENTAL CHANGES

a) *Daily changes.*—The physiological responses of animals to these changes have an important bearing on their relation to each other. Some forms are diurnal, others nocturnal, others crepuscular. Some are probably active all the time, but move into different positions in the day and in the night. For example, some *pelagic* animals (which float or swim freely in water and are independent of bottom) are numerous near the surface at night, but migrate to considerable depths during the day, as a response to light. Many animals bear relations to day and night, which in some cases may be of an adaptive character. Some forms are active during the day, and hide themselves during the night, either in burrows or under suitable objects. Those which simply crawl under loose objects during the day frequently appear at artificial lights in the evening.

It is not impossible that there are structures in the bodies of many animals *which are the product of the different conditions of day and night during critical periods of growth.* Thus Riddle (46) has found that the barrings of the feathers of certain birds are due to low blood pressure at night (feeble circulation) during the growth of the feathers.

b) Seasonal changes.—These involve great changes in the physiological states. Inactivity is the rule in winter; growth and activity in the other seasons. The plants and animals of a locality do not all reach sexual maturity or the greatest growth activity at the same time during the growing season, but different species succeed each other as the season advances (47). The food and enemies of a given species, which is present in an animal community for a large part of the growing season, differ from time to time.

c) Weather changes.—These constitute fluctuations of conditions calling forth special types of behavior. Some animals hide when the wind begins to blow; some burrow into the ground on cool and cloudy days.

4. HABITAT PREFERENCES

By virtue of being unlike or possessing different properties, the various animal species require different conditions for the best adjustment of their internal processes. For example, the carp lives in shallow and muddy ponds and rivers, while the brook trout lives only in clear swift streams. These two organisms are able to move about and find places to which they are suited. The differences between them are clearly indicated by the differences in the habitats which they prefer.

By observation and by experimentation it has been shown that animals select their habitats. By this we do not mean that the animal reasons, but that selection results from regulatory behavior (p. 29). The animal usually tries a number of situations as a result of *random movements*, and stays in the set of conditions in which its physiological processes are least interfered with. This process is called selection by trial and error. If animals are placed in situations where a number of conditions are equally available, they will almost always be found living in or staying most of the time in one of the places. The only reason to be assigned for this unequal or local distribution of the animals is that they are not in physiological equilibrium in all the places. However, some animals move about so much that it is with some difficulty that we determine what their true habitats are.

5. THE MOST IMPORTANT ACTIVITIES OF ANIMALS

Animal activities are classified as feeding, breeding, hiding, sleeping, etc. The strength of a chain is the strength of its weakest link; the activity which determines the range of conditions under which a species will be successful is the activity which takes place within the narrowest limits. This is usually the breeding activity. The breeding instincts are the center about which all other activities of the organism rotate, and the breeding-place is the axis of the environmental relations of the organism (6, 48, 49, 50). Migratory birds are our most striking motile forms. They may migrate great distances, but always come back to the same kind of area to breed.

Failure to recognize the relative importance of the different activities is in part responsible for the general unorganized state of our knowledge of natural history. Investigators have often failed to interpret the relations of animals to their environments because they have regarded the records of the occurrence of all stages of the life history as equally important. They have considered the occurrence of the most motile stage in the life history as significant, for example the occurrence of an adult butterfly. Plant ecologists would have met with equal success if they had studied only the environmental relations and distribution of wind-disseminated seeds.

We have noted reasons for not putting primary emphasis on structure and form as a basis for the organization of ecology. The above discussion shows that activities are actually most important, and accordingly may be used in ecological study. However, since structure and activity (function) are always correlated, we should never lose sight of the former.

IV. Scope and Meaning of Ecology

1. SPECIES AND ECOLOGY

In practice, species are diagnosed in terms of structures, such as number and arrangement of bristles, hair, form, color, size, etc. Such characters are commonly called *morphological*. In ecology, the morphological characters of species are of little or no significance. Still, since habitat preferences are commonly closely correlated with the characters used to separate species, some progress in ecology can be made by the study of the distribution and environmental relations of species, but if this is not carefully checked by experimentation one may constantly fall into error.

2. MORES—PHYSIOLOGY THE BASIS OF ECOLOGY

As we have already seen, ecology[1] is that branch of general physiology which deals with the organism as a whole, with its general life processes as

[1] The unorganized phases of ecology are sometimes called natural history, biology, ethology, or bionomics, but usually by men having little understanding of plant ecology or who for some reason object to the word ecology (see 35a, pp. 18–21). The term ecology is applied to those phases of natural history and physiology which are organized or are organizable into a science, but does not include all the unorganizable data

distinguished from the more special physiology of organs (13, 51). With these limitations upon the term physiology, what may be termed physiological life histories (52) covers much of the field. Under this head fall matters of rate of metabolism, latency of eggs, time and condition of reproduction, necessary conditions for existence, and especially behavior in relation to the conditions of existence. Reactions of the animal maintain it in its normal environment; reactions are dependent upon rate of metabolism (53 and citations), which may be modified by external conditions. Behavior reactions throughout the life cycle are a good index of a physiological life history.

If we knew the physiological life histories of a majority of animals, most other ecological problems would be easy of solution. The chief difficulty in ecological work is our lack of knowledge of physiological life histories. With elaborate facilities these may be worked out in a laboratory. Ecology, however, considers physiological life histories primarily in nature, and for this reason the central problem of ecology is the *mores* (13) problem. This may be defined as the problem of physiological life histories in relation to natural environments together with that of the relations of organisms in communities. The latter is not a part of physiological life histories, the *mores* conception being the broader. An ecological classification is a classification upon a physiological basis, but since structure and physiology are inseparable, we must not forget the relations of structure to ecology and to ecological classification.[1]

V. COMMUNITIES AND BIOTA

1. BASIS

Animals select their habitats probably by trial and error. The simple fact of selection is, we believe, familiar to all naturalists. A given

of natural history. There has never been any attempt to organize natural history and physiological data into a science under the head of ethology, biology, or bionomics, and the use of these terms will not seem justified until the materials to which they have been applied are organized into a science.

[1] *Mores* (Latin singular *mos*), "behavior," "habits," "customs"; admissible here because behavior is a good index of physiological conditions and constitutes the dominant phenomenon of a physiological life history and of community relations. We have used this term just as *form* and *forms* are used in biology, in one sense to apply to the general ecological *attributes* of motile organisms; in another sense to *animals* or *groups* of animals *possessing* particular ecological attributes. When applied in this latter sense to single animals or a single group of animals the plural is used in a singular construction. This seems preferable to using the singular form *mos* which has a *different* meaning and introduces a second word. The organism is viewed as a complex of activities and processes and *mores* is therefore a plural conception.

environmental complex is selected by a number of species. All of the animals of a given habitat constitute what is known as an *animal community;* all the life (plant and animal) is a *biota.* It follows that there is often a certain physiological or ecological similarity in the species which select the same or similar habitats. When not ecologically similar, animals living in the same or similar habitats are usually *ecologically equivalent,* i.e., they meet the *same* conditions in *different* ways. For example, in a swift stream, the small fishes known as darters maintain themselves against the swift current by their strong swimming powers and by orienting against the current (positive rheotaxis). The snails (*Goniobasis*) are able to maintain themselves because of the strength of their foot and positive rheotaxis. The darters and snails are ecologically equivalent with respect to current.

There is a marked agreement of all the animals of this community in their reactions to the factors encountered in the stream. This agreement is due (*a*) to the *selection of* the habitat through innate (instinctive) behavior (40, 49, 54, 55), and (*b*) to the *adjustment of behavior to the conditions* through the effects of physical factors and through formation of habits and associations (44, 53).

Animals of the same species show behavior differences in different habitats (44, chap. xxi; 55, p. 584; 53). Bohn found that the sea anemones living near the surface of the sea, where the wave and tide action are strongest, showed more marked rhythms of behavior in relation to tide than those living lower down where the action of the tide and waves is less marked (53*a*, p. 156; 53 *b*, p. 155). These rhythms disappeared slowly when the animals were removed from the tide to the aquarium. Many such cases are probably to be found in the natural-history literature. For example, the chipmunk differs in behavior under different conditions (21, p. 523). Abbot (53*c*, p. 104) makes a similar statement about fish. It is apparent then that one species may have *several mores.* Different species may sometimes have *identical mores;* these cases are usually separated geographically (55, p. 604). In addition to these relations, the relation of ecology to species is largely a matter of language, names being necessary as a means of referring to animals.

The physiological and behavior relations of animals in the same community are of much importance and are included under (*a*) *inter-physiology* or *psychology* and (*b*) *inter-mores physiology* or *psychology.* (*a*) *Inter-physiology.*—Tarde (55, citations) is the author of the idea of inter-psychology—the psychology of the relations of individuals of the same species (man). He suggests that the social psychology of man may be traced to the inter-psychology and physiology of the lower animals. If this is true, then we can be more certain that the inter-psychology of the higher forms has developed from the inter-physiology of the lower forms (55 and citations). To this should be added the behavior between different species, while acting or living together as one. In the steppes ecologically similar animals frequently act as one species. Mr. Roosevelt has said that one of the most interesting features of African wild life is a close association and companionship often seen between totally different species of game (3). Mr. Roosevelt shows the zebra and hartebeest herding together. (b) *Inter-mores physiology* (between ecologically dissimilar forms, or antagonistic forms).—The relations of animals of different size, habits, etc., to one another involve some of the most striking features of behavior. Much of the behavior which tends to protect animals from enemies falls under this head.[1]

In all cases of modification of behavior by the physical environment or by relations to other animals of the community and in all cases where the habitat is selected, the habitat is the *mold* into which the organism fits. The study and analysis of the habitat is a necessity *as soon as the selection of habitat and the adjustment of behavior and physiological makeup to the environment are shown to be general facts.* Since habitats are different, animal communities occupying different habitats are physiologically different for the reasons just given.

The relations of the animals which make up communities are relations of life histories. The life histories of the different species are so adjusted to conditions that all animals do not reach maturity and greatest abundance at the same time. Some species continue throughout the season; for example, mammals because of their long lives, and some species of aphids or copepods because of their great fecundity and peculiar physiological makeup. There is a succession of mature or breeding animals with the change of season. A similar phenomenon is noticeable in plants. Such succession is called seasonal succession (47, 56). Different species of the same community come into relation at different seasons of the year.

Communities are systems of correlated working parts. Changes are going on all the time as a sort of rhythm much like the rhythm of activity in our own bodies related to day and night. In addition to this, communities grow by the *addition of more species*, decline, and finally

---

[1] It is at this point that ecology comes into contact with the theories of natural selection, adaptation, mimicry, etc.

disappear from the locality with changes in environment produced either by themselves or by physiographic or climatic changes (57, 58). The general growth or evolution of environmental conditions and the communities which belong to them, are included under *succession*. The word succession is used in three distinct senses. We speak of (*a*) geological succession, (*b*) seasonal succession, and (*c*) ecological succession.

a) *Geological succession* is primarily a succession of species throughout a period or periods of geological time. It is due mainly to the dying-out of one set of species and the evolution of others which take their places, or in some cases to migration.

b) *Seasonal succession* is the succession of species or stages in the life histories of species over a given locality, due to hereditary and environic differences in the life histories (time of appearance) of species living there.

c) *Ecological succession* of animals is succession of *mores* over a given locality as conditions change. If species have relatively fixed *mores* we have succession of species. When *mores* are flexible we may have the same species remaining throughout, with changes in *mores*. It is on the basis of *ecological succession* that we arrange the data presented in chaps. iv to xiv and proceed with discussion. The response of the organism to the condition of the environment is only occasionally or partially dependent upon ecological succession, but this is the only notable phenomenon about which habitats and animal communities can be arranged into a natural order.

2. CLASSIFICATION OF COMMUNITIES

Ecological classification of animals must be based upon community or similarity of physiological makeup, behavior, and mode of life. Those natural groups of animals which possess likenesses are the communities which we must recognize. One community ends and another begins where we find a general more or less striking difference in the larger *mores* characters of the organisms concerned. These communities usually occupy relatively uniform environments (58*a*).

a) *Ecological terminology* (13).—Terminology in ecology is still unsettled and changing. Groupings have thus far been based upon similarity of habitat. Habitat likenesses have in general been based upon general resemblances. General resemblances have not always been accompanied by similar physical conditions. In general there has been an agreement in the recognition of strata, of associations as communities based upon minor differences in habitats, and formations based upon larger major differences in habitats.

We give the communities of different orders below with taxonomic divisions of corresponding magnitude opposite for comparison. With the exception of the first, these taxonomic groupings do not bear the slightest relation to the ecological groupings, but are added to indicate magnitude.

Ecological Groups	Taxonomic Groups
(*Mos*) *mores*	Form (forms) (species)
Consocies	Genus
Stratum or story	Family
Association or society	Order
Formation	Class
Extensive formation	Phylum
(Aquatic and terrestrial)	(Vertebrates and invertebrates)

Mores, in the technical sense in which the term is used here, are groups of organisms in full agreement as to physiological life histories as shown by the details of habitat preference, time of reproduction, reactions to physical factors of the environment, etc. The organisms constituting a *mores* usually belong to a single species but may include *more* than one species as *specificities of behavior* are not significant (13).

Consocies are groups of *mores* usually dominated by one or two of the *mores* concerned and in agreement as to the main features of habitat preference, reaction to physical factors, time of reproduction, etc. Example: the prairie aphid *consocies*. The aphids dominate a group of organisms which for the most part prey upon them, as, for instance, certain species of lacewing, lady beetles, syrphus-flies, etc. (13).

Strata are groups of *consocies* occupying the recognizable vertical divisions of a uniform area. Strata are in agreement as to material for abode and general physical conditions but in less detail than the *consocies* which constitute them (13).

For example, a forest-animal community is clearly divisible into the subterranean-ground stratum, field stratum (zone of the tops of the herbaceous vegetation), the shrub stratum (zone of the tops of the dominant shrubs), the lower tree stratum (zone of the shaded branches of the trees), and the upper tree stratum. A given animal is classified primarily with the stratum in which it breeds, as being most important to it, and secondarily with the stratum in which it feeds, etc., as in many cases most important to other animals. The migration of animals from one stratum to another makes the division lines difficult to draw in some cases. Still, the recognition of strata is essential but a rigid classification undesirable. *Consocies* boring into the wood of living trees probably should be considered as *consocies* relatively independent of stratification phenomena (13).

Associations are groups of strata uniform over a considerable area. The majority of *mores, consocies,* and *strata* are different in different associations. A minority of strata may be similar. The term is applied in particular to stages of formation development of this ranking. The unity of associations is dependent upon the migration of the same individual and the same *mores* from one stratum to another at different times of day or at different periods of their life histories. Migration is far more frequent from stratum to stratum than from one association to another (13).

Formations are groups of physiologically similar associations. Formations differ from one another in all strata, no two being closely similar. The number of species common to two formations is usually small (e.g., 5 per cent). Migrations of individuals from one formation to another are relatively rare (13).

Extensive formations are groups of formations clearly influenced by a given climate in the case of land formations and by the topographic age of a large area and by climate in the case of aquatic formations (13, 58a).

A sub-formation is an association or a poorly developed phase of a formation. The term is used in comparing communities of the ranking associations when viewed from the standpoint of physiological differences but without reference to genetic history. Accordingly the same community is referred to as an association in the genetic sense and a sub-formation when the point of view is that of physiological difference or resemblance.

b) Animal communities of the forest-border region.—The forest-border region is the western line of demarcation between forest and steppe (see prairie area of Fig. 8, p. 51). The following is a list of the chief animal communities of the area about the south end of Lake Michigan. It is not intended to be complete, but rather to illustrate the use of the terms with particular reference to the communities to be mentioned later on. The term community is used in the general sense. Association is applied to stages in genetic development, with sub-formation as an alternative as defined above. The classification here presented in outline is artificial and attempts to combine the historical or genetic with the purely physiological points of view. Accordingly some communities are mentioned more than once. Others have two names.

I. Stream Communities
  1. Intermittent stream communities
    a) Intermittent rapids *consocies*
    b) Intermittent pool *consocies*
    c) Permanent pool, or horned dace association
  2. Permanent stream communities
    a) Spring dominated stages
      1) Spring *consocies*
      2) Spring brook associations
  3. Creek and river communities
    a) Pelagic sub-formation
    b) *Hydropsyche,* or rapids formation (turbulent-water formations)
    c) *Anodontoides ferussacianus,* or sand or gravel-bottom formations
    d) Sandy-bottomed stream sub-formation (shifting-bottom sub-formations)
    e) Silt or sluggish-stream communities
      1) Sluggish-creek sub-formations
      2) Pelagic formations
      3) *Hexagenia,* or silt-bottom formation
      4) *Planorbis bicarinatus,* or vegetation formation

II. Large Lake Communities
  1. Pelagic formations
  2. Eroding rocky-shore sub-formations (turbulent-water formations)
  3. Depositing sandy-shore sub-formations (shifting-bottom sub-formations)
  4. Lower-shore formations
  5. Deep-water formations

III. Lake-Pond Communities
  1. Pelagic sub-formations
  2. *Pleurocera subulare,* or terrigenous-bottom formation
  3. Vegetation formation
    a) *Leptocerinae,* or submerged vegetation association
    b) *Neuronia,* or emerging vegetation association
  4. Temporary pond formations

IV. Prairie or Grassland Formation of the Savanna Climate
  1 *Xiphidium fasciatum,* or grassland association of moist ground and marsh vegetation in the savanna and forest climates
  2. Prairie chicken, or high-prairie association of the savanna climate

V. Thicket or Forest Margin Sub-Formations of the Savanna and Forest Climates. Physiological group in the main though the "candlehead" sub-formation may develop into V, 2, or VI, 7, *d*

   1. Wet-ground thicket sub-formations (lower strata periodically submerged)

     a) River deposit or stream-margin thicket sub-formations. Association in the development of flood-plain forest

     b) Marsh and pond-margin thicket sub-formations (first association in the development of forests in marshes)

     c) Candlehead or moist forest margin sub-formation of the savanna and deciduous forest climates

   2. *Straussia longipennis*, or high forest margin sub-formation of the savanna climate (a climatic sub-formation of considerable permanency, probably not usually a genetic type)

VI. Forest Communities of the Deciduous Forest Climate

   1. Elm-ash series of communities

     a) Low prairie associations (see IV, 1)

     b) Marsh-margin thicket associations (see V, 1, *b*)

     c) Elm-ash associations

   2. Tamarack or floating-bog series of communities

     a) Low prairie or floating-bog association (pitcher-plant *consocies*)

     b) Marsh-margin thicket associations (see V, 1, *b*)

     c) Tamarack-forest formations

   3. Flood-plain series of communities

     a) Terrigenous river-margin associations

     b) Stream-margin thicket associations (see V, 1, *a*)

     c) Elm and river-maple associations (not studied)

   4. Clay series of communities

     a) *Cicindela purpurea limbalis*, or bare clay association

     b) Sweet-clover association

     c) High forest margin associations (see V, 2)

   5. Rock series of communities (little studied)

     a) Bare rock *consocies*

     b) Thicket association (probably high forest margin, V, 2). Later stages not well represented near Chicago

   6. Sand series of communities

     a) Lake-margin association

     b) *Cicindela lepida*, or cottonwood association

     c) *Cicindela lecontei*, or the pine association

     d) Ant-lion or black-oak association

   7. Climatic forest formation of the deciduous forest climate

     a) Birch-maple association of the tamarack-forest series

     b) *Panorpa*, or oak-elm-basswood association of the flood-plain and marsh-forest series

c) *Hyaliodes vitripennis*, or black-oak, red-oak association of the sand and other sterile soil series

d) *Cicindela sexguttata*, or red-oak, hickory association (climax, or final forest association of the savanna climate)

e) Wood-frog or beech-maple association (the climax or final association of the forest climate)

The evidence for the relations of the different formations and associations here suggested is presented from time to time in the following pages, and on the basis of this, is graphically represented in Diagram 9, on p. 312, where both physiological and genetic relations are indicated.

CHAPTER III

THE ANIMAL ENVIRONMENT: ITS GENERAL NATURE AND ITS CHARACTER IN THE AREA OF STUDY

I. NATURE AND CLASSIFICATION OF ENVIRONMENTS (35a, 55, 58)[1]

The environment is a complex of many factors, each dependent upon another, or upon several others, in such a way that a change in any one effects changes in one or more others. The most important environmental factors are water, atmospheric moisture, light, temperature, pressure, oxygen, carbon dioxide, nitrogen, food, enemies, materials used in abodes, etc. In nature the combinations of these in proportions requisite for the abode of a considerable number of animals are called "environmental complexes" (55). It is our purpose to consider animals as inhabiting environmental complexes, rather than to isolate their responses to various single factors. The consideration of environmental complexes in any comprehensive way would consume much space and require extensive and special knowledge of many fields. Accordingly, we can present here only the briefest outline of some of the principles of classification, and the important features.

If one is to understand the most elementary principle of the classification of environments, he must recognize the distinction between local and (55, 58a) climatic environmental complexes. Local complexes are often referred to as secondary or minor conditions or as edaphic or soil conditions. The climate, and such features as types of vegetation covering large areas, e.g., steppe, deciduous forest, etc., are commonly regarded as *climatic*. Opposed to these, and lying within them, are the local conditions, such as streams, lakes, soils, exposure, etc, which are only indirectly dependent upon climate. The idea can be better illustrated by the desert than by our own region. For example, in the Mohave Desert, the climatic conditions may be characterized as hot and arid. Within this desert are a few streams fed by mountain rainfall. These streams are local conditions in themselves, and produce others, such as moist soil, and types of vegetation which do not belong to the desert. Within the area about Chicago are represented two geographic complexes, the savanna and the deciduous forest, and lying

[1] Numbers in the text in parentheses refer to references in the Bibliography (pp. 325–36).

in and among these are various local complexes. The history to follow applies particularly to the local complexes. The analysis into factors applies to both local and climatic.

II. THE IMPORTANT FACTORS AND THEIR CONTROL IN NATURE

Little experimentation has been conducted with a view to determining the relative importance of different factors in the control of animals within an environmental complex. It is known, however, that moisture (evaporating power of the air), light, and materials for abode are factors important in the life of land animals; carbon dioxide, oxygen, materials for abode (including bottom), and current, in the life of aquatic animals. The evidence for these statements cannot be presented here, but will be given in appropriate places throughout the discussion which follows.

I. THE CONTROL OF FACTORS

This is related to *physiography, surface geology,* and *vegetation.*

a) Physiography.—In streams, current and oxygen content are determined very largely by physiographic conditions. Current is a function of volume of water and slope of stream bed. Oxygen content is largely determined by the rate of flow, and therefore is influenced by physiography. In lakes, oxygen content is determined by the depth, the temperature, and winds—physiographic factors are again important. On land, moisture and light are in a measure controlled by physiographic features. Slope and direction of facing profoundly affect vegetation, moisture, and light.

b) Surface materials and vegetation.—Materials for abode are largely the surface soil or rock or the vegetation. Surface soil or rock influences the moisture. Both moisture and surface materials influence the kind and amount of vegetation. All are interdependent (35a).

Physiographic features change with time. Erosion changes the gradient of streams, the width of valleys, the steepness of valley walls and cliffs, the ground-water level, etc. The weathering of rock is a process familiar to all. It is the aggregate of processes by which the coarse and hard or massive materials are reduced to clay and soil. This requires time.

The fact that vegetation grows upon the so-called sterile, coarse rough-surface materials, usually scattered or ephemeral at first, but increasing in denseness with each generation, is also familiar (58). Plants add organic matter to the soil. This organic matter holds the water so that moisture increases and plants may increase. With such changes it is obvious that an area of sterile soil will support more animals as time goes on, than at the outset, when the conditions were such that only a few hardy species could live. Here again, then, *time* is the important factor in determining the change of the area, so as to be suitable for more species (because more species are adapted to live in the resulting than in the initial conditions). The length of time which has elapsed since a given set of surface and physiographic conditions became exposed to the atmosphere is very important in governing the number, kind, and distribution of animals in a given area.

c) The value of physiographic form.—Physiographic features are classified according to their form and their mode of origin. What is the importance of their forms and modes of origin to the animal ecologist? Has a kame or an esker or a valley train any significance so far as animals are concerned? So far as anyone has been able to observe, the fact that they possess their particular form is of no significance whatever. Their relations to present ground-water level, their slope, relation to the sun, etc., are significant. The amount of surface soil and the denseness of vegetation are also of very great importance, and conditions in these respects are usually closely correlated with the length of time that the structures have been exposed to the atmosphere.

Since age is important, we turn at once to the history of an area in order to *learn the relative age of the various features* present. We have parted company with the physiographer and his discussion of mode of origin, and are interested in origins only in point of time.

III. HISTORY OF THE REGION ABOUT LAKE MICHIGAN (59)

I. PHYSIOGRAPHIC HISTORY

We will give the briefest possible account of the history of the Chicago area, following Leverett (59), Salisbury (57, 60), Alden (61), Atwood (62), Goldthwait (62, 63, 64), and Lane (65).

The most important features of our area were shaped during and since the glacial epoch. To us, the only important movement of the ice was that of the last Wisconsin ice sheet. This came to us mainly from the east and north. It spread out over the great basins now occupied by the Great Lakes and thence pushed on to the higher rock to the south of them and reached its southernmost extent in Southern Illinois.

In retiring from here (Fig. 1) one of the positions in which the edge of the ice halted corresponded to the present Valparaiso Moraine. The crest of this moraine extends from the Fox Lake region (see map)

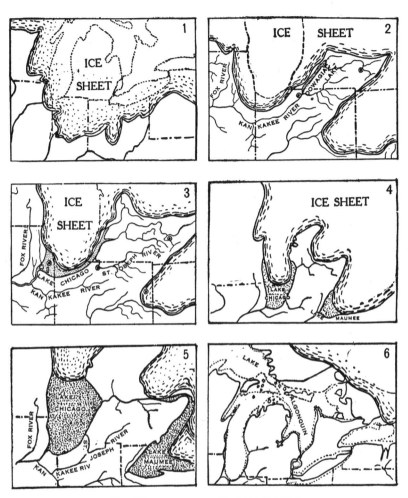

THE HISTORY OF THE CHICAGO REGION

FIG. 1.—Showing the region of the Great Lakes when the Wisconsin ice sheet was retreating from its maximum extent (after Atwood and Goldthwait).

FIG. 2.—A part of the same area, showing the drainage of the ice sheet by the Kankakee and Huron rivers through Dowagiac Lake (from Lane after Leverett).

FIG. 3.—Showing a later stage of the retreat of the ice sheet—the Glenwood stage (from Lane after Leverett).

FIG. 4.—A later stage of the same—the Calumet stage of Lake Chicago (from Goldthwait after Leverett and Taylor).

FIG. 5.—A still later stage—probably the Tolleston stage (from Lane after Leverett).

FIG. 6.—A post-Tolleston stage (from Goldthwait and Atwood after Leverett and Taylor).

southward around the head of Lake Michigan, nearly parallel with the shore, then northward into Michigan, there turning somewhat more to the east (Fig. 2). Beyond the edge of the ice, early lines of drainage were established and temporary lakes came into existence. All of our southward flowing rivers bore the sediment-laden waters from the melting ice. The results of this may be seen in the gravel and sand outwash, valley trains, etc., along the DuPage and other rivers, the more sandy portion usually being farthest downstream.

In Southwestern Michigan, these early lines of drainage were by the St. Joseph and the Dowagiac valleys. In the latter a small lake is believed to have existed (Fig. 2). These waters did not flow into the south end of the lake, as at present, but united and flowed down the present course of the Kankakee River. The Kankakee marsh area and the region at the mouth of the Kankakee (Morris Basin) are believed to have been occupied by a lake. These basins are surrounded by sand areas which are probably the oldest in our area of study. Dunes are said to be present to the south and east of "Lake Kankakee," a few being present on the moraine in the extreme southeast corner of our map (frontispiece).

The next stage was marked by the retirement of the ice from the position of the Valparaiso Moraine to the present basin of Lake Michigan. The drainage of glacial waters down the Fox, DuPage, and Upper DesPlaines rivers stopped (Fig. 3). The lakes to the south and east probably began to disappear. Later, the St. Joseph and Dowagiac changed their lower courses and flowed directly into Lake Michigan, which found an outlet by way of the lower DesPlaines.

Now begins the history treated in the first bulletin of the Geographic Society (60), and *Bulletin 7* of the Illinois Geological Survey and elsewhere (61, 62, 63, 64). The predecessor of Lake Michigan stood at a level 55 to 60 feet above the present lake. The stage is known as the Glenwood stage of Lake Chicago. Cliffs were cut, beaches of sand and gravel were deposited, and dunes were formed. These are our second oldest sand and gravel areas. Their position is shown on the map (facing p. 52).

The water then fell to a level of 35–40 feet above the present lake. This is known as the Calumet stage (Fig. 4). Here again cliffs and beaches of sand and gravel were formed, and constitute our third in point of age. These beaches have not been indicated on the map because their distribution within the state of Michigan has not been studied by physiographers. In the vicinity of Waukegan they are very close to the Glenwood beach.

The lake again receded, probably to a low level, and readvanced to a 20-foot level known as the Tolleston stage (Fig. 5). Here the development of beaches continued and the cutting of new cliffs was inaugurated. From these beaches, dunes were developed which are fourth in point of age. The position of these beaches is not indicated on the map. The lake is believed to have fallen after this to a level of 60 feet below the present level of Lake Michigan (60–62), which is known as the Champlain stage. At this time the sea came up the Gulf of St. Lawrence as far as Lake Ontario. Since the cliffs and beaches of this stage were again submerged, they are no doubt of some importance to the aquatic life in Lake Michigan, because they affected slope and bottom locally. The water rose again to a level 12 to 15 feet above the present lake, known as the Algonquin or post-Tolleston stage (Fig. 6), which was followed by a retreat to the present level.

2. THE FORMER CLIMATE AND ANIMALS (66)

During the ice age, the entire region about Chicago was overridden by the ice, and plants and animals migrated southward. There are at present a few animal species which inhabit glaciers and ice fields, and probably such were the only regular inhabitants at that time. The tundra and coniferous forest were crowded to the southward, and with them the caribou, musk ox, and other northern animals. As the ice retreated north of the southern end of the basin of Lake Michigan and the Lake Chicago stage was inaugurated, a tundra climate no doubt prevailed in the Valparaiso Moraine. It was probably the breeding-place of the present tundra species of birds; the home of the musk ox, the caribou, the snow grouse, and other northern animals. The ponds grew aquatic plants and probably supported hordes of mosquitoes (2) and other aquatic insects in summer. Early Lake Chicago is said to show no evidence of life. If we may judge from Arctic lakes at present, it had a summer fauna, especially of small crustaceans and probably some fishes.

As the ice retreated still farther northward, the coniferous forest displaced the tundra, and the musk ox and caribou were presumably only winter visitors; the woodland caribou and the moose were probably regular residents. Conditions in the lake were similar to those of the preceding stage. By this time a relatively rich flora and fauna probably existed. Organic material accumulated in the soil, shade was produced, etc. With the further retreat of the ice, the coniferous forest continued for a long time, but the plants and animals became gradually more and more like those of the southern portion of the coniferous forest (67), and gradually gave way through processes of ecological succession to the species of the present day. Just preceding our period, the mastodon roamed over the site of Chicago. The skeleton of one of these was found in a marsh near Crown Point, Ind., another at Cary, Ill.

IV. Extent and Topography of the Area Considered[1]

The area which we shall consider has its center at a point 18 miles east of Lincoln Park. It extends 67 miles (108.1 kilometers) to the east and to the west and 40 miles (64.4 kilometers) to the north and 40 miles to the south from this point. Measured from the mouth of the Chicago River it extends 85 miles (137 kilometers) eastward, 49 miles (79 kilometers) westward, 38 miles (61 kilometers) southward, and 42 miles (68 kilometers) northward. It is 80 by 134 miles (128.8 by 216 kilometers) and contains over 10,700 sq. miles (27,820 sq. kilometers).

The range of altitude in the Chicago area is not great. The lowest part of the bottom of the lake included in our map is about 80 feet above sea-level. The highest point on the Valparaiso Moraine is 900 feet above sea-level, which gives a range of altitude of 820 feet. The surface of the lake is 581 feet above sea-level. The plain of Lake Chicago is

---

[1] See frontispiece map. The term "Chicago Area" has been applied to regions varying in extent and direction, according to the points of view and interests of various authors. Chicago biologists have as yet written but little concerning the ecology of areas to the east of Millers, Ind. It becomes necessary to go farther from Chicago every year. The areas in Michigan and Northern Indiana offer the only substitute for those nearer to Chicago which are being so rapidly destroyed.

The following maps covering the area have been published:
1. Lake Michigan
    a) U.S. Hydrographic Office, Maps Nos. 1467–75.
    b) U.S. Lake Survey Maps, Custom House Bldg., Detroit, Mich.
2. Land
    a) County surveyors often publish maps covering particular counties, e.g., LaPorte Co., Ind.
    b) Illinois Internal Improvement Committee, *The Water-Way Report*, Springfield, 1909.
    c) Topographic sheets of the U.S. Geological Survey (prepared for much of the region covered by our map).
    d) The U.S. Land Office has maps of the original land surveys which are said to give roughly the distribution of prairies, forests, and marshes.
    e) Rand McNally & Co. publish maps of all local counties.
    f) Brown & Windes' (Chicago) map of the Fox Lake Region.
    g) Davis, "Peat" (map of marshes), *Ann. Rept. Mich. Geol. Surv.*, 1906.

chiefly between 581 and 600 feet, and presents very little relief. The lowest point of land on our map is in the valley of the Illinois River below the entrance of the Kankakee. This is 480 feet above tide, or 101 feet below the level of Lake Michigan. In passing from the lowest point in the lake shown on our map to the vicinity of Lake Zurich, which is the location of one of the high points on the moraine, one would travel 64 miles and make an ascent of only 12 feet per mile on the average. Indeed, if Lake Michigan were to become dry and its bottom a prairie, it would appear an undulating plain.

V. CLIMATE AND VEGETATION OF THE AREA

1. METEOROLOGICAL CONDITIONS AFFECTING ANIMALS (68)

The table (I) illustrates the fact that there are some notable differences between the different parts of our area. Extreme points would

TABLE I

STATION	TEMPERATURE				MEAN RAINFALL		SUNSHINE		RATIO OF RAINFALL TO EVAPORATION
	April to September			Year	April to September	Year	April to September	Year	July, 1887, to July, 1888
	Mean	Mean of Maxima	Mean of Minima	Mean					
Chicago...	62.6	70.0	55.6	48	19.3	33.4	1695 hrs.	2616 hrs.	95%
South Bend	65.3	76.3	54.3	49	18.3	34.5	....	....	105%

show greater differences. The evaporating power of the air is probably one of the best indices of conditions which affect animals. The ratio of rainfall to evaporation is the only expression of the evaporating power of the air which has been mapped. Fig. 7 shows this phenomenon in Central North America, with our area indicated.

2. VEGETATION (69, 70)

Those features of the vegetation which are called climatic must be discussed first. The two main climatic divisions of vegetation represented in the Chicago area are savanna including the prairie vegetation, and deciduous forest. The prairie or savanna, as distinguished from steppe, is a strip of country (the forest-border area) a few hundred miles wide, from Athabaska to Texas, where trees, chiefly oak, hickory, basswood, and elm, occur in groves and along streams. It has the general form of a bow, with its central and most eastern point at Chicago (Fig. 8). To the east of Valparaiso, Ind., the forest is chiefly beech and maple (see frontispiece). The types are believed to stand in close relation to climate, especially to ratio of rainfall to evaporation (Fig. 7).[1]

The vegetation of local conditions, as indicated on p. 42, is different from that of the region as a whole and we are concerned in part with

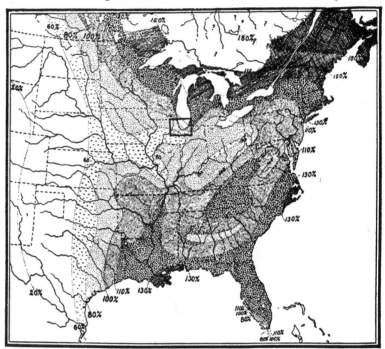

FIG. 7.—Map showing ratio of rainfall to evaporation in percentages, with area of special study inclosed in rectangle (after Transeau). Compare with Sargent's map of the "Forests of North America" (10th Census Report and, Fig. 8 below).

the relations of the animal communities of local conditions to animal communities of the climatic vegetation.

VI. LOCALITIES OF STUDY

In beginning the investigation of any biological subject from the point of view of general principles, the most important step is the selection of the material (animals to be studied). In ecological work we

[1] A glance at the map shows us that our area of study is in the center of the *Forest-Border Region.*

have not only this, but we must make a still more important choice, namely, that of the locality of study. To make this selection one must possess a good knowledge of animal environments, such as we have touched upon in the preceding pages.

1. BASIS OF SELECTION AND SUBDIVISION

Such knowledge can be acquired from texts of physiography and plant ecology, and from special works on the area at hand. The basis

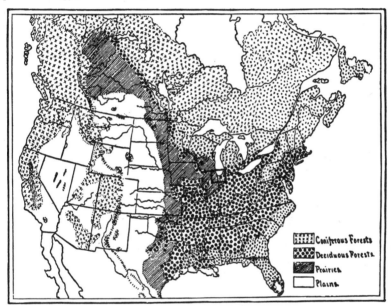

Fig. 8.—Map showing the location of the plains, savanna (prairie), and forest regions of North America, with area of special study inclosed in rectangle (from Transeau after Sargent).

of selection is either that of age or of present conditions, or both. The points selected for study are called *stations*. *Stations* are subdivided on the basis of plant and animal habitats into substations. The substations may represent either formations or divisions of formations. For example, a station like Wolf Lake may be divided into sandy shore substation, vegetation of open-water substation, and embayment substation.

2. ENUMERATION OF STATIONS—GUIDE

In the study at hand we have made use of a large number of stations which are enumerated below and are referred to in the text. The list of stations and accompanying remarks with the Guide Map may serve as a guide to the region about Chicago for field students.

<small>LIST OF STATIONS WITH DIRECTION AND DISTANCE BY RAIL FROM THE MOUTH OF THE CHICAGO RIVER, AND TRANSPORTATION</small>

A. *Aquatic Communities*

I. Large Lake Communities (chap. v).

Station 1. The open water, piers at Jackson Park, 6 miles south.

Station 1a. The eroding shore, Jackson Park, introduced rocks.

Station 2. The eroding shore, Glencoe, Ill., C. & N.W. R.R., 20 miles north.

Station 3. The depositing shore, Buffington, Ind., L.S. & M.S. R.R., and P. R.R., 22 miles southeast. Pine, L.S. & M.S. R.R., 24 miles southeast. Boats and launch from fishermen.

II. Stream Communities (chap. vi).

Station 4. Youngest ravines, Glencoe, Ill., C. & N.W. R.R., 20 miles north.

Station 5. Youngest brooks, Glencoe, Ill., C. & N.W. R.R., 20 miles north.

Station 6. County Line Creek, Glencoe, Ill., 21 miles north.

Station 7. Pettibone Creek, North Chicago, Ill., C. & N.W. R.R., 34 miles north.

Station 8. Bull Creek, Beach, Ill., C. & N.W. R.R., 41 miles north.

Station 9. Dead River, Beach, Ill., 41 miles north.

Station 10. Spring-fed streams and springs, Cary, Ill., C. & N.W. R.R., 40 miles northwest.

Station 11. Spring-fed streams and springs, Suman, Ind., B. & O. R.R., 52 miles southeast.

Station 12. Rock ravine stream, the Sag, Joliet Electric, 22 miles southwest.

Station 13. Intermittent headwaters, Butterfield Creek, Matteson, Ill., I.C. R.R., 28 miles south.

Station 14. Small swift permanent stream, Butterfield Creek, Flossmoor, I.C. R.R., 24 miles south.

Station 15. Larger swift stream and effect of rock outcrop, Thornton, Ill., C. &. E.I. R.R., 23 miles south.

Station 16. Permanent headwaters and pre-erosion stream, Hickory Creek, Alpine to Marley, Wabash R.R., 28 to 31 miles southwest.

Station 17. Permanent swift stream, Hickory Creek, Marley to New Lenox, Marley (Wabash R.R. only). New Lenox, C.R.I. & P. R.R. or Wabash R.R., 31 to 34 miles southwest.

Station 18. Sluggish small stream, North Branch of the Chicago River, Schermerville, C.M. & St.P. R.R., 21 miles northwest.

Station 19. Moderately swift, medium-sized stream, North Branch of the Chicago River, Edgebrook, C.M. & St.P. R.R., 12 miles northwest.

Station 20. Fine gravel bottom, DuPage River, Winfield, C. & N.W. R.R., 28 miles west.

Station 21. Gravel bottom, DesPlaines River, Wheeling, Ill., W.C. R.R., 33 miles northwest.

Station 22. Sandy bottomed streams, headwaters of the Calumet, Otis, Ind., L.S. & M.S. R.R., 50 miles southeast.

Station 23. Larger sandy stream, Little Calumet, Chesterton, Ind., L.S. & M.S. R.R., 42 miles southeast.

Station 23a. Deep river, E. Gary, Ind., M.C. R.R., 36 miles southeast.

Station 24. Small and intermittent sandy streams, South Haven, Mich. (4 miles south), steamer, 80 miles northeast.

Station 25. Small sandy stream, Deep River at Ainsworth, Ind., G.T. R.R., 46 miles southeast.

Station 26. Medium sandy stream, Black River, South Haven, Mich., steamer, 80 miles northeast.

Station 27. Large drowned sandy stream with marsh border, Deep River, Liverpool, Ind., P. R.R., 31 miles southeast; boats at saloon.

Station 28. Sandy large drowned stream, Grand Calumet, Clark, Ind., P. R.R. (destroyed by industrial waste), 25 miles southeast.

Station 29. Sluggish stream of the base-level type, Fox River, Cary, Ill.; boats near railroad bridge, C. & N.W. R.R., 40 miles northwest.

III. Small Lake Communities (chap. vii).

Station 30. Wolf Lake (a) Roby, Ind., L.S. & M.S. R.R., P. R.R., electric railway from 63d St., and Sheffield boathouse, 15 miles southeast; (b) Hegewisch, L.S. & M.S. R.R., P. R.R., or South Shore Electric R.R., boats from Delaware House (not practicable at low water).

Station 30a. Small lake, Lake George, Ind. Electric railway from Hammond or to Hammond from 63d St., or from Robertsdale, L.S. & M.S. R.R., P. R.R., 18 miles southeast; boats near south end of lake. For information regarding Indiana lakes, boats, etc., see *Report of the Indiana Fish and Game Commission* for 1907.

Station 31. Fox and Pistakee lakes, Fox Lake, Ill., C.M. & St.P. R.R., 50 miles northwest; boats at all hotels.

IV. Pond Communities (chap. viii).

Station 32. Young ponds, Pond 1, Buffington, Ind., L.S. & M.S. R.R. or P. R.R., 22 miles southeast (1 mile east from station).

Station 33. Middle-aged pond, Pond 5, Pine, Ind., L.S. & M.S. R.R., 24 miles southeast (pond at rear of station).

Station 34. Middle-aged pond, Pond 7, Pine, Ind., L.S. & M.S. R.R., 24 miles southeast (pond to the right in front of station).

Station 35. Mature pond, Pond 14, Clark Junction, Ind., P. R.R., 23 miles southeast (the fourth pond south of bridge over P. R.R. tracks).

Station 36. Late mature pond, Pond 30, Clark, Ind., P. R.R., 25 miles southeast (pond parallel with main street and east of school)

Station 37. Senescent pond, Pond 52, Cavanaugh, Ind., South Shore Electric R.R., 27 miles southeast.

Station 38. Prairie ponds, Roby, Ind., 26 miles southeast, east side of Wolf Lake, between second and third icehouses.

Station 39. Morainic pond or small lake, Butler's Lake, Libertyville, Ill., C.M. & St.P. R.R., 36 miles northwest.

B. *Temporary Pond and Swamp Communities.*

AQUATIC PHASES (CHAPS. VIII AND X)

Station 40. Young artificial temporary ponds, Pine, Ind., L.S. & M.S. R.R., 24 miles southeast (ponds 1 mile northwest of station).

Station 41. Middle-aged temporary ponds, Pine, Ind., L.S. & M.S. R.R., 24 miles southeast (ponds 1 mile northeast of station).

Station 42. Prairie temporary ponds, south of Jackson Park, I.C. R.R., South Chicago Branch to Bryn Mawr, 10 miles south.

Station 43. Prairie temporary ponds, 81st St. and Stony Island Ave., electric railway from 63d St. and Jackson Park Ave., south.

Station 44. Temporary pond of prairie type, but being captured by shrubs, Pond 90 or 93, Ivanhoe Station, L.S. & M.S. R.R., to Gibson, Ind., and G. & I. R.R. to Ivanhoe (1 mile south of Ivanhoe), 36 miles southeast.

C. *Marsh, Forest Margin, and Prairie Communities*

Station 45. Low forest margin (see Station 30).

Station 46. Intermediate forest margin, Beverly Hills, C.R.I. & P. R.R., 12 miles southwest.

Station 47. High prairie, Chicago Lawn, 63d St. electric railway, 11 miles southwest.

Station 48. High prairie (some low prairie), Riverside, Ill., C.B. & Q. R.R. or LaGrange electric railway, 12 miles west.

Station 49. Temporary forest pond of early stage, Pond 93, near Station 44.

Station 50. Strictly temporary forest pond, Pond 92, near Station 44.

Station 51. Spring-fed marsh, Cary, Ill., C. & N.W. R.R., 40 miles northwest.

Station 52. Swamp forest, elm, and ash, Wolf Lake, Roby, Ind., southeast (same as Station 30).

Station 53. Swamp forest, wood west of Dempster St., Evanston, Ill., C. & N.W. R.R., elevated, or surface cars, 12 miles north.

Station 54. Tamarack swamp, Mineral Springs, Ind., South Shore Electric R.R, 46 miles southeast. (For other tamarack swamps, see map.)

Station 54a. Tamarack swamp, Pistakee, Ill., 4 miles south of Fox Lake (see Station 31).

D. *Dry Forest Communities*

I. EARLY STAGES (CHAP. XII)

Station 55. On rock, Stony Island, L.S. & M.S. R.R., 12 miles south on suburban loop. Also Pullman electric car from 63d St. and Jackson Park Ave.

II. ON CLAY (CHAP. XII)

Station 56. Bluff at Glencoe, Ill., C. & N.W. R.R., 20 miles north.

Station 57. On sand, moving dunes. Mineral Springs, Ind. (near Lake Mich. and Station 54).

Station 58. Lower beach, cottonwood and pine, Pine, Ind. (near Station 40).

Station 59. Pine and oak, Miller, Ind., near bridge over the Calumet, L.S. & M.S. R.R., 31 miles southeast.

Station 60. Black oak (same as Station 59 but near village).

Station 61. Clark, Ind., near Station 28.

Station 62. Cavanaugh, Ind., near Station 37.

Station 63. Black oak, white oak, red oak, near Station 44.

E. *Moist Forest Communities*

(CHAPS. XI AND XII)

Station 64. White oak, red oak, hickory, upland forest, near Station 56.

Station 65. Forest on Blue Island, Beverly Hills, C.R.I. & P. R.R., 12 miles southwest.

Station 66. Youngest flood-plain forest, New Lenox, Ill., C.R.I. & P. R.R., also Wabash R.R., 35 miles southwest.

Station 67. Early flood-plain forest, near Station 15.

Station 67a. (Near station 71a).

Station 68. Mature flood-plain forest, near Station 48.

Station 69. Elm, basswood, oak, hickory forest, Gaugars (near New Lenox), 37 miles southwest, Joliet So. Electric R.R. from Joliet or New Lenox.

Station 70. Oak, hickory, beech, maple, Suman, Ind., near Station 11.

Station 71. Beech and maple, Otis, Ind., L.S. & M.S. R.R., 50 miles southeast.

Station 71*a*. Beech and maple, Sawyer, Mich., P.M. R.R., 73 miles east (4 miles southwest).

Station 71*b*. Beech, maple, and hemlock, Sawyer, Mich., P.M. R.R., 73 miles east (1½ miles northwest).

F. *Secondary Communities*

Station 72. Roadsides, Flossmoor, Ill., near Station 14.

Station 73. South Haven, Mich. (see Station 24).

Station 74. Stream contamination, Riverdale, Ill., I.C. R.R., 17 miles south.

Station 75. Pasturing of forests, Beatrice, Ind., C.C. & L. R.R., 45 miles southeast.

Station 76. The growth of a modern city, Gary, Ind.; many lines of transportation; 27 miles southeast.

VII. Legal Aspects of Field-Study

The student must recognize that legally, when he leaves the public highway, he usually becomes a trespasser, even though he walks in a stream bed or along a lake margin. Public property is scarce. Still, since the cost of prosecution is far greater than the remuneration secured by it in the way of damages, etc., even the most unreasonable owners are not inclined to insist upon the enforcement of the laws concerning trespassing. It should be borne in mind, however, that owners or tenants are entitled to respect, and that as a usual thing they will not object to the student's working on their property if they be treated with courtesy. Damaging gates, fences, etc., should be carefully avoided, and gates should be left as they are found.

Small wild animals such as insects, snails, etc., are not property, in the eyes of the law, and an owner would probably not be able to prevent their removal from his land except by trespass procedure. Many of the larger animals are considered as public property and are therefore protected by law. In most states nearly all birds are protected by law. It is usually legal to kill certain game birds in season, and certain condemned birds at all times. Game mammals are protected in accordance with a similar plan. It is usually necessary that a license to shoot be obtained before shooting of any sort be carried on. This would apply even to the shooting of snakes, lizards, and such animals, as well as game.

Fishes, turtles, and fresh-water mussels are protected in Illinois, as are fishes in nearly all states. The use of seines and nets of all sorts, including hand dip-nets, dynamite, and all other devices for securing fishes, is usually forbidden. The hook and line is the only exception in some states. Forbidden equipment is nearly always confiscatable, and the fines for illegal fishing are usually very heavy.

In some states it is possible to obtain licenses or permits to take birds, birds' eggs, and sometimes fishes for scientific purposes. For specific information one should consult the state fish and game warden.

CHAPTER IV

CONDITIONS OF EXISTENCE OF AQUATIC ANIMALS

I. Introduction: Comparison of Land and Aquatic Animals

The conditions of existence of aquatic plants and animals are very different from those of land plants and animals. Some of the most important differences are as follows:

a) Water, the surrounding medium, is about 768 times as heavy as atmospheric air at the sea-level.

b) The necessary gases are in solution in the water and their diffusion is much less rapid than in the atmosphere.

c) The necessary inorganic salts are in solution in the surrounding medium.

d) The necessary organic food substances for plants and some of the carbon compounds necessary for animals are in solution in the water and are taken directly by the plants and animals (47).

e) Vegetation rooted to the bottom is important in most bodies of water. In large lakes like Lake Michigan, however, there are very few attached or rooted plants, and therefore nothing comparable to the vegetation of the land, or to the plant-eating animals which live on it, is to be found. Most of the plants float freely in the water. Such plants are present also, however, where rooted vegetation occurs.

II. Chemical Conditions

1. Dissolved Content of Water

In order to support animals and plants, water must contain certain minerals and gases in solution (71). Salts (carbonates, sulphates, and chlorides) of magnesium, calcium, and sodium and salts of potassium, iron, and silicon are practically always present in solution in water, and their presence in definite proportions is essential to the life of the animals (72). Water without these has been shown to kill fish (71). Dissolved gases in definite proportions are also necessary.

Gases.—The chief facts regarding the occurrence of gases in nature and their solubility under experimental conditions are shown in Table II. The standard method of expressing quantity of gas in solution is in cubic centimeters per liter at o° C. and 760 mm. of mercury (73). All values are therefore given in these terms.

58

TABLE II

SHOWING THE DISTRIBUTION AND SOLUBILITY OF ATMOSPHERIC GASES

GAS	COMPOSITION OF AIR IN PERCENTAGES	GAS VALUES IN CUBIC CENTIMETERS PER LITER AT 0° C AND 760 MM. MERCURY			KIND OF WATER HAVING GAS CONTENT GIVEN IN PRECEDING COLUMN
		At Temperature 20° C. 760 mm.		Maximum Amounts Found in Natural Fish Waters, Springs Excepted	
		Water Absorbs from Air	Water Absorbs Pure Gas		
Nitrogen, argon, etc..	79.02	12.32 c.c.	15.00 c.c.	19.00 c.c.	Lakes (74, p. 152)
Oxygen......	20.95	6.28 c.c.	28.38 c.c.	24.00 c.c.	Streams, lakes, winter, with green algae
Carbon dioxide....	0.03	0.27 c.c.	901.00 c.c.	30.00 c.c.	Ponds
Ammonia....	Small traces locally	....	Very large quantities	14.00 c.c.	Sewage contaminated
Methane.....	Small traces locally	....	34.00 c.c.	10.00 c.c.	Bottom of lake in September (74, p. 101)

Nitrogen has little effect upon animals except when present in excess. Under these conditions in the laboratory, bubbles of the gas accumulate in the tissues and blood-vessels of fishes and cause death. It is not certain that such conditions exist in nature (Fig. 9).

Oxygen is usually necessary to the life of animals. Most animals that have been studied select water with a rather high oxygen content instead of water with little or no oxygen. The resistance of animals to lack of oxygen varies in different groups. It has been found that water with about 6 c.c. of oxygen and 14 c.c. of nitrogen per liter is suitable for brook trout. Mackinaw trout have been taken in water containing but 1 c.c. of oxygen per liter (6).

In general, carbon dioxide is a narcotic in its action upon animals. In small quantities it is a stimulant, especially to respiratory action. In large quantities it produces anesthesia and death. Several workers have shown that carbon dioxide is very toxic to fishes. Most aquatic animals that have been studied turn back when they encounter water containing large amounts of the gas. This turning away from carbon dioxide is much more decided than it is in the case of corresponding differences (2.4 c.c. per liter) in oxygen content. Fishes, for example, turn away when they encounter as small an increase as 5 c.c. per liter of carbon dioxide. Since a large amount of dissolved carbon dioxide is commonly accompanied by a low oxygen content as well as other important factors, the carbon dioxide content of water (strongly alkaline waters excepted) is probably the best single index of the suitability of the water for fishes.

Fishes do not turn away from ammonia. Ammonia is rarely present in any great amount in nature. The effect of dissolved methane is unknown. Oxygen and nitrogen go into solution from the atmosphere and oxygen is also produced by green plants. The other gases are produced chiefly by organisms as excretory and decomposition products.

9

FIG. 9.—A marine fish affected with gas-bubble disease causing protrusion of the eyes, due to excess of dissolved nitrogen in aquarium water (after Gorham).

III. Physical Conditions

1. CIRCULATION

The distribution of dissolved salts and gases is dependent upon the circulation of the water, as their diffusion is too slow to keep them evenly distributed. The circulation of water in streams is probably such as to keep all dissolved gases and salts about equally distributed. The water of streams has been found to be supersaturated with oxygen (74). Oxygen is taken up by the water near the surface. Nitrogen and carbon dioxide are produced especially near the bottom, and if the water did not circulate they would be too abundant in some places and deficient in others for animals to live.

In lakes, during strong winds (74), there is a piling-up of water on the leeward side and a lowering of the level on the windward side. This is usually compensated for by a downward flow of the waters along the bottom, as shown in Fig. 10. Small lakes with little exposure to the wind and with considerable depth frequently develop a summer circulation, such as is shown in Fig. 11. Such lakes are without oxygen in the deeper water in summer (74), and will not support the fishes which are known to inhabit the deeper water of Lake Michigan; hence we conclude that Lake Michigan must have a deep circulation at all times.

We have been able to find no record of the amount of lowering of the waters of Lake Michigan at a given point, by the wind, nor any discussion of the relations of the surface currents to the effects of winds and the vertical circulation. The waves of large lakes rise to considerable heights, as is familiar to all. They are of much importance in keeping a large amount of gas in solution in the lake waters.

The current in streams differs from that in lakes in that it is for the most part in one definite direction, while the lake currents often alternate. There are backward flows and eddies at various points in streams, in front of and behind every object encountered in the current (57, p. 124). On the basis of the current, streams are classified as intermittent, swift,

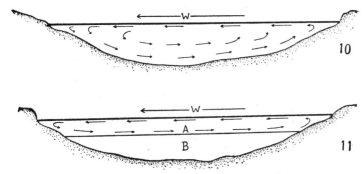

FIG. 10.—Showing the circulation of the water in a lake of equal temperature. W represents the direction of the wind (after Birge).

FIG. 11.—The circulation of the waters of a lake of unequal temperature (after Birge).

moderately swift, sluggish, and stagnant or ponded. The current within the same stream differs at different times, and in different places. As we pass across a stream we find the current swiftest near the surface in the middle, and least swift at the bottom near the sides.

2. TEMPERATURE

Temperature has always been regarded as of great importance in the direct control of the distribution of life in water. The tendency of modern investigation is to show that its influence is of great indirect importance, and the belief in its direct importance is correspondingly weakened.

The temperature in a stream is probably about the same at the various points in any cross-section. The extent to which daily, seasonal, and weather fluctuations in atmospheric temperature affect a lake is determined by the depth. Small lakes with incomplete circulation in summer are cold at the bottom, being heated at the surface only (Fig. 11). Lake Michigan is a deep lake and none of these fluctuations is felt throughout (see Table III below and Table IX, p. 74). In summer the water of the surface is warmed, but if the vertical circulation is what we suppose it to be, all the heat in the waters flowing downward at the leeward side (Fig. 10) must be absorbed above 110 meters. Table III shows the temperatures recorded by Ward (75); these were evidently taken at the bottom and do not therefore represent the temperatures at the same level in the open water, especially those records made in the shallower situations where the sun's rays can reach the bottom essentially undiminished in intensity.

TABLE III

TEMPERATURE OF LAKE MICHIGAN

Date	Hour P.M. Unless Stated	Sky	Temperature of Air	Temperature at Surface	Temperature at Depth in Next Column		Depth	
							Meters	Feet
Aug. 16	4:05	Clear	16.7° C.	18.3° C.	18.3° C.	64.0° F.	5.66	18.6
Aug. 18	9:00 A.M.	Cloudy	18.9° C.	17.2° C.	16.7° C.	62.0° F.	11.32	37.1
Aug. 18	12:25	Clearing	16.7° C.	17.5° C.	7.2° C.	44.9° F.	22.63	74.1
Aug. 16	5:10	Clear	16.7° C.	18.3° C.	7.5° C.	45.5° F.	32.06	105.2
Aug. 25	3:25	....	20.0° C.	19.4° C.	7.2° C.	44.9° F.	43.38	142.3
Aug. 16	12:05	Clear	15.6° C.	18.3° C.	5.2° C.	41.3° F.	55.93	183.5
Aug. 11	10:30 A.M.	Hazy	....	18.9° C.	5.1° C.	41.1° F.	108.22	355.0
Aug. 16	1:50	Clear	16.7° C.	18.3° C.	4.2° C.	39.5° F.	112.00	367.5
Aug. 18	4:30	Scattered clouds	18.9° C.	18.3° C.	4.2° C.	39.5° F.	132.66	436.0

3. LIGHT (76)

Light is an important factor in controlling the distribution and activities of animals. The depth to which light penetrates water is therefore of importance. Forel found that in Lake Geneva, Switzerland, during the period when the water was clearest, light diminished gradually from 25 to 65 meters, and then decreased rapidly to 115 meters where there was not sufficient light to affect the photographic plate. No doubt future investigation with more accurate means of measuring light will show that very faint light penetrates much farther. The depth of light penetration in fresh water is usually determined by the amount of sediment in the water. Forel found that in Lake Geneva the depth of light penetration decreased with the melting of the mountain snows and the beginning of the rainy season. The drainage area of Lake Michigan is very small and has little relief, and the amount of sediment carried in is small at all times. The depth of light penetration is therefore not so much influenced by these factors as in Lake Geneva. Wave-action is also important in stirring the bottom materials near shore. We would expect the light penetration in Lake Michigan to be least during the rainy and windy seasons, and greatest in calm, dry weather— late summer and autumn.[1] All of the surrounding physiographic conditions are factors controlling light. Table IV shows the seasonal distribution of rainfall and light penetration in Lake Geneva, and the seasonal distribution of winds and rainfall at Chicago.

TABLE IV

SHOWING DEPTH OF LIGHT PENETRATION IN LAKE GENEVA AND CONDITIONS AFFECT-
ING THE SAME IN BOTH LAKE GENEVA, AFTER FOREL (76, Vol. II,
p. 439), AND LAKE MICHIGAN

In the eighth column the results are given in seconds, in terms of the effect on the photographic plate, of equivalent exposures to the sun.

MONTH	LAKE MICHIGAN				LAKE GENEVA, SWITZERLAND (AFTER FOREL)			
	Rainfall		Velocity of Wind at Noon		Rainfall and Light		Light and Depth	
	Inches	Centi-meters	Miles per Hour	Meters per Second	Prec. in Cm.	Light Limit at Depth in Meters	Intensity of Light (March) at Depth in Next Column	Depth in Meters
January.....	2.0	5.1	17.8	8.0	4.87	....	500 sec.	0.0
February....	2.3	5.2	20.0	9.0	3.65	....	500 sec.	19.6
March......	2.5	6.4	20.4	9.1	4.72	110	500 sec.	25.2
April.......	2.7	6.9	19.4	8.7	5.68	....	400 sec.	45.5
May........	3.5	8.9	18.3	8.2	7.91	75	360 sec.	55.5
June........	3.7	9.4	14.4	6.5	7.59	....	120 sec.	65.6
July........	3.6	9.2	14.6	6.6	7.08	45	60 sec.	75.6
August......	2.8	7.2	13.4	6.0	8.04	....	25 sec.	85.7
September...	3.0	7.7	16.7	7.5	9.42	50	10 sec.	95.8
October.....	2.6	6.6	17.6	7.9	10.10	....	2 sec.	105.4
November...	2.6	6.6	19.0	8.5	7.4	85	0 sec.	115.6
December...	2.1	5.3	19.9	8.9	5.11	....	....	....

4. PRESSURE (76)

Pressure in water increases with depth. The results given by Forel are shown in Table V.

[1] The Lake Michigan Water Commission has reported greatest turbidity in January, February, March, and April.

TABLE V (76)

Pressure in Atmospheres	1	2	3	5	8	10	20
Depth in meters.	10.328	20.6	30.9	51.5	82.4	103.27	206.49

It will be noted that there is a little less than one atmosphere increase in pressure for each 10 meters (33 feet) in depth because water is very slightly compressible. According to this, animals in the deepest parts of Lake Michigan are living under a pressure of about 375 pounds to the square inch.

5. BOTTOM

The character of materials and topography of the bottom are very important to animals living on the bottom, but it has its effect also on free swimming animals as a determining factor in the amount of sediment.

The kind of bottom is important because many animals are dependent upon solid objects for attachment and are absent from bottoms made up of fine materials. Others must burrow into mud or creep on sand and gravel. This will be discussed later in special cases, particularly in streams.

Topography of the bottom in shallow water is important in lakes locally in affecting wave-action and currents, and through these, bottom vegetation and temperature. Ward (75) noted such effects but did not carry the work far enough to solve any of the problems involved, which are usually local. In lakes, bottom materials are most important in shallow water, because of their effect in connection with wave-action, the amount of sediment in suspension, and the stability of the bottom. The bottom materials of lakes vary greatly locally. Taking Lake Michigan as an example, if we were to see the region about Chicago denuded of all vegetation, we would be able to appreciate the fact that there are bowlder deposits, gravel deposits, sand, clay, and bare rock. Evidently the ice sheet left the same kind of bottom materials strewn with the same irregularity in the bottom of the lake as on the land. Apparently wave-action has not affected them below 25 meters (85 feet). The waves of Lake Michigan are believed not to move sand below 9 meters (30 feet). It is thought that, during the Champlain stage, the lake stood at a level 60 feet below its present level. Along the north shore there is a cliff at this level with sand deposits lying on the side toward the deeper water. Inside of this is an area of clay and then, next to the present shore, sand and gravel again. It is seen that this lower level of the lake influenced both the topography and bottom material locally, both of which probably have an influence on the occurrence of certain animals.

6. VEGETATION

The amount and kind of rooted vegetation is very important to animals. Of all the aquatic situations with which we have to deal Lake Michigan has fewest attached plants, and these are all algae. *Cladophora, Chara,* and filamentous algae are the most important. These do not appear to have been recorded below about 25 meters; some of them require solid bodies for attachment, and are probably most abundant on the rock outcrops of shallow water.

The vegetation of the younger streams consists largely of holdfast algae like those along the rock shores of the lake. These are of importance to animals. The more sluggish streams have rooted aquatic vegetation.

The vegetation is used as breeding-places. Eggs are stuck into plant tissues by the predaceous diving beetles (*Dytiscidae*) and by the water scorpions (*Ranatra*). Eggs are attached to plants by the electric-light bugs (*Belostomidae*), back-swimmers, May-flies, caddis-flies, water scavengers (*Hydrophilidae*), long-horned leaf beetles (*Donacia*), snails, and many fish (*Umbra,* and probably *Abramis*). Young animals are often dependent upon plants for shelter, to escape from enemies, etc. Many animals must use plants as a means of reaching the surface for oxygen. The most important of these are the *Dytiscidae* (adults and larvae), the *Hydrophilidae* (adults and larvae), the back-swimmers, *Zaitha, Belostoma, Donacia,* snails, *Ranatra,* and *Haliplidae*. Some, for example *Zaitha* and dragon-fly nymphs, lie in the vegetation and wait for their prey.

Different kinds of vegetation have different values for animals. The bulrush is barren for the following reasons: (1) hardness makes it a bad place for eggs; (2) there are no clinging-places; (3) there is little shade; (4) it gives a high temperature in summer; (5) there is no great addition of oxygen by vegetation; (6) it does not afford a suitable place for securing food. *Equisetum* is unfavorable for similar reasons. *Elodea* is excellent; *Myriophyllum,* good; water-lilies and *Chara,* only fair.

IV. ELEMENTARY FOOD SUBSTANCES (47)

Nitrogen, in the form of nitrates, is necessary for the growth of the plants of a pond, lake, or stream, and an insufficient quantity is secured from mineral soil. Nitrogen can be taken from the air only by nitrogen-fixing bacteria, such as *Azotobacter*, an aerobe, and *Clostridium*, an anaerobe. These bacteria occur on the outside of plants and animals, in the mud of the bottom, etc. Plants and animals provide carbon for the bacteria; bacteria provide the nitrites or nitrates for the plants.

Ammonia, resulting from the decomposition of proteid of the dead bodies of plants and animals, is oxidized to nitrous acid; nitrous acid is oxidized to nitric acid by the bacteria (*Nitrosomonas, Nitrobacter, Nitrococcus*). This acid unites with bases to form nitrates and nitrites. There are accordingly two sources of nitrate and nitrite. Working against these are the denitrifying bacteria (*Bacterium actinopelte* [Baur]) which reduce nitrogen compounds to free nitrogen. Their work is influenced by temperature. Baur placed a standard quantity of nitrate infected with *Bacterium actinopelte* at several temperatures (47, p. 271) with results as follows:

1. Temperature 25° C.: Denitrification began 24 hours after inoculation; in 7 to 11 days later the solution was nitrate-free.
2. Temperature 15° C.: Denitrification began 4 days after inoculation; in 27 days the solution was nitrate-free.
3. Temperature 4–5° C.: Denitrification began 20 days after inoculation; process incomplete 112 days after.
4. Temperature 0° C.: Denitrification not initiated.

The quantity of life in water has been held by some to be in proportion to the available nitrogen. The amount of plankton in the sea is greatest in the polar regions in summer. It has been suggested that the greater retarding effect of low temperature on the denitrifiers, as compared with the producers of nitrates, is a cause of the greater quantity of life in colder waters. Atmospheric nitrogen in solution is important in the building of nitrogen compounds by nitrogen-fixing bacteria. Oxygen is necessary for the life of most organisms, though a few can live for considerable periods in its absence. Carbon dioxide is necessary for starch building by chlorophyll-containing plants and animals. These organisms form the principal (food) basis of all other organisms.

Complex foodstuffs, such as proteids, are necessary for most animals. It is only animals which contain chlorophyll in the form of algae living symbiotically in their bodies, or otherwise, that can live without taking in proteid from the outside. Proteids are made only when light for the production of starch, nitrates, and several other inorganic foods are present. Light is then indirectly necessary to animals which can live in darkness.

The smaller aquatic animals are commonly either alga-eaters or predatory. The larger aquatic animals are commonly predatory or scavengers. The rooted vegetation is eaten only to a small extent. Small floating or swimming plants and animals, called plankton (Figs. 12–18, pp. 75, 76) are the basis of the food supply of larger animals. We could probably remove all the larger rooted plants and substitute something else of the same form and texture without greatly affecting the conditions of life in the water, that is, so far as the life habits of the animals are concerned. The aquatic plants are commonly covered with a coating of green algae, protozoa, and other small organisms, so that animals such as small snails may rasp the surface of the plants and secure food without eating the plant tissues themselves. Plants in water are of particular use to animals as clinging- and nesting-places.

V. Quantity (47) of Life in Water

The quantity of living matter in water, so far as it is plankton or floating organisms, has been much studied. The quantity is usually expressed in one of two ways: number of organisms per liter or cubic meter of water, determined by counting a part of a collection; or in cubic centimeters per cubic meter of water. In Lake Michigan (August) Ward (75) found an average of 11.5 c.c. per cubic meter in water from the surface to 2 m.; from 2–25 m., 3.9 c.c.; 25 m. to bottom, 0.4–1.5 c.c. He found that Pine Lake (a small lake) contained relatively less plankton than Lake Michigan, the surface stratum of Pine Lake containing more and the deeper strata much less than the larger lake. Lake St. Clair contains only one-half as much plankton as Lake Michigan. Lake Michigan contains only about one-tenth as much plankton as some of the small European lakes (Dobersdorfer See). Kofoid (77) found 71.36 c.c. per cubic meter the maximum record for the Illinois River. The average for the year is 2.71 c.c. per cubic meter. The largest amount recorded by Kofoid is 684.0 c.c. per cubic meter (Turkey Lake, Ind.).

Small streams and lakes with large inflow and outflow have but little plankton. Large amount of plankton is commonly associated with high CO_2 content, low oxygen content, and a large amount of carbonate in solution.

The amount fluctuates from season to season. Kofoid (77) found the maximum for the Illinois River in April to June. The amount gradually decreases until December and January, when the minimum is reached. He also found evidence that the light of the moon increases photosynthesis and the amount of plankton. The maximum of *Crustacea* was found by Marsh (78) to fall in July, August, and September, differing in different years. The maximum in Lake Michigan probably is usually in late summer or early autumn. Smaller bodies of water are similar in this respect.

I. LAW GOVERNING QUANTITY (47)

Liebig's Law of Minimum, as applied to plants, is stated as follows: ".A plant requires a certain number of foodstuffs if it is to continue to live and grow, and each of these food substances must be present in a certain proportion. If one of them is absent, the plant will die; if one is present in a minimal proportion, the growth will also be minimal. This will be the case no matter how abundant the other foodstuffs may be. Thus the growth of a plant is dependent upon the amount of the foodstuff which is presented to it in minimal quantity" (47, p. 234). The amount of plankton is determined by the same law. All food substances must be present in the correct proportions. The amount of plankton may be determined by one substance which is deficient in amount.

2. AGE AND QUANTITY (6 and citations)

In bodies of water with small outlet, the quantity of plant and animal life probably increases with the age of the water body. This is because the foodstuffs are washed in by the inflowing water, and because rooted plants absorb food from the soil in which they grow, and when they die and decay these foodstuffs are added to the water. Accordingly, the older the pond and the longer rooted vegetation has grown, the greater the quantity of life. This principle is illustrated by an age-series of ponds at the south end of Lake Michigan to be discussed in detail later. The numbers used indicate relative age. Ponds 1, 5, 7, 14, 30, 52, 89, and 95 were studied, but especially 1, 5, 7, and 14 (6). Tables VI–VIII give a summary of the results.

TABLE VI

SHOWING QUANTITATIVE RESULTS OF EXAMINATION OF FACTORS RELATED TO QUANTITY OF PLANKTON

	POND NUMBERS—AGE-SERIES			
	1	7	14	No. of Collection
Total carbonates in parts per million	138.800	160.200	160.300	1
CO_2, c.c. per liter*	0.0	3.4	2.7	2
Oxygen, c.c. per liter*	6.28	3.47	2.78	4
Bacteria per c.c.	779	2450	3550	2

* Average of collections, April, May, June, July, taken over sandy bottom (pond 1) or at the top of submerged vegetation (ponds 7 and 14).

We note that on the whole the carbonates, CO_2, and bacteria are greater in quantity according to age. Oxygen is on the whole less.

TABLE VII

Showing the Number of *Entomostraca* in Approximately 90 Liters of Water

Body of Water	September 3, 4	April 30, 1910	Average of Collections in Parentheses	Relative Age
Wolf Lake............	213	2,900	1,556 (3)	½
Prairie Pond I.........	232	9,333	4,781 (3)	3
Prairie Pond II........	4,115	19,866	11,991 (3)	14
		Aug. 28, 1912		
Pond 1..............	556	104	874 (6)	1
Pond 7..............	539	.....	927 (6)	7
Pond 14.............	2,773	133	2,680 (6)	14
Pond 30.............	1,039	.....	....	30
Pond 52.............	351	2,600	....	52
Pond 89.............	2,870	11,400	....	89
Pond 95.............	....	2,480	....	95

TABLE VIII

Showing Ratio of Number or Quantity of Different Organisms When the Maximum Is 100

	Pond Numbers—Ecological Age-Series		
	1	7a	14b
Rooted vegetation....................	20	60	100
Entomostraca.........................	32	35	100
Midge larvae.......................	80	80	100
Sphaeridae.........................	0	50	100
Gilled snails.......................	20	50	100
Lunged snafls.......................	10	50	100
Amphipoda.........................	50	90	100
Crayfishes	10	50	100
Insects.............................	40	90	100
Fish................................	100	87	87

The *Entomostraca* are rated on the basis of actual count of six collections. The other figures are estimates (6).

Here we note that the number of *Entomostraca* was greater in the older ponds though some irregularities occur, dependent upon the amount of rainfall. In rainy seasons the increase with age appears almost throughout.

As we pass from younger to older ponds we note an increase in the number of animals, excepting fish. These appear to decrease, probably because of the increasing unsuitability of the ponds as fish breeding-places. The oxygen content decreases, particularly on the bottom. The distribution of the fish present in these ponds, and whose breeding habits were known, was found to be correlated with the distribution of the bottom upon which they breed. This becomes less and less in amount as the ponds grow older.

3. EQUILIBRIUM

Each animal prefers certain food. The food relations of pond animals are shown in Diagram 3, below. For purposes of illustration let us suppose the existence of a community composed of the species named *only*.

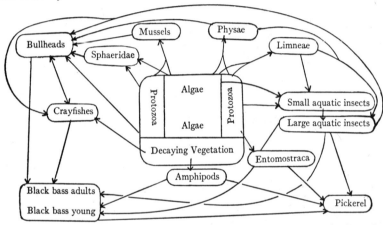

DIAGRAM 3.—Showing food relations of aquatic animals. Arrows point from the organisms eaten to those doing the eating. For explanation see text.

Any marked fluctuation of conditions is sufficient to disturb the balance of an animal community (see chap. i, p. 18). Let us assume that because of some unfavorable conditions in a pond during their breeding period the black bass (79) decreased markedly. The pickerel, which devours young bass, must feed more exclusively upon insects. The decreased number of black bass would relieve the drain upon the crayfishes, which are eaten by bass, crayfishes would accordingly increase and prey more heavily upon the aquatic insects. This combined attack of pickerel and crayfishes would cause insects to decrease and the number of pickerel would fall away because of the decreased food supply. Meanwhile the bullheads, which are general feeders and which devour aquatic insects, might feed more extensively upon mollusks because of the decrease of the former (see chap. i, p. 15), but would probably decrease also because of the falling-off of their main article of diet. We may thus reasonably assume that the black bass would recover its numbers because of the decrease of pickerel and bullheads, the enemies of its young. A further study of the diagrams shows that a balance between the numbers of the various groups of the community would soon result.

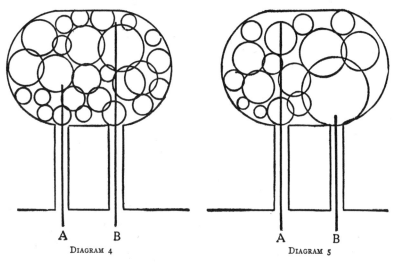

A   B   A   B

DIAGRAM 4   DIAGRAM 5

DIAGRAM 4.—Showing the life histories of the animals of the pond community in the form of circles. The heavy, vertical, black lines represent the animals which are dependent upon the most elementary food substances. *A* represents dead animal matter; *B*, the protozoa, rotifers, and *Entomostraca*, the smallest animal food. The black lines come into contact with different numbers of life cycles, but are indirectly connected with all so that any change in the position or rate of movement (meaning number or rate of reproduction and growth) of the rod must effect the entire community; compare with Diagram 3.

DIAGRAM 5.—Showing the food relations in the brook community. *A* represents algae which grow upon the stones. *B* represents the floating animal bodies and other organic matter. The latter are of small importance because of their small number and the swift current.

Under other circumstances, such as the extinction of the black bass, the resulting condition would be entirely different from the original one, but a balance between supply and demand would nevertheless finally be established. The community is said to have *equilibrated* when such a condition is reached; that is, a new equilibrium is established which *may or may not be like the old.*

The causes of fluctuations of numbers of organisms are numerous. Cold winters often destroy aquatic vertebrates. Large rainfall dilutes the plankton in streams and carries it away. Too little sunshine causes a poor production of the chlorophyll-bearing organisms which are the food basis of all the others. High temperature favors denitrification. From Diagram 3 and brief discussion above it will be seen that there are in a pond community, close interrelations traceable to certain groups which are closely dependent upon the more elementary food substances A representation of these relations is given in Diagrams 4 and 5.

CHAPTER V

ANIMAL COMMUNITIES OF LARGE LAKES (LAKE MICHIGAN)

I. CONDITIONS

1. GENERAL (75)

Lake Michigan lies between 41°-40' and 46°-5' N. latitude. Its total length is about 350 miles and its greatest width is approximately 85 miles. Its area is about 25,000 sq. miles. Its greatest depth is nearly 275 meters (900 ft.) and its average depth is approximately 122 meters (400 ft.).

Within the area covered by our map (frontispiece) there are about 3,200 sq. miles. The maximum depth is about 152 meters (500 ft.). It has been estimated that the lake contains 262,500,000,000,000 cubic feet of water. It becomes obvious at once that the lake constitutes one of the most uniform and extensive environments with which we have to deal.

2. CIRCULATION

The level of the lake fluctuates from season to season with the amount of rainfall, but we have been unable to find a statement as to the amount of such fluctuation. Changes in atmospheric pressure over part of the lake cause various fluctuations in level, called seiches. In Lake Michigan there is a definite circulation of the surface waters. Here the current moves southward along the west shore (57), around the head of the lake, and northward along the east shore. The rate of flow is 4 to 90 miles per day.

II. COMMUNITIES OF THE LAKE[1] (80, 81, 82, 83, 84)

One of the recognizable animal communities of Lake Michigan is made up of the animals which live freely in the water, either swimming or floating. This community is called the Pelagic or Limnetic community. Other communities are governed directly or indirectly by depth

---

[1] The only published account of the invertebrate fauna of the Great Lakes is that of Lake Superior. From this account and from incidental scattered notes found in various publications cited we have been able to bring together enough data to give an idea of the conditions and life which we may expect future investigations to show. The attempts to study Lake Michigan have been ill-fated. In 1871, the Chicago Academy of Sciences and the United States Fish Commission co-operated in an attempt to study the fauna of the lake. The work on the vertebrates was published

and bottom. Accordingly the conditions on the bottom at various depths are roughly shown in Table IX.

TABLE IX

| PHYSICAL CONDITIONS | DEPTH | | VEGETATION |
	Meters	Feet	
Limit of sand-moving waves......	8	26	
Limit of daily temperature fluctuations; limit of wave action; beginning of light decrease; pressure about 2½ atmospheres.........	25	82	Lowest record of *Chara* and (75) *Cladophora*
Pressure 4 atmospheres; light reduced to ⅞..................	39	128	Scanty filamentous algae (75)
Seasonal temperature fluctuations less than 1°; light reduced to ¾; pressure 5⅜ atmospheres........	54	177	*Nostoc* and diatoms (75)
Light ⅛; pressure 7 atmospheres...	70	230	No bottom plants recorded
No light; pressure 11½ atmospheres; no change in temperature; uniform conditions..............	115	377	No plants recorded
Greatest depth in the area considered; pressure 15 atmospheres	153	500	No plants recorded
Greatest depth in lake; pressure 27½ atmospheres................	274	900	No plants recorded

I. THE LIMNETIC COMMUNITY

(Station 1; List I)

Chicago is famous for its good water supply. However, if one fastens a small sack of miller's bolting-cloth under an open water tap for an hour in summer and examines the contents of the sack with the naked eye and then with the microscope, he will be of the opinion that he has *not* been straining drinking water but stagnant ditch water. He finds small microscopic plants in great numbers (75), as well as large numbers of small animals, most of the larger ones dead. Every person drinking water from a lake or river drinks the small plants and animals. If every one of the 2,000,000 persons in Chicago drank a quart of unfiltered

by the United States Fish Commission, and Doctor Stimpson of the Academy published a brief note on the invertebrate forms found in the lake, but never gave more than a hint of the work, as the collections were all burned with the Academy's building. Subsequently, collections were made by the State Laboratory of Natural History, and later by the Fish Commissioners of Michigan. In the summer of 1902, the University of Chicago and the Academy of Sciences made a single-day excursion, but no report was ever published.

city water in a day in August, all together they would be consuming about 10 quarts of solid plant and animal substance—enough to make a meal for about forty people.

One does not think of the lake as an area of luxuriant vegetation, teeming with animal life, but rather as a barren waste of water. However, if one's vision for small objects were only better, he would see as he passes over the water in a boat, thousands of small animals and plants such as are shown in Figs. 12–18 together with about fifty other forms of

protozoa, wheel animalcules, crustaceans, insects, and small fish. Most of these spend their entire existence freely floating or freely swimming. With the exception of the fish and insects they constitute the plankton which is the basis of the food of the millions of pounds of fish taken from Lake Michigan every year.

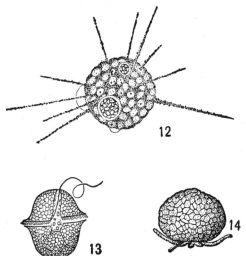

From the standpoint of our economic·interests, the limnetic formation is of great importance. It deserves comment also because of its scientific interest, and the aesthetic value of the various forms of which it is composed.

Fig. 12.—A sun animalcule (*Actinophrys sol* Ehrbg.); 330 times natural size (after Leidy).

Fig. 13.—Protozoan (*Peridinium tabulatum* Ehrbg.); 400 times natural size (after Kent).

Fig. 14.—A shelled protozoan (*Difflugia globulosa* Duj.); 130 times natural size (after Leidy).

a) Its composition (85, 86, 87, 88, 89).—The minutest animals of this formation are the protozoa. About thirteen species have been found to inhabit the open waters of the lake. Of these the sun animalcule (*Actinophrys sol*) (Fig. 12) and the shelled protozoan (*Difflugia globulosa*) (Fig. 14) are easiest to recognize. Nine of the thirteen common species are mixotrophic in their nutrition (i.e., contain chlorophyll and manufacture their own food) (Fig. 13) and share with the algae and diatoms the important function of furnishing food for the rotifers (wheel animalcules) and the crustaceans.

About a dozen species of crustaceans are common in the lake. They feed chiefly on the protozoa, diatoms, desmids, and possibly the rotifers (85). *Such crustaceans constitute almost the sole food of young fishes and are the first food of the young whitefishes* (79). They are divided into copepods and *Cladocera* (and ostracods, rare). This division of the crustaceans is known as the *Entomostraca*. The smallest and most

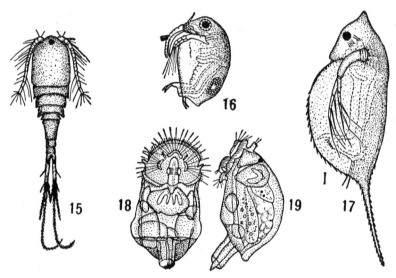

REPRESENTATIVE CRUSTACEANS AND ROTIFERS OF THE LIMNETIC COMMUNITY OF LAKE MICHIGAN

FIG. 15.—A common copepod (*Cyclops bicuspidatus*); 25 times natural size (after Forbes).

FIG. 16.—A cladoceran (*Bosmina*); enlarged (from Forbes after Gerstaecker).

FIG. 17.—A cladoceran (*Daphne hyalina galeata*); enlarged as indicated (after Smith).

FIG. 18.—A pelagic rotifer (*Notops pelagicus* Jen.); 180 times natural size (after Jennings).

FIG. 19.—The same, side view.

abundant of the *Entomostraca* of the lake is only 1.1 mm. in length and is slender and colorless. It is the slender *Cyclops bicuspidatus*, shown in Fig. 15.

The commonest *Cladocera* of the lake are *Bosmina* (Fig. 16), *Daphne retrocurva*, and *Daphne hyalina* (Fig. 17). One other small species (*Leptodora hyalina*) belonging to this group is a very interesting creature.

"When in its native element it is almost perfectly transparent and consequently invisible—a true microscopic ghost" (Forbes, 89).

The wheel animalcules are as a rule larger than the protozoa and are of a much higher structural organization, capable of making more complex movements. About thirteen species of these may be found in the waters of the lake in midsummer. *Notops pygmaeus* Calm. (see Figs. 18-19) is a characteristic member of the group.

In addition to these forms there are also worms, such as round worms, planarians, leeches, etc., found in the limnetic formation either incidentally or habitually.

None of the adult fishes of the lake belong strictly to the limnetic formation. Fishes such as the whitefish, lake herring, and lake trout are sometimes found in the open water, and the young of some lake fishes may belong there strictly (90).

b) Characters.—Specialists in the various groups of animals might be able to pick out some structural characters which would distinguish the forms of such open-water situations from the forms living in among the vegetation or on the bottoms of this or smaller lakes. The only striking structural character is the transparent or translucent color of most of the forms.

A large number, if not all, of the limnetic crustaceans are in deep water during the day and come to the surface at night. The behavior of the rotifers is somewhat different. Jennings (87) says: "During the day the limnetic rotifers are found in much greater numbers near the surface than near the bottom, reversing the condition commonly observed for the crustaceans. At night the distribution seems not to be materially changed. The immense numbers of crustaceans obscure the rotifers; but there was no greater number of rotifers near the bottom in the few towings made at night than in the day time."

The most striking characteristic of the limnetic formation is that it is independent of bottom and in its reactions is indifferent to the bottom. Jennings (44) states that pelagic forms have a more simple type of behavior than the attached and bottom forms.

2. BOTTOM COMMUNITIES

Forms inhabiting the bottom of lakes and also of the sea in a general way bear the same relation to the water that the terrestrial animals do to the surface of the land. Usually they do not leave it to rise to any considerable height above the bottom. The fishes of lakes correspond to the birds of the land.

Other relations are, however, different. As has been stated, there are no truly rooted plants in the bottom of Lake Michigan. Those attached to the bottom are not rooted in the way that land plants are. The things which land plants get from the soil are supplied to the aquatic plants by the water itself. The same is true of the bottom animals; food is floating in the water in quantities and can accordingly be secured without effort, and some animals have the form of plants and simply depend upon the food which may be brought within reach by accident.

Classification of bottom formations: Bottom formations are determined by depth (and associated phenomena) and bottom. Bottom is of greatest importance in shallow water (less than 8 meters). Its importance is inversely proportional to depth.

Within the zone of wave-action conditions are somewhat different than below it. Here the kind of animals is determined by (1) strength of wave-action, (2) erosion and kind of material eroded, and (3) deposition, and animal communities may be classified as those of (1) eroding—rocky or stony—shores, (2) depositing or sandy shores, and (3) protected situations.

a) *Eroding rocky shore sub-formation* (80, 81, 82, 83, 84) (Stations 1a, 2; Table XV).—There are a considerable number of rock outcrops in the bottom inside the 8-meter (26 ft.) line, between Gross Point and the mouth of the Calumet River at South Chicago (61). As we shall see later, these are of great importance to the animals of the lake. However, the communities of such situations are known to us only through the study of the very shallow water in the vicinity of Glencoe. Here, attached to the rocks by their silk, are caddis-worms (*Hydropsyche*). (Mr. W. J. Saunders has given me specimens of *Parnidae* (*Psephenus*) and stone-fly nymphs (*Perla*) taken from Lake Ontario at Kingston, Ontario.) All these ordinarily live in swift streams. Under the stones and among the algae attached to them are amphipods (*Hyalella knickerbockeri*) and May-fly nymphs (*Ephemeridae*), but so far as we have been able to record these are the only forms common here. The animals avoid the waves by creeping under stones or are attached to withstand wave-action. The lake trout (Fig. 20) is known to breed on the rocks off Lincoln Park. These rocks are then of considerable importance to the fish. Some species of small fish may be common here, but they have not been studied.

b) *Sandy depositing shore sub-formation*, o–8 meters (26 ft.), *shifting sand bottom* (Station 3; Table XII).—On the open shore inside of 1.5 meters (5 ft.) of water we have found *nothing on the bottom*. From this depth to 4 meters (13 ft.) *Sphaerium vermontanum,* which occurs rarely in Hickory Creek also, and midge larvae (a red and a white species) appear characteristic. A number of species of small fish such as the blunt-nosed minnow, the straw-colored minnow, and shiners are likely to be found in from 4–8 meters (13–26 ft.) of water. An occasional *Lymnaea woodruffi* is found at this depth.

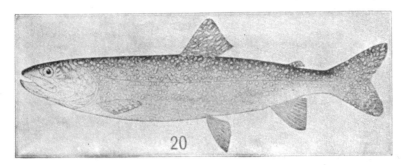

REPRESENTATIVE FISHES BELONGING MAINLY TO THE TRANSITION BELT OF LAKE MICHIGAN (25–54 M.)

FIG. 20.—Great Lakes trout (*Cristivomer namaycush*); length 3 feet (after Jordan and Evermann).

FIG. 21.—The long-jaw whitefish (*Argyrosomus prognathus*); length 15 inches; from the depth of 74 meters (after Smith).

c) Communities of protected situations (Table X).—Near Chicago, bays and inlets are rare. Doubtless the mouths of some of the larger rivers, before they were modified for navigation, were of this character. Such places have been studied in Lake Superior (80, 83) and the Grand Traverse Bay region. Out of 21 species recorded here, 16 are definitely recorded below 9 meters and not on the open shores. All are found in small lakes and sluggish streams.

d) *Lower shore formation* (8–25 meters) (Station 3; Tables XI, XIII, XV).—The belt immediately below the shore belt is characterized by wave-action sufficient to move only the finest material. Its lower limit is the limit of wave-action; the beginning of light diminution; the lower limit of daily fluctuation in temperature; and the lower limit for most of the species of *Mollusca* (75, appendix). Practically all the forms that have been recorded here are inhabitants of still, shallow water also. Notable among these are the common still-water amphipod *Eucrangonyx gracilis*, the little bivalve *Sphaerium striatinum*, and several species of *Amnicola* and *Valvata* which, together with *Lymnaea woodruffi*, are more characteristic of Lake Michigan than of shallow waters. While a large number of *Mollusca* are recorded from the lake above 25 meters only the *Sphaeridae* are found below this limit. Small annelids, midge larvae, and leeches are very abundant north of Gary, Ind., in 11 meters of water.

This belt is the principal breeding-ground of the whitefish. The eggs are deposited on the bottom and left unguarded. It appears that the young fish stay in the shallow waters for a considerable time. Wherever the bottom is firm the lake trout breeds also. Nearly all the fish traps are set in the upper edge of this belt and in the lower boundary of the one above.

e) *Belt of overlapping: upper deep-water belt* (25–54 meters) (Tables XIV, XV).—This belt is characterized as below wave-action, below daily fluctuations of temperature, with seasonal fluctuations not exceeding 3° C. It is intermediate between the belt above and the deep belt, and is the characteristic feeding-ground of the whitefish and the regular home of the long-jaw (*Argyrosomus prognathus*, Fig. 21). On the other hand, it is the upper limit for some of the deeper-water forms, such as the well-known *Mysis relicta* and *Pontoporeia hoyi* (Figs. 22, 23), the deep-water crustaceans which are the chief food of the whitefish.

f) *Deep-water formation* (54 meters to bottom) (Table XV).—This belt is characterized by weak or no light and by seasonal changes in temperature less than 1 degree. Below 115 meters there are no light and no seasonal changes, and the temperature is 4° C. throughout the year. Off Racine in 82 meters (265 ft.) the bottom is of reddish-brown sandy mud (82); in 95–125 meters (311–410 ft.) dark-colored impalpable mud, depressions with decaying leaves (82*a*). In the Grand Traverse Bay region, Milner found decaying sawdust in 183 meters (600 ft.) (81). Except for unimportant variation in bottom, conditions are practically uniform throughout. Milner (81) states that the invertebrates are abundant and evenly distributed throughout the deep-water belt. The principal invertebrates are *Pontoporeia hoyi*, *Mysis relicta*, water-mites, midge larvae, and a species of *Pisidium*.

The fish, however, show some noteworthy peculiarities of distribution. The lake trout rarely leaves this belt, except during the breeding season. The blackfin (*Argyrosomus nigripinnis*) is below 70 meters, except in December, when it has been recorded in 60 meters. Hoy's whitefish

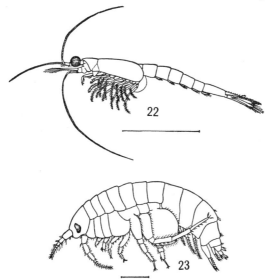

REPRESENTATIVE CRUSTACEANS OF THE DEEP-WATER COMMUNITY OF
LAKE MICHIGAN

FIG. 22.—A schizopod (*Mysis relicta*); enlarged as indicated (after Smith).
FIG. 23.—An amphipod (*Pontoporeia hoyi*) (after Smith).

(*Argyrosomus hoyi*) is rare, and *Triglopsis thompsoni* has not been recorded above 115 meters; all accordingly live under uniform conditions—no day, no night, no seasons.

III. SUMMARY

The available data on the conditions and life in the lake are of such a nature as to justify few conclusions of weight. We find only hints here and there which may be useful to those who shall investigate the lake in the future.

1. Bottom forms are the most abundant on the open shores which are rocky, and which form good substrata for the attachment of algae and the holdfast organs of animals.

2. The sand-depositing shores are without animals, at least to a depth of 1.5 meter, and life is scanty to 8 meters, on account of the shifting character of the bottom.

3. Animals are abundant in protected bays; the species inhabiting these situations are commonly found in sluggish streams and small lakes, and a few of them have been recorded below 8 meters also, which is relatively quiet water.

4. The animals of the upper shore belt, o-8 meters, are found also in swift streams.

5. The animals of the lower shore and upper deep-water zone are below effective wave-action and are those found in still waters.

6. The animals of the deep-water zone are not found outside of deep lakes, and cannot be compared with any others of our Chicago area.

7. We have, then: swift-water animals in the upper belt, still-water animals in the middle belt, and deep-water animals in the lowest.

8. The fish are migratory and deserve special comment.

DISTRIBUTION OF WHITEFISH AND DEEP-WATER FISH IN LAKE MICHIGAN (75)

Argyrosomus artedi, the lake herring, is near the surface.

Coregonus clupeiformis, the whitefish, lives most commonly between 21 and 36 meters; it spawns in water between 3 and 28 meters, most commonly between 15 and 19 meters. It makes migrations into the 9-meter belt in summer, supposedly on account of bad aeration; has disappeared where breeding-grounds have been destroyed.

Argyrosomus prognathus, the long-jaw, is found mainly in from 36–66 meters.

Argyrosomus nigripinnis, the blackfin, is found in from 70–80 meters, coming up to 60 in December.

Argyrosomus hoyi, Hoy's whitefish, is usually recorded below 115 meters.

Triglopsis thompsoni is confined below 115 meters.

Cristivomer namaycush, the lake trout, is confined below 25 meters, except during the breeding season. It breeds between 2 and 25 meters on rock or other hard bottom.

Lota maculosa, the lawyer, appears to be distributed throughout, but no account is to be found regarding its movements or their causes.

An interesting truth is illustrated by the species of whitefishes (*Argyrosomus* and *Coregonus*). If a group is to be successful and become extensive in its distribution, it must so differentiate in habits as to bring the different races out of competition with each other. We usually find that different species which are closely related have different habitats. Here we have these species of fish arranged one above the other. The separation in such cases is usually horizontal.

ANIMALS RECORDED FROM LAKE MICHIGAN[1]

LIST I

Common *Entomostraca*

Copepods: *Cyclops leuckarti* Claus, *C. bicuspidatus* Claus, *C. prasinus* Fischer, *Epischura lacustris* Forbes, *Diaptomus ashlandi* Marsh, *D. oregonensis* Lil.; Cladocerans: *Daphne hyalina* Ley., and *D. retrocurva* Forbes.

TABLE X

Animals occurring in protected situations (bays, harbors, etc.) in Lake Superior in from 0–2 meters of water, and known also to occur in Lake Michigan where habitats are not recorded:

Common Name	Scientific Name	Literature
Mussel..................	*Anodonta grandis* Say................	(75, 83, 91)
Mussel..................	*Anodonta marginata* Say..............	(75, 83, 91)
Snail....................	*Amnicola lustrica* Pils...............	(75, 83, 91)
Snail....................	*Valvata tricarinata* Say..............	(75, 83, 91)

TABLE XI

Animals of the lower shore belt. Those definitely recorded from 8–15 meters of water are marked * and **, the latter indicating that the records are original from 11 meters of water north of Gary, Ind. (Station 3); † indicates that the animals are recorded from protected bays in 0–2 meters of water (Lake Superior), and ¶ that they occur in inland waters, especially ponds:

Common Name	Scientific Name	Literature
†¶ Snail................	*Lymnaea stagnalis* Linn..............	(75, 83, 91)
†¶ Snail................	*Planorbis bicarinatus* Say............	(75, 83, 91)
†¶ Snail................	*Planorbis exacutus* Say...............	(75, 83, 91)
†¶**Snail...............	*Amnicola limosa* Say.................	(91)
†¶**Snail...............	*Amnicola limosa porata* Say..........	(91)
†¶**Snail...............	*Amnicola emarginata* Küster.........	(91)
†¶**Snail...............	*Amnicola lustrica* Pils..............	(91)
**Snail...............	*Valvata bicarinata perdepressa* Walk.....	
†¶ Snail...............	*Valvata sincera* Say..................	(91)
† **Bivalve............	*Pisidium idahoense* Roper	(75, 83)
†¶**Bivalve............	*Pisidium scutellatum* Sterki..........	(83, 91)
†¶**Bivalve............	*Pisidium compressum* Prime..........	(91)
†¶* Bivalve............	*Pisidium variabile* Prime............	(75, 83, 91)
†¶* Bivalve............	*Pisidium ventricosum* Prime..........	(75, 83, 91)
†¶* Bivalve............	*Pisidium punctatum* Sterki..........	(75, 91)
†¶**Bivalve............	*Sphaerium striatinum* Lamarck........	(80)
†¶**Bivalve............	*Calyculina transversa* Say...........	(91)
¶* Midge larva.........	*Metriocnemis* sp....................	‡
¶* Leech..............	*Glossiphonia stagnalis* Linn..........	(91a)
¶**Worm..............	*Limnodrilus claparedianus* Ratzel......	

‡ See citation 98.

[1] The numbers in parentheses in the column headed "Literature" refer to references in the Bibliography at the end of the book.

TABLE XII

Animals on depositing shores in from 0–8 meters of water, * indicating that records are original.

Common Name	Scientific Name	Literature
*Bloodworm	Chironomid larvae	
*Bivalve	Sphaerium vermontanum Prime (characteristic)	
*Midge larvae	Metriocnemus sp	
*Snail	Lymnaea woodruffi Baker (rarely)	
Long-nosed sucker	Catostomus catostomus Fors	(81, 84)
Common sucker	Catostomus commersonii Lac	(81, 84)
Hog sucker	Catostomus nigricans LeS	(81, 84)
Red-horse	Moxostoma aureolum LeS	(81, 84)
*Trout perch	Percopsis guttatus Ag	(81)
Minnow	Notropis hudsonius DeW. Clin	(84)
Straw-colored minnow	Notropis blennius Gir	(84)
*Shiner	Notropis atherinoides Raf	(84)
*Blunt-nosed minnow	Pimephales notatus Raf	(84)
Top minnow	Fundulus diaphanus menona J. and C	(84)
Johnny darter	Boleosoma nigrum Raf	(84)
Least darter	Microperca punctulata Put	(84)
Lake herring	Argyrosomus artedi LeS	(75, 84)
Pumpkinseed	Eupomotis gibbosus Linn	(81)
Bluegill	Lepomis pallidus Mitch	(81)
Mud minnow	Umbra limi Kirt	(81)
Eel	Anguilla rostrataLeS	(81, 84)

TABLE XIII

Animals occurring in from 15–25 meters of water:

Common Name	Scientific Name	Literature
Snail	Amnicola walkeri Pils	(75, 83)
Polyzoan	Plumatella sp	(81, 82)
Snail	Pleuroceridae	(81, 82)
Snail	Lymnaea sp	(81, 82)
Leech	Clepsine sp	(81, 82)
Larvae	Neuropteroid insects	(81, 82)
Rotifer	Rotifer elongatus Weber	(75)
Rotifer	Dinocharis tetractis Ehrbg	(75)

TABLE XIV

Animals occurring in from 25–54 meters of water:

Common Name	Scientific Name	Literature
Bivalve	Pisidium sp	(82)
Polyzoan	Paludicella ehrenbergii van Ben	(75)
Polyzoan	Fredericella sultana Blum	(75)

TABLE XV

Showing the recorded distribution of animals occurring in several of the vertical belts of Lake Michigan. The star indicates that the animal is present at the depth indicated at the head of the column in which the star occurs. B indicates that it breeds, and F that it feeds, at the indicated levels. The numbers in the column headed "Literature" refer to the Bibliography at the end of the book. The lower depth limit of many of the fishes listed is somewhat uncertain, as Milner does not indicate their exact distribution inside of 35 meters, but implies that they may occur at the depths indicated in the table. Other records bear out Milner's implications.

Common Name	Scientific Name	Depth in Meters						Literature
		0–8 m.	8–15 m.	15–25 m.	25–54 m.	54–70 m.	70 and Below	
Sturgeon	*Acipenser rubicundus* LeS.	B	F					(75, 81)
Crayfish	*Cambarus propinquus* Gir.	*	*	*				(75,p.15)
Crayfish	*Cambarus virilis* Hag.	*	*	*				(75,p.15)
Long-nosed gar	*Lepisosteus osseus* Linn.	*	*	*				(81, 84)
Lake catfish	*Ameiurus lacustris* Wal.	*	*	*				(81, 84)
Croaker	*Aplodinotus grunniens* Raf.	*	*	*				(84)
Perch	*Perca flavescens* Mitch.	*	*	*				(81, 84)
Wall-eyed pike	*Stizostedion vitreum* Mitch.	*	*	*				(81)
Large-mouthed black bass	*Micropterus salmoides* Lac.	*	*	*				(81)
Small-mouthed black bass	*Micropterus dolomieu* Lac.	*	*	*				
Northern moon-eye	*Hiodon alosoides* Raf.	*	*	*				(81, 84)
Toothed herring	*Hiodon tergisus* LeS.	*	*	*				(81, 84)
Tadpole cat	*Schilbeodes gyrinus* Mitch.	*	*	*				(81, 84)
Carp	*Carpiodes* sp.	*	*	*				(81)
Pike	*Esox lucius* Linn.	*	*	*				(81, 84)
Brook silverside	*Labidesthes sicculus* Cope.	*	*	*				(81)
Stickleback	*Eucalia inconstans* Kirt.	*	*	*				(75, 81)
Whitefish	*Coregonus clupeiformis* Mitch.	B	B	F				(75, 81)
Rock bass	*Ambloplites rupestris* Raf.	*	*	*				(81)
Amphipod	*Eucrangonyx gracilis* Smith		*	*				(80)
Snail	*Lymnaea lanceata* Gld.			*	*			(75, 80)
Long-jaw	*Argyrosomus prognathus* Smith				*			(75)
Lawyer	*Lota maculosa* LeS.	*	*	*	*	*		(75, 81)
Lake trout	*Cristivomer namaycush* Wal.	B	B	B	*	*	*	(75, 81)
Hoy's whitefish	*Argyrosomus hoyi* Gill (MSS)	*	?	?	*	*	*	(75, 81)
Amphipod	*Pontoporeia hoyi* Smith				*	*	*	(82, 75)
Schizopod	*Mysis relicta* Loven				*	*	*	(82, 75)
Blackfin	*Argyrosomus nigripinnis* Gill						*	(75, 81)
Small cottoid	*Triglopsis thompsoni* Gir.						*	(75, 81)

CHAPTER VI

ANIMAL COMMUNITIES OF STREAMS

I. Introduction

The conditions in streams from headwaters to mouth have many features in common with lakes, like Lake Michigan. It is therefore appropriate that they follow the discussion of such a lake. The streams belong to two drainage systems—the Mississippi and the Saint Lawrence. All are tributary either to Lake Michigan or to the Illinois River. The principal tributaries of the lake near Chicago are the Chicago River, the Calumet River, Trail Creek, the Galien River, the St. Joseph River, and the Black River. The principal tributaries of the Illinois River, with which we are concerned, are the Fox River, the DesPlaines River, the DuPage River, the Kankakee River, Salt Creek (Ill.), Hickory Creek.

The factors of greatest importance in governing the distribution of animals in streams are current and kind of bottom. They influence carbon dioxide, light, oxygen content, vegetation, etc.

These factors are controlled by age (physiographic), length of stream, and elevation of source above the mouth, all of which are physiographic. The typical stream begins as a gully and works its way into the land (Fig. 68, p. 112). The importance of some of the factors is greater in some stream stages than in others. For example, in the younger stages (*a*) material eroded, (*b*) relation to ground water, and (*c*) slope of stream bed play a more important rôle than they do in later stages.

II. Communities of Streams

1. Classification

The classification of stream communities is based upon physiographic history and physiographic conditions. In the early stages of stream development there are two types to be distinguished: (*a*) the communities of intermittent streams, and (*b*) spring-fed streams. As soon as the intermittent stream cuts below the ground-water level, it becomes much like the spring-fed stream. Permanent streams are divided into brooks, swift and moderate, and rivers, sluggish and moderate, with communities named accordingly. We undertake a discussion, first, of the history of the communities of streams developing in materials easily weathered and eroded, containing bowlders, gravel, and occasional strata of hard rock.

2. THE INTERMITTENT STREAM COMMUNITIES

(Stations 4–8; Tables XVII, XVIII)

There are two types of these—intermittent rapids and pool communities.

An Intermittent Stream

Fig. 24.—The young stream at Glencoe in spring at high water, showing the leaf-barren trees.

Fig. 25.—The same in summer, showing the stream entirely dry.

a) *Temporary rapids consocies* (Figs. 24, 25).—Small gullies in which water runs only when it is raining do not have any aquatic residents. As soon as such a gully has cut a channel deep enough to stand below ground-water level during a few days or weeks of the rainy season, aquatic insects make their appearance. The species which is usually found in the smallest trickle of water is the larva of the black fly, *Simulium* (Figs. 27–32). As the stream grows a little larger, and per-haps even at such a young stage also, we sometimes find the nymphs

of May-flies.   Such streams have, however, no permanent aquatic residents.   These aquatic forms are not aquatic during their entire lives.
They require water only during their early stages.   If the water is
running at the time the female is ready to deposit eggs and if she is
properly stimulated by the conditions, she deposits them without regard
to future conditions.   If the wet weather continues long enough, the
larvae will mature and the other adults will appear, otherwise they die.
This type of animals continues after the stream becomes large enough

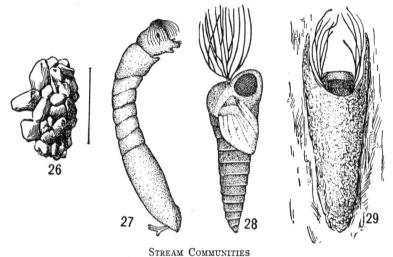

STREAM COMMUNITIES

FIG. 26.—The pupal case of one of the caddis-worms (*Rhyacophila*) from the
rapids of the temporary stream at Glencoe; enlarged as indicated (original).

FIG. 27.—The larva of the black fly (*Simulium*); about 15 times natural size
(after Lugger).

FIG. 28.—Pupa of the same (after Lugger).

FIG. 29.—Pupa of the same in the pupal case (original).

to have permanent pools.   At such a stage the number of species is
increased, but no two collections are alike (see Table XVII).   Clinging
to the upper surface of the stones are black-fly larvae, caddis-worms
(*Rhyacophilidae*) (Fig. 26); under stones, May-fly nymphs, those collected as different times often belonging to different species.   On some
occasions there are great numbers of unidentifiable dipterous larvae
and caddis-worms without gills or cases.   Such a stream may possess
any or all of these on one occasion, and none or only a few of them on
another.

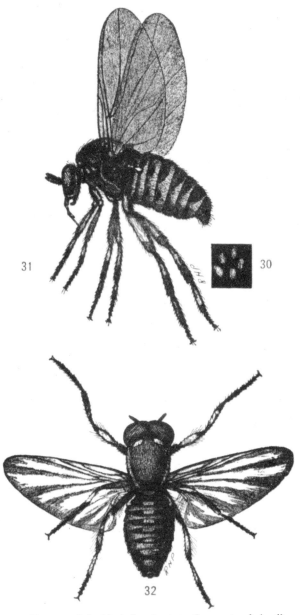

FIG. 30.—The eggs of the black fly, about 15 times natural size (from Williston after Lugger). FIG. 31.—Side view of the adult fly (from Williston after Lugger). FIG. 32.—The same from above (from Williston after Lugger).

b) *Temporary pool consocies.*—As a young stream grows deeper it often reaches some depression or marsh at its headwaters of which it forms the outlet in the early spring. It is now permanent for a longer period each season of normal rainfall, and small pools usually alternate with the rapids just described. In these pools aquatic insects, crustaceans, and snails which belong primarily to stagnant ponds make their appearance. The first resident species are the crayfishes. They are found in the pools in the early spring when the water is high. The drying of the stream calls forth behavior suited to the conditions, and in summer their *burrows* are common in the stream bed. They come out at night and are preyed upon by raccoons, the tracks of which are commonly seen.

c) *The horned dace, or permanent pool communities.*—The first permanent parts are permanent pools. In these, conditions such as current, sediment, oxygen content, etc., are intermittent or spasmodic. The current in the rapids is distinctly spasmodic and conditions in these rapids are similar to those in the stream before even temporary pools were developed. Streams with permanent pools are represented in the Chicago region by many which enter the lake where high bluffs are present. County Line Creek (Figs. 24, 25) has been studied as an illustration of this type (Table XVII).

The larger pools possess a practically permanent fauna. The characteristic forms are the crayfishes (*Cambarus virilis* and *propinquus*). The young are to be found in the pools at all seasons of the year. Water-striders, back-swimmers, and water-boatmen are common. Occasionally one finds dragon-fly nymphs (*Aeshna constricta* and *Cordulegaster obliquus*), dytiscid beetles (*Hydroporus* and *Agabus*), crane-fly larvae, the brook amphipod (*Gammarus fasciatus*), and the brook *mores* of the sow-bug (*Asellus communis*) (Fig. 55, p. 98). These are common among the lodged leaves. They move against water current.

The species of fish (Table XVIII) which is most commonly found in the smallest streams (92) and nearest the headwaters of the larger streams is the horned dace or creek chub (*Semotilus atromaculatus*) (Figs. 33, 34). It possesses certain noteworthy physiological characters. Like many other species of fish, it goes farthest upstream for breeding (50). Its nest is made of pebbles. Often after the breeding season is over, and the adults have gone downstream, the water lowers so that young fishes are left in large numbers in small drying pools. Here they swim about, with their mouths at the top of the water, which is constantly being stirred up by the many tails, and which often contains much blackened, oxygen-consuming excreta and decaying plant materials. This would cause death to less hardy fishes. Allee (53) found very little oxygen in the waters of such pools. As it is, the pools often dry up, and the fish die. The second fish to enter a small stream appears to have many of the characters of the first. It is usually the red-bellied dace (*Chrosomus erythrogaster*), which breeds on sandy or gravelly bottom (93) but tolerates standing water, being found also in some of the stagnant ponds at the south end of Lake Michigan. In some streams, the black-nosed dace (*Rhinichthys atronasus*) (Fig. 35) is second from the source. These fishes go against the current, but avoid the places where it is most violent.

 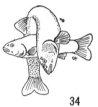

33   34

BREEDING HABITS OF A PIONEER STREAM FISH

FIG. 33.—Showing, in longitudinal section, the nest of a horned dace (*Semotilus atromaculatus*), with male and female fish in the nest. The stream flows in the direction indicated by the arrow at the upper left-hand corner of the picture; ⅛ natural size (after Reighard).

FIG. 34.—Male and female horned dace during the spawning act. Each time the male clasps the female she deposits 25 to 50 eggs in the nest. Note pearl organs on the head of the male (after Reighard).

This one also breeds on gravel bottom, and can withstand the stagnant conditions of the summer pools.

As the stream lowers its bed, this type of formation passes gradually into a later one. The beginning of the succeeding formation is heralded by the coming of the Johnny darter (*Boleosoma nigrum*), the common sucker (*Catostomus commersonii*) (Fig. 36), and the blunt-nosed minnow (*Pimephales notatus*) (Fig. 37) (79).

d) Characters of the communities.—The intermittent-stream communities are made up of animals which are dependent upon water during only a part of their lives and which possess a means of attachment and move against current (94) (positive rheotaxis). The pool communities are made up of animals tolerating great extremes of conditions and being also positively rheotactic. The fish are able to meet the current and to withstand the conditions of the stagnant pools. The crayfishes live in the water in the spring and burrow in the

PIONEER STREAM FISHES

FIG. 35.—Black-nosed dace (*Rhinichthys atronasus*) (from Forbes and Richardson).

FIG. 36.—Common sucker (*Catostomus commersonii*); length 18 in. (from Meek and Hildebrand after Forbes and Richardson).

FIG. 37.—Blunt-nosed minnow (*Pimephales notatus*); length 2 to 3½ in. (from Forbes and Richardson).

dry weather; adults of the aquatic insects creep into moist places when the stream dries. Allee (53) has found that isopods are positively rheotactic and that they can be acclimated to extreme conditions.

3. SPRING BROOK COMMUNITIES

(Stations 10 and 11; Table XIX)

In glaciated areas many of the streams are fed by springs which have not been produced by erosion, but are the result of porous and impervious layers of till arranged as in regions possessing artesian wells. The presence or absence and numbers of animals in a spring depend largely upon the chemical content of its water. Spring waters commonly have insufficient oxygen to support animals and at the same time may contain sufficient nitrogen and carbon dioxide to be detrimental if not fatal to animals. The mineral matter in solution may be large in quantity and in some cases poisonous also. As the water flows away from the spring it becomes aerated and diluted with surface water so that the animals of the spring brook can live in it. *Spring consocies* differ in different springs because of variations in the character of the water.

In an area where there are springs, they are usually numerous. The little brooks unite to form larger streams. Typically, such streams may not be larger than intermittent streams, but a nearly constant flow at all times of the year is one of the characteristic conditions. Pools and riffles are not so well defined, but contain some small fishes. The watercress grows abundantly at the sides of the stream and affords a lodging-place for aquatic animals not furnished so abundantly by young streams of other types. The water is colder in summer and warmer in winter than in other streams.

Spring brook associations.—Among the watercress are the amphipods (*Gammarus fasciatus*), the larvae of *Simulium* attached to the leaves, beetles, dragon-fly nymphs, and young crayfishes. Here are also found occasional snails (*Physa gyrina*). The species of the cress association are nearly all found under stones or on stones in the riffles. On the stones are *Simulium* larvae and *Hydropsyche* (95), the net-building caddis-worm (Figs. 39, 40, p. 96). Under the stones are the nymphs of the May-fly (*Baetis* and *Heptagenia*), the larvae of flies and midges (*Chironomus, Dixa,* and *Tanypus*), the brook beetles (*Elmis fastiditus*) (Fig. 47, p. 98), and occasional amphipods and crayfishes.

4. THE SWIFT-STREAM COMMUNITIES

As the spring brooks and the intermittent streams continue to erode their beds, they increase the extent of their drainage systems and become larger streams. Springs tend to disappear in connection with the spring brook and the intermittent stream reaches the ground-water level and becomes permanent. The two sets of conditions converge toward the *larger swift stream* (Fig. 38). While the conditions in these are like those of the spring brook, the watercress is absent and there are few rooted plants. Pools and riffles are well developed and the flow of water is constant, but fluctuates in volume. These streams differ in size, but the formation *mores* are practically the same, although larger species commonly inhabit the larger stream.

a) Pelagic sub-formation is very poorly developed in the smaller streams and will be discussed in connection with sluggish streams.

Fig. 38.—The permanent swift stream showing the stones in the rapids, and the stiller places below (New Lenox, Ill., Gaugars Station) (original).

b) Hydropsyche or rapids formations (Stations 14, 15, 17, 19, 20, 21; Tables **XX, XXI, XXII**).—These are usually due to the presence of coarse material or an outcrop of rock. They are typical in streams with large bowlders and stones of all sizes. Here current is probably the controlling factor. In these streams, we find the best expression of the riffle formation, which we have seen is poorly developed in the smaller streams. This formation includes three ecologically equivalent modes of life, each meeting the current in a different way. These are (i) clinging to stones in the current, (ii) avoiding the current by creeping under stones, (iii) self-maintenance by strong swimming powers.

Upper surface of stones (stratum 1): Here again we find the black-fly larvae, particularly in the smaller streams. They are provided at the posterior end of the body with a sucker surrounded with hooks (Figs. 27–32). The salivary glands are, as is common in insects, modified into silk glands and the silk is of such a nature that when it is brought into contact with a stone it adheres. The animals are usually found attached to the rock by the sucker, with the head downstream. The fans are extended and serve to catch diatoms and other floating algae. If for any reason the sucker gives way, the animal starts to float downstream. If the mouth can be brought into contact with a stone, the silk is exuded and the animal is held until it can make the sucker fast again. The pupae of this fly are also attached to the stones. They are surrounded with a cocoon. We have removed them from the stream and have found that they *cannot make this cocoon in the absence of the current, but make a shapeless tangle* instead. The adults deposit their eggs at the sides of the streams (96).

On the tops of stones caddis-worms (*Hydropsyche* sp.) usually have cases made of pebbles stuck together with silk (Figs. 39, 40). They also have a net for catching floating food. The net faces the current (usually upstream) (Fig. 40). The river snail (*Goniobasis livescens*) (Fig. 54) is common on the upper surfaces of the larger rocks and is distinguished by a strong adhesive foot. These snails are usually headed upstream. When placed in a long piece of eave-trough into which the tap water was running at one end, they nearly all made their way to the upper end within a short time. *They are ecologically equivalent to the caddis-worms and the black-fly larvae.*

Among the stones (stratum 2): Of the animals living among stones, the darters are most important. Of these the banded darter (*Etheostoma zonale*) (Fig. 44), the fan-tailed darter (*E. flabellare*), and the rainbow darter (*E. coeruleum*) (97) (Fig. 45) live among and under the stones or in the algae which cover the rocks (especially the fantail). With them are sometimes found the Johnny darter (*Boleosoma nigrum*), the black-sided darter (*Hadropterus aspro*) (Fig. 46), and the small bullhead or stonecat (*Schilbeodes exilis*). These fish are all positively rheotactic. They apparently orient because of unequal pressure on the two sides of the body when it is not parallel with the direction of the current.

Under the stones (stratum 3): There are many more forms living under and among the stones than on the tops of them. Here are the

May-fly nymphs, the flattened *Heptageninae*, and the very awkward damselfly nymph, *Argia*, evidently succeeding well together. This fact makes the value of the flattening as an adaptation appear nil. There are also the larvae of midges (*Chironomus* sp.) (98) and of horse-flies (*Tabanus*) (Figs. 51, 52). The adults of the latter deposit their eggs in great masses on the tops of the stones which protrude from the water. The stone-fly nymphs, similar to the *Heptageninae* May-fly

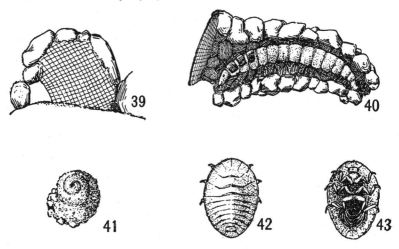

REPRESENTATIVE AQUATIC INSECTS OF A RAPIDS COMMUNITY

FIG. 39.—The net of the brook caddis-worm (*Hydropsyche*) seen from the front. Drawn from a specimen which made its case against the side of an aquarium (original).

FIG. 40.—The same in its case with the net adjoining the opening which faces upstream (original).

FIG. 41.—The larva of a caddis-fly (*Helicopsyche*) with a case made from pebbles, in the form of a spiral; 2½ times natural size (original).

FIGS. 42, 43.—The water-penny larva of the brook beetle (*Parnidae*) seen from above and below (43); 2½ times natural size (original).

nymphs in form and appearance, are found here also. Perhaps the most bizarre of all are the water-pennies. These are round flat objects adhering to the under sides of stones, and not looking like animals at all. They are the larvae of a parnid beetle (*Psephenus*). Figs. 42 and 43 show two views of a larva. The old larval back becomes the cover for the pupa. The adults live under the stones also and their general appearance is like that of the parnid in Fig. 47. Sessile or attached animals are common in the brooks, but their numbers vary greatly from year to year. On one occasion the surface of the rocks and stones in Thorn Creek was almost covered with sponge, and while some sponge is always to be found, we have not seen it so abundant again. Polyzoa

44

45

46

REPRESENTATIVE FISHES OF A RAPIDS COMMUNITY

FIG. 44.—The banded darter (*Etheostoma zonale*); length 2 in. (from Forbes).

FIG. 45.—The rainbow darter (*Etheostoma coeruleum*); length 2 in. (from Forbes).

FIG. 46.—Black-sided darter (*Hadropterus aspro*); length 3-4 in. (from Forbes).

are usually present under the stones. Such animals depend upon foods in solution and small floating plants and animals.

In addition to those rapids which have large rocks, are those in which the bottom is of coarse sand and gravel, with only a few small stones.

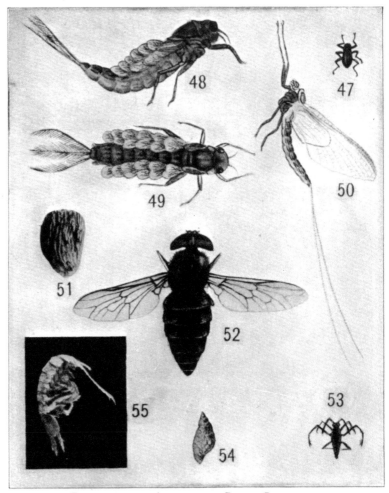

REPRESENTATIVE ANIMALS OF A RAPIDS COMMUNITY

FIG. 47.—An adult brook beetle (*Parnidae*); twice natural size (original).

FIGS. 48–50.—Different views of the nymph and adult of the May-fly (*Siphlurus alternatus*); 3½ times natural size (after Needham).

FIG. 51.—The eggs of a tabanid fly taken from a protruding stone; twice natural size (original). FIG. 52.—Adult fly.

FIG. 53.—A water-strider (*Rhagovelia collaris*), from the margin of the swift brook (New Lenox, Gaugars); twice natural size.

FIG. 54.—The common river snail (*Goniobasis livescens*), covered with calcium carbonate secreted by algae; natural size (original).

FIG. 55.—An intermittent stream sowbug (*Asellus communis*); twice natural size (original).

Here we find the caddis-worm (*Helicopsyche*) (Fig. 41, p. 96), which has a spiral case made of sand grains. These are most abundant where some sand and swift current are both found. There is from time to time some vegetation in such situations and on it we find the brook damsel-fly nymph (*Calopteryx maculata*), the adult of which is the black-winged damsel-fly.

Characters of the formation: The swift-stream formation has a striking behavior character, namely, strong positive rheotaxis. Other physiological characters, such as the toleration of only low temperatures and high oxygen content, and the necessity for current for the successful carrying-on of their building operations, are probably common to the animals. So far as the fishes of the rapids are known, they breed on coarse gravel bottom or under stones. The *mores* of the formation are, then, current resisting and current requiring, dependent upon large stones or rock bottom for holdfast and building materials.

c) *Sandy and gravelly bottom formation* (*pools*) (Stations 15–22; Tables XVII–XXV).—The pools of streams with characteristic formations are usually 2 or 3 to 10 feet deep, depending upon the size of the stream. The bottom is sand or coarse gravel. In these we find conditions very different from those in the rapids. The pools are the home of the rock bass (*Ambloplites rupestris*), the small-mouthed black bass (*Micropterus dolomieu*), the sunfishes (*Lepomis pallidus* and *megalotis*), and the perch (*Perca flavescens*), together with a number of interesting small fishes whose distribution is shown in Tables XXI and XXII (79, 92).

With these are also the mussels (91), frequently as many as nine or ten species, among which are *Lampsilis luteola*, *ventricosa*, and *ligamentina*, the little *Alasmidonta calceola* (Figs. 57, 58), and *Anodontoides ferussacianus* (Figs. 59, 60), the last-named being perhaps the most characteristic of them all. They are often found beneath the roots of willows along the sides of the pools. Mr. Isely found that mussels migrate to shallow water during flood time. Mussels are dependent upon fish for a part of their lives. The young are carried by the adult until ready to attach to the body of the fish (99). When they leave the fish they are able to take care of themselves. Burrowing in the gravel are bloodworms (*Chironomus* sp.) (95, 98), the burrowing dragon-fly nymph (*Gomphus exilis*), a burrowing May-fly (Fig. 64*a*, p. 107), a caddis-worm, and occasionally snails, *Campeloma* (Fig. 61 or 64*c*) and *Pleurocera* (Fig. 64*d*). There are a few plants that grow on the sandy bottom in such places, and among these one finds the snail (*Amnicola limosa*),

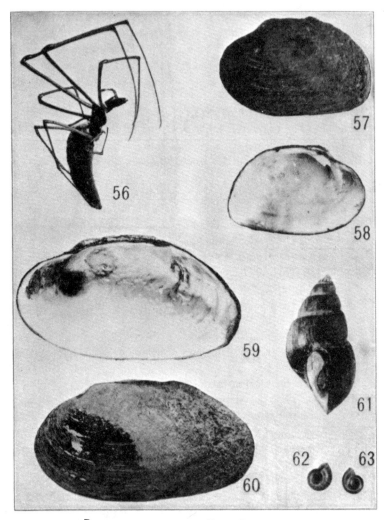

REPRESENTATIVES OF THE POOL COMMUNITY

FIG. 56.—A long-legged spider taken from a stone out of water in a stream (*Tetragnatha grallator*); twice natural size (original). FIG. 57.—Outside of shell of a small mussel from Hickory Creek (*Alasmidonta calceola*); natural size (original). FIG. 58.—Inside of the same. FIG. 59.—Inside of shell of mussel from Hickory Creek (*Anodontoides ferussacianus*, subspecies *subcylindraceus* Lea); natural size (original). FIG. 60.—Outside of the same. FIG. 61.—A snail from the still water of Thorn Creek (*Campeloma subsolidum*); natural size (original). FIG. 62.—A snail from the still water of Hickory Creek (*Planorbis bicarinatus*), seen from the left; natural size (original). FIG. 63.—The same seen from the right.

May-fly nymphs (*Siphlurus*) (Figs. 48–50), and hair-worms (*Gordius*). In some localities bivalved mollusks (*Sphaeridae*) and leeches are numerous. Under primeval conditions beavers are associated with the pool formation. They build dams which contribute to the deepening of the water of the pools. For a good account of their habits see citation 99b. An old beaver dam is supposed to have turned the waters of the DesPlaines out of the Chicago River and down the Chicago outlet.

Characters of the formation: The *mores* of the pool formation are distinctly those of partially burying the body just beneath the surface of the fine gravel and moving against the current. The few animals that make cases usually use gravel or sand grains. A single caddis-worm makes its case from small sticks such as commonly lodge in eddies. Some of the fishes breeding in these situations cover their eggs (50). Some fishes orient the body and swim upstream as a result of seeing the bottom apparently move forward below as the fish floats down (94). They behave the same if put into a trough with a glass bottom and the trough drawn forward. Some orient also when their bodies rub against the bottom when floating downstream.

5. THE COMMUNITIES OF SANDY BOTTOMED STREAMS (SHIFTING BOTTOM SUB-FORMATIONS)

(Stations 22–26; Table XXIV)

We have studied the upper course of the Black River, the upper course of the Calumet River, and the Deep River, and two or three tributaries of Lake Michigan near South Haven. The kind of material eroded is of the greatest importance in determining the *mores* present in a stream. The streams of the eastern part of our area are in till which is sandy and their bottoms are sandy. This material is always slipping and moving downstream. There are few large stones. The bottom is not suitable for animals. The swift-water animals are almost entirely absent. The forms present are those which belong to moderately swift water.

Composition and subdivisions.—Such streams are poorly populated. Their *mores* resemble those of the formations of the pools of streams eroding coarse material, but the shifting is so much more general and the species found so different, that it has been thought wise to separate the two. In the Michigan streams there are in summer a few scattered plants, which support a considerable number of insects; some of the brook beetles (*Parnidae*) are found attached to them. The logs and roots that happen to be in the water are important; they are the only places that support any amount of life. From these logs I have taken hundreds of specimens of small *Parnidae*, and with them predaceous diving beetles (*Dytiscidae*) which were found hiding in the cracks, also a few scattered caddis-worms (*Hydropsyche*). The fauna of the bottom is made up of burrowing and semi-burrowing forms. The little dytiscid (*Hydroporus mellitus* Lec.) (99c) is characteristic: it has the habit of burying itself in the sand. The bivalved mollusks, especially mussels, are present. From the Deep River (upper course) we have taken nearly a dozen species. The only snail found is a burrowing form also.

Animals of such a stream are subject to severe conditions. Many of them burrow. The substratum is very unstable and the logs and parts of trees to which many of them are attached are free to float downstream with every flood. We know nothing of the reactions of these animals to various stimuli. They are distinctly subjects for investigation.

6. THE SLUGGISH STREAM COMMUNITIES

(Stations 19, 27, 28, and 29; Tables XVII, XVIII, XX–XXV)

There are several phases or types of sluggish stream formations. The most important of these are the *sluggish or base-level creek*, the *sluggish river*, and the *drowned river*. These are all illustrated in the Chicago area.

The sluggish creek type is illustrated by the west branch of the DuPage River and its tributaries; the upper course of the west branch of Hickory Creek, Dune Creek, some parts of the Little Calumet south of Millers, and the Kankakee and some of its tributaries.

The sluggish rivers are the Upper Fox, the lower St. Joseph, the Grand Calumet, the lower Galien, the lower Black, and others. These constitute a group of streams representative of the sluggish type about the Great Lakes.

a) Sluggish creek sub-formations (Stations 16, 18).—The west branch of Hickory Creek has been studied in a cursory manner. The fish are a strange mixture of semi-temporary stream and *pond* forms. The black bullhead (*Ameiurus melas*) (79) is probably the most characteristic fish. The golden shiner (*Abramis crysoleucas*) and sunfish (*Lepomis cyanellus*) are also found.

Baker (100) studied the upper portion of the east-north Chicago River. He recorded the same species of *Mollusca* as were taken in the upper part of Hickory Creek. He records also the black bullhead. The insects which he mentions are those commonly found in ponds. This community is distinctly of the pond type in its general *mores*. Stagnation and low oxygen content and the partial drying of the stream are tolerated by all the residents.

b) *Sluggish river formations.*--The conditions in sluggish rivers are different from those in smaller swift streams in many respects. The bottom is for the most part of fine materials; there are no rocks. The difference between pools and rapids no longer exists. The river is a gently flowing mass with relatively little distinction as to different parts. The margins of such streams are lined in summer with typical rooted and holdfast aquatic plants. The small bays and out-of-the-way spots, out of the current, support bulrushes and sometimes cattails. We can distinguish several formations in the Fox River: (1) The pelagic formation, (2) the formation of sand and silt bottom (association of sandy bottom where the current drags in midstream or beats against the shore; association of silt bottom where least current is present), and (3) the formation of the zone of vegetation.

Pelagic formation: This is well developed in the larger rivers, e.g., the Illinois River (77). While the Illinois no doubt differs from the Fox in many respects, doubtless the general features are much the same. It does not differ greatly from that of Lake Michigan.

Burrowing May-fly or sand and silt bottom formations: On the bottom in ten feet of water we have found mussels (*Anodonta grandis* and *Quadrula undulata*), the snail (*Goniobasis livescens*), bloodworms (*Chironomidae*), green midge larvae (*Chironomidae*). On the old mussel shells were large colonies of the bryozoan *Plumatella* and occasional caddis-worms (*Hydropsyche*) (Figs. 39, 40, p. 96). On sandy bottom, conditions near the margin are similar to those on the bottom. We find here also an occasional snail (*Goniobasis, Pleurocera,* and *Campeloma*), the midge larvae and bloodworms, occasional burrowing May-fly nymphs, and a number of mussels (*Unio gibbosus* and *Quadrula rubiginosa* being the most characteristic). There is also an occasional specimen of the long-legged dragon-fly nymph (*Macromia taeniolata*) and the black-sided darter. A considerable number of these species occur in the stillest pools of Hickory Creek, indicating the types that will dominate later. Silt is often found in particular spots. The most characteristic animals in this are the large mussel (*Quadrula undulata*), the burrowing May-fly nymph (*Hexagenia* sp.), and the bloodworms (*Chironomidae*). There are also the worms (*Annelida*) which burrow in the mud and protrude their posterior ends, often also the common mussel (*Lampsilis luteola*), the *Sphaeridae*, and the mud leech (*Haemopis grandis*). All of the animals of the silt formation burrow and probably require little oxygen.

Planorbis bicarinatus formation, or formation of the vegetation: Here we have for the first time the conditions which we find in ponds— a dense rooted vegetation. With such a growth of vegetation we have a very different fauna: a large number of aquatic insects and pulmonate (lunged) snails. Of these there are a considerable number of species which must come to the surface for air, both in the adult and the young stages. The most important of these are the bugs: water scorpions (*Ranatra fusca*), the creeping water-bugs (*Pelocoris femoratus*), the small water-bug (*Zaitha fluminea*), the water-boatmen (*Corixa* sp.), the still-water brook beetles or parnids (*Elmis quadrinotatus*), several species of predaceous diving beetles (*Dytiscidae*) (99c), and water scavengers (*Hydrophilidae*). The pulmonate snails are *Physa integra, Planorbis bicarinatus* (Figs. 62, 63), and often species of *Lymnaea*.

Where the bottom is not too soft we often find numbers of viviparous snails (*Campeloma*) and an occasional mussel (*Anodonta grandis*). The crustaceans are distinctly clear-water forms: the crayfish (*Cambarus propinquus*) (101), the amphipod (*Hyalella knickerbockeri*), and the brook amphipod (*Gammarus fasciatus*) (102).

The gilled aquatic insects are the May-fly nymphs (*Caenis* and *Callibaetis* sp.) and the damsel-fly nymphs (*Ischnura verticalis*) and dragon-fly nymphs (*Aeschnidae* and *Libellulidae*). To practically all of these the vegetation is necessary as a resting-place or clinging-place, or a place to enable them to creep to the surface to shed the larval skin and become adult.

Variations of the formation: The Fox is fairly representative of base-level rivers beyond the reach of tide-water except perhaps that the presence of gravel and sand in this stream may not seem fully in accord with this statement. There are, as has been noted, rivers near Chicago in which these conditions, which go along with old age in a stream, are still more marked. The lower Deep River is perhaps a good example of this. It is very sluggish and the bottom in the vicinity of Liverpool, Ind., is, so far as we have been able to ascertain, entirely covered with silt, with considerable humus mixed with it. The margins are peaty. The Calumet and the lower Black are similar. In these, sand and gravel areas, and animals which inhabit them, are reduced to a minimum and the silt and vegetation associations are better developed.

Characters of the formation: The vegetation formation is distinct and clearly marked off from all others. The animals are dependent upon the vegetation for support. The adult aquatic insects must creep to the surface of the water to renew their air. The forms that have gills are, at least many of them, dependent upon the vegetation for crawling to the surface to molt the old skin. The crustaceans are forms that cling to the vegetation and the snails must come to the surface for air. Doubtless this formation should be divided into strata, but our data do not justify such division.

III. SPECIAL STREAM PROBLEMS (103, 92)

The first special problem is that of the relations of animals to seasonal changes, to changes in volume of water, amount of silt, shifting of bottom materials, and the seasonal aspects of the vegetation. The second problem of streams is the historic or genetic, which includes the phenomena of the origin of the animals of the stream, their mode of entrance, and the effect of rejuvenation, drowning, etc.

I. SEASONAL CHANGES

Streams are more strikingly affected by rainfall and drought than are any other of the aquatic habitats. In extremely dry years streams dry up in the rapids where they have perhaps not been dry for a century. Floods change all the landmarks of the stream bottom and often scatter the animals of the stream over the flood-plain.

a) Floods.—We found at the side of the high bank of the stream where the water is quiet at low water, the Johnny darter (*Boleosoma nigrum*), the little pickerel (*Esox vermiculatus*), the tadpole cat (*Schilbeodes gyrinus*), the crayfish (*Cambarus virilis*), and an occasional *Hydropsyche*. Here were also an occasional sphaerid mollusk and one or two leeches.

Caught in a mass of driftwood behind the roots of a tree were casebearing caddis-worms (*Phryganeidae*), the black-winged damsel-fly nymph (*Calopteryx maculata*), the larvae of the black fly (*Similium* sp.), and two species of May-fly nymphs (one *Heptageninae*). The last two belong to the swift water, the others to the still water or the pools. During floods the still-water fauna and the swift-water fauna become mixed in the still places.

At the time of our study there was a growth of rank weeds on the flood-plain. While the stream had been swollen for a long period and had stood higher than at the time of observation, little or no invasion of these weeds by aquatic animals had occurred. Animals evidently react negatively to such bottom and vegetation.

We have had but little opportunity to study the swift-water formation during floods, though some of the riffles in Butterfield Creek have been studied when the stream was bank full, but no marked changes were noted. It is obvious that the extreme floods which move large stones crush large numbers of swift-water animals.

b) Droughts.—There was an unusual drought in the autumn of 1908. The data on the distribution of fishes in Glencoe Brook and County Line Creek were collected before this date (Fig. 67, p. 111). Table XVI shows the arrangement after the drought.

TABLE XVI

SHOWING THE EFFECT OF DROUGHT ON FISHES

The localities 1, 2, 3, 4 are indicated on the maps of the North-Shore Streams (Fig. 67, p. 111). P=before drought. *P=after drought.

Name of Stream and Common Name of Fish	Scientific Name	1	2	3	4
Glencoe Brook............	*Semotilus atromaculatus*..	P	......	......	......
County Line Creek					
Horned dace	*Semotilus atromaculatus*...	P	P	*P	*P
Black-nosed dace	*Rhinichthys atronasus*....	......	......	P	*P
Common sucker........	*Catostomus commersonii* ..	......	......	......	*P

County Line Creek was entirely dry except the pool nearest its mouth in September, 1908. This is locality 4 in Fig. 67, p. 111. The following spring was one of normal rainfall. The fish *proceeded upstream a distance of only three rods.* This partially restored the usual arrangement. If this represents the rate, the fish proceed upstream slowly. Glencoe Brook has not recovered its fish.

As evidence of upstream migration of *Mollusca*, the following seems to be important. Frequent examination of a section of the North Branch of the Chicago River at Edgebrook, between 1903 and 1907, showed that *Pleurocera elevatum* and *Campeloma* occur in this stream. *Pleurocera* was not found during this period (ending November, 1907) above a certain point. *Campeloma* was found only sparingly above this point. The spring of 1908 was one of heavy rainfall and the streams were in flood from April to June. On July 6 the snail *Pleurocera* was found in numbers one-fourth of a mile farther upstream than formerly. *Campeloma* had gone nearly as far. The season from November to April was not different from other seasons and there is no reason to assume that the migration began before the spring floods. If this is true the snails could make their way toward

the headwaters at the rate of at least a mile per year, if they were introduced into a large stream.   This must be a response to both water
pressure and current.   The small value of such single observations is
recognized but they are presented here because the opportunity to secure
such data is small.   In this river there are also notable relations between
especially dry seasons and the distribution of other animals.   The
season in which the riffles were dry (October 31, 1907) the pools presented

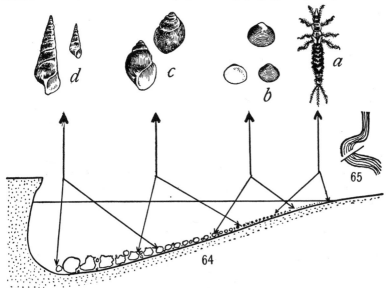

THE TRANSVERSE DISTRIBUTION OF STREAM ANIMALS

FIG. 64.—Shows the form of bottom and size of bottom materials in a cross-
section of the North Branch of the Chicago River. *a–d*, natural size (original).

a, a burrowing May-fly nymph (*Hexagenia* sp.).
b, small bivalve (*Sphaerium stamineum*), two individuals, two views.
c, viviparous snail (*Campeloma integrum*), seen from two sides.
d, the long river snail, young and full grown (*Pleurocera elevatum*).

FIG. 65.—Cross-section of the stream with reference to a curve.

an unusual aspect.   The standing pools were choked with water-net.
The minuter forms, such as protozoa and flatworms, were present in the
greatest profusion.   *Hydra* was abundant.   All this is in marked contrast to the conditions which one finds when the stream is running.
The season following the dry riffles, we found small *Hydropsyche*
larvae, and a few young stone-fly nymphs.   *The only forms present were
those that could be introduced by terrestrial, egg-laying females.*

In the autumn of 1906 Professor Child found that the May-fly and stone-fly nymphs were not present in the riffles but were present in the moderately swift and more quiet parts below. The spring of 1906 was a dry spring and the females probably laid their eggs in the moderately swift instead of the preferred swift water. The distribution is determined by the conditions *at the time of egg laying.*

We note that even in the larger streams the weather conditions affect the presence and absence and abundance of animals. The *mores*, however, remain essentially the same.

2. TRANSVERSE STUDIES

Cross-section studies of streams are of interest as showing a horizontal arrangement of forms belonging properly to different formations. This is best illustrated in the cross-sections of curves where there is a horizontal gradation of current and in the size of material of the bed. Figs. 64 and 65 illustrate this. The burrowing May-fly nymph, belonging to the silt, is in the finest materials of the inside of the curve; passing toward the center of the stream we next encounter the sphaerid (*Sphaerium*) and a little farther in the snail (*Campeloma integrum*), with it often mussels (*Anodontoides ferussacianus*); and still farther into the stream we find, clinging to the larger stones, the long snail (*Pleurocera elevatum*). While depth of water may be a factor here, the size of bottom material is of first importance.

3. LONGITUDINAL STUDIES

(Figs. 66, 67, 68, 69)

If one passes from the headwaters of a stream to its mouth, he will usually find either the spring brook formation or the intermittent formation in the upper course, the swift-water formations in the middle course, and the sluggish stream or river formations in the lower course. There are very numerous variations of this and several of them deserve comment. Large streams with a large drainage area and much sediment, and with much of the upper part in a young stage, are subjected to many changes in the lower courses, such as silting-up at the end of the flood periods and washing out later. *This often prevents the development of the vegetation formation and favors the shifting sand and gravel formations.*

a) Rejuvenation, ponding, and retarding of erosion.—Streams are often dammed by some obstruction in their mid course, or erosion is checked at a point by a hard stratum, or the stream which has reached base-level is rejuvenated by a lowering of the water level at the mouth.

The obstruction of the hard layer encountered always produces local swift water. Above this the water may be sluggish and the area reduced to the general level of the obstruction. In the case of rejuvenation the head of erosion proceeds upstream; the part of the stream above the point to which erosion has reached is sluggish and is sometimes called the pre-erosion stream.

Of the rivers and creeks which we have considered, nearly all the larger ones are sluggish or pre-erosion in their upper courses. This is true of the DesPlaines, which is held in this condition largely by rock at Riverside. Hickory Creek (Fig. 66) is also of this type, the head of erosion being at Marley. In passing from source to mouth of such a stream we find formations arranged as follows: In the upper sluggish courses of all the streams mentioned we find (1) sluggish creek or river formations, (2) chiefly swift-water formations below the sluggish, (3) chiefly gravel bottom formations below the swift-water formation,

Fɪɢ. 66.—Diagrammatic profile of Hickory Creek: *A*, source; *B*, mouth; *C*, head of erosion; *D*, rock outcrop. The figures below refer to the columns in Table XXI and represent parts from which fish were collected.

and (4) typical sluggish river formations farthest downstream where the vegetation, silt, and sand formations are arranged much as in the Fox River.

Tables XVIII, XXI, and XXII and Figs. 67–69 show the longitudinal distribution of fishes in six streams. A few moments' study and comparison of these tables will make the following facts evident:

a) The only species in the youngest stream of the North Shore series is at the headwaters of all the others.

b) The species found in County Line Creek are found in the same order in the upper courses of Pettibone Creek and Bull Creek; additional species are found farther downstream in the larger streams.

c) The same species are at the headwaters of Thorn-Butterfield and Hickory creeks and in the upper courses of the North Shore streams. Other species are with them. The species of the North Shore streams are crowded together in these large streams which have permanent deeper water at their sources (due to springs) and in which the graded series of conditions found in the North Shore streams is wanting.

d) The swift-water fishes begin markedly at the head of erosion in Hickory Creek.

e) The fish communities differ as to species where the conditions are very similar, for example, in Thorn-Butterfield and Hickory creeks. The general habits of the fishes are the same.

f) Larger fishes are found in the larger water course and in the downstream portions of the smaller streams.

g) Fish, when entering a stream, go upstream to a point suited to their physiological constitution, regardless of its physiographic mode of origin.

4. GENETIC ECOLOGY OF STREAMS

Several years ago Adams (103) pointed out that the dispersal of aquatic animals is determined by the shifting backward of the headwaters and other conditions in streams as erosion proceeds. The forms that are in the young streams are moved back as the headwaters are moved back and as the river system spreads out into the usual fan shape, the animals that belonged in or near the headwaters move backward as the conditions migrate backward. In a broad geographic way this is unquestioned but details may be studied in the small streams of the bluff between Glencoe and the Wisconsin state line.

Fish are the only strictly aquatic forms in these streams that might not have entered by some other method than through the mouth of the stream. We have made a study of the fish of these streams for the purpose of determining whether the fish in the headwaters of the large streams are the same as the fish that are found in streams that are just large enough to have a single fish species, and the relation of the animals to stream development. The changes in animal communities which take place at one point are called succession.

a) *Ecological succession.*—Ecological succession is the succession of ecological types (physiological types, modes of life) over a given point or locality, due to changes of environmental conditions at that point. From this point of view *we have nothing to do with species, except that names are necessary.* However, we may speak of the succession in terms of species whenever their life habits (*mores*) are not easily modifiable.

Succession always involves all the animals of a community but it is often easier to discuss the changes which take place with respect to one group, such as the fishes. It is always to be understood that with changes in the fish communities there are similar changes in the communities of

other animals living with them. To illustrate the succession of fish in
streams we shall consider succession of fish in the North Shore streams.

b) Statement of ecological succession.—Succession is a reconstruction.
Here it is based on the superposition of all the fish communities (Fig. 67)
over the oldest part of the oldest and largest stream. To make this
clearer we will state, with the aid of the diagram (Fig. 69), the succession
of fish in Bull Creek. This succession will be considered as taking place

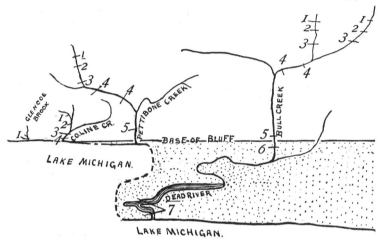

FIG. 67.—Diagrammatic arrangement of the North Shore streams. The streams
are mapped to a scale of one mile to the inch, and the maps are placed as closely
together as possible in the diagram. The intermediate shore-lines are shown in broken
lines which bear no relation to the shore-lines which exist in nature. Toward the top
of the diagram is west. Each number on the diagram refers to the pool nearest the
source of the stream which contains fish, as follows: 1, the horned dace (*Semotilus
atromaculatus*); 2, the red-bellied dace (*Chrosomus erythrogaster*); 3, the black-nosed
dace (*Rhinichthys atronasus*); 4, the suckers and minnows; 5, the pickerel and blunt-
nosed minnow; 6, the sunfish and bass; 7, the pike, chub-sucker, etc. The bluff
referred to is about 60 ft. high. The stippled area is a plain just above the level of the
lake (see Table XVIII).

over the oldest part of the portion of Bull Creek which lies back of the
bluff and higher levels of Lake Michigan. This is the point designated
as 5. (Table XVIII and Figs. 67 and 69 should be before the reader.)

When Bull Creek was at the stage represented by the first stage in
our diagram (which is represented by the present Glencoe Brook), its
fish, if any were present, were ecologically similar to those now in Glencoe
Brook in their relations to all factors except climate. This ecological
type is represented by the horned dace alone. As Bull Creek eroded its

bed and became hypothetical stage C of the diagram, the fish community of stage 1 was succeeded by a fish community ecologically similar to the fish communities at the localities marked 2 in Fig. 67. The fish now ecologically representing this community are the horned dace and the red-bellied dace. The community of the single species, the horned dace, had at such a period moved inland to the point where line 1–1 (Fig. 69) crosses the curved line representing the profile of hypothetical stage C. As erosion continued, the fish community ecologically represented by the horned dace and red-bellied dace moved gradually inland and was succeeded by a fish community occupying the mouth of hypothetical

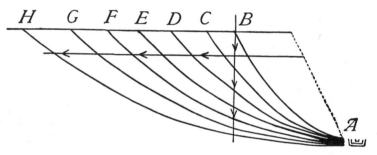

Fig. 68.—A diagram showing the successive stages in the profile (general shape of the bottom) of a very young stream, curved lines, *A–B, A–C, A–D, A–E, A–F, A–G, A–H* representing the successive profiles. The uppermost horizontal line represents the surface of the land into which the stream is eroding. The horizontal line with the arrowheads indicates the migration of the source of the stream and accordingly of similar stream conditions. The vertical line with arrowheads when followed downward passes through a succession of stream conditions and represents physiographic succession at the locality *B*. The point *A* is the mouth of the stream. Opposite this are shown three successive sizes of the stream, and therefore succession at that point.

stage D, ecologically similar to that now found at the point 3. This is represented by the three daces and the Johnny darter.

As the hypothetical stage D eroded its bed and became stage E, which is represented by County Line Creek, fish community 3 was then succeeded by a fish community ecologically similar to the fish community now present at point 4. This is ecologically represented by the three daces, the Johnny darter, and the young of the common sucker. The fish communities designated as 1, 2, 3 have meanwhile moved inland and are arranged in the order which their ecological constitution requires.

The continuation of the process resulted in displacing a fish community ecologically similar to the fish community 4 by a fish community ecologically similar to the present fish community 5. This is represented in the lower waters of Bull Creek—stage F.

Ecological succession is one of the few biological fields in which prediction is possible. We may carry this discussion a little farther. We have noted that the developing streams continue to erode their beds, grow larger, and bring down the surface of the land. These processes have not stopped in Bull Creek; it will become larger, contain a larger volume of water at the locality 5, and the fish community of locality 5

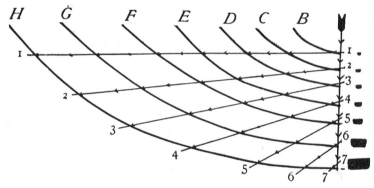

Fig. 69.—This figure is based on Fig. 68. The profiles of the streams shown here are separated vertically at the mouth. The curved lines represent seven stream stages as follows: B, Glencoe Brook; C, hypothetical stage; D, hypothetical stage; E, County Line Creek; F, Pettibone Creek; G, hypothetical stage; H, Bull Creek-Dead River. The hypothetical stages could, no doubt, be found along the shore of Lake Michigan; the difficulty arises from the introduction of sewage into so many streams.

The comparative size of the mouth of each stream stage is represented by a stream cross-section at the right. The direction of reading in succession is indicated by the vertical line with the arrowheads pointing downward. The oblique lines marked 1-1, 2-2, 3-3, etc., pass through points in the stream profiles which are in the same physiographic condition and occupied by similar fish communities.

will be succeeded by a fish community ecologically similar to that now at locality 6. This stage has been designated as hypothetical stage G in the diagram. With a further continuation of the process, the fish community of stage G, locality 6, will be succeeded by a fish community ecologically similar to that now found at the locality 7 (Dead River)— stage H. The communities of every stream have some such history as we have reconstructed, but the details may be modified by conditions. That branch of ecology which deals with such histories is called *genetic ecology*.

TABLE XVII

DISTRIBUTION OF INVERTEBRATES IN NORTH SHORE STREAMS

The meaning of the numbers is shown in Figs. 67 and 69. *a*=Temporary pool (*consocies*); *b*=Very young stream and intermittent riffles (ephemeral *consocies*).

Common Name	Scientific Name	a	b	1	2	3	4	5	6	7
Caddis-worm	*Phryganeidae*	*								
Mosquito larva	*Anopheles*	*								
Amphipod	*Eucrangonyx gracilis* Smith	*								
Isopod	*Asellus communis* Say	*								
Snail	*Lymnaea modicella* Say	*								
Crayfish	*Cambarus diogenes* Gir	*								
Black-fly larva	*Simulium* sp			*	*	*	*	*		
May-fly nymph	*Heptageninae*			*	*	*	*	*		
Crayfish	*Cambarus blandingi acutus* Girard				*	*				
Burrowing dragon-fly	*Cordulegaster obliquus* Say				*	*	*	*		
Dragon-fly nymph	*Aeschna constricta* Say				*	*	*	*		
Amphipod	*Gammarus fasciatus* Say				*	*	*	*		
Snail	*Physa gyrina* Say					*	*	*		
Crayfish	*Cambarus virilis* Hag				*	*	?	?	?	?	*
Crayfish	*Cambarus propinquus* Gir							*	*	
Crane-fly larva	*Pedicia albivitta* Walk (rarely)									*
Amphipod	*Hyalella knickerbockeri* Bate									*
Snail	*Planorbis campanulatus* Say									*
Dragon-fly nymph	*Tetragoneuria cynosura* Say									*

TABLE XVIII

SHOWING THE DISTRIBUTION OF FISH (NOMENCLATURE AFTER 79) IN THE NORTH SHORE STREAMS AT THE TIMES INDICATED

(The numbers refer to Figs. 67 and 69)

Name of Stream and Common Name of Fish	Date and Scientific Name	1	2	3	4	5	6	7
Glencoe Brook...........	August, 1907							
Horned dace..........	*Semotilus atromaculatus*...	*						
County Line Creek.......	1907–8							
Horned dace..........	*Semotilus atromaculatus*...	*	*	*	*			
Black-nosed dace	*Rhinichthys atronasus*....			*	*			
Johnny darter..........	*Boleosoma nigrum*.......			*	*			
Blackhead minnow......	*Pimephales promelas*.....			*				
Blunt-nosed minnow	*Pimephales notatus*.......			*				
Common sucker........	*Catostomus commersonii* ..			*				
Pettibone Creek†........	September, 1909, and April, 1910							
Horned dace..........	*Semotilus atromaculatus* ..	?	*	*	*			
Red-bellied dace........	*Chrosomus erythrogaster*...		*	*	*			
Black-nosed dace	*Rhinichthys atronasus*....			*	*			
Johnny darter..........	*Boleosoma nigrum*.......			*	*			
Common sucker........	*Catostomus commersonii*...			*				
Bull Creek–Dead River....	September, 1909							
Horned dace..........	*Semotilus atromaculatus*...	*	*	*	*	*		
Red-bellied dace........	*Chrosomus erythrogaster*...		*	*	*			
Black-nosed dace	*Rhinichthys atronasus*....			*	*			
Common sucker........	*Catostomus commersonii*...				*	*		
Blunt-nosed minnow	*Pimephales notatus*.......					*	*	
Little pickerel..........	*Esox vermiculatus*					*	*	*
Bluegill	*Lepomis pallidus*........						*	*
Large-mouthed black bass	*Micropterus salmoides*....						*	*
Pike..................	*Esox lucius*.............							*
Crappie...............	*Pomoxis annularis*.......							*
Red-horse.............	*Moxostoma aureolum*.....							*
Chub-sucker...........	*Erimyzon sucetta*........							*
Golden shiner..........	*Abramis crysoleucas*......							*
Common shiner........	*Notropis cornutus*........							*
Cayuga minnow........	*Notropis cayuga*.........							*
Tadpole cat...........	*Schilbeodes gyrinus*......							*

† The lower part of Pettibone Creek has been destroyed by the United States Naval School, otherwise the table would include the records for a point 5 and perhaps a point 6, but probably not 7.

NOTE.—Table XIX follows Table XX.

TABLE XX (Table XIX follows Table XX)

HICKORY CREEK LONGITUDINAL AND HABITUDINAL DISTRIBUTION

(For meaning of Roman numbers see Table XXI and Fig. 66.)

Common Name	Scientific Name	STILL			MODERATE			SWIFT			LONGITUDINAL				
		Veg.	Silt	Water-Lilies	Sides Gravel	Pools	Riffles	Among Stones	Under Stones	On Stones	V	IV	III	II	I
Snail	Planorbis trivolvis Say	*	*		*										*
Snail	Planorbis parvus Say	*	*		*										*
Snail	Physa gyrina Say	?			*									*	*
Crayfish	Cambarus propinquus Gir.					*			*		*			*	*
Bug	Zaitha fluminea Say	*	*	*	*	*					*			*	
Dragon-fly nymph	Plathemis lydia Drury			*	*	*									
Bivalve	Sphaerium striatinum Lam.												*	*	
Crayfish	Cambarus blandingi acutus Gir.												*		
Burrowing dragon-fly	Gomphus exilis Selys							*	*				*		
Mussel	Alasmidonta calceola Lea									*					
Bivalve	Sphaerium stamineum Con.														
Crayfish	Cambarus virilis Hag.											*	*		
Black-wing nymph	Calopteryx maculata Beauv.										*		*		
Snail	Goniobasis livescens Mke.										*		*		
Black-fly larvae	Simulium sp.										*	?			
Midge larvae	Chironomus sp.										*	?			
Dobson	Corydalis cornuta Linn.										*				
Stone-fly nymph	Perla sp.								*		*				
Caddis-worm	Chimarrha sp.								*		*				
Sponge	Spongilla sp.								*		*				
Damsel-fly nymph	Argia putrida Hagen								*		*				
Water-penny	Psephenus lecontei Lec.								*		*				
Horse-fly larvae	Tabanus sp.							*	*		*				
May-fly nymph	Siphlurus alternatus Say							*	*		*				

TABLE XX—Continued

Common Name	Scientific Name	I	II	III	IV	V	On Stones	Under Stones	Among Stones	Riffles	Pools	Gravel Sides	Water-Lilies	Silt	Veg.
Caddis-worm	Helicopsyche sp.					*				*		*			
Mussel	Lampsilis iris Lea.					*				⌐					
Snail	Pleurocera elevatum levisii Lea.					*									
Mussel	Lampsilis ellipsiformis Conrad.					*									
Mussel	Quadrula undulata Barnes.					*									
Mussel	Anodontoides ferussacianus Lea.					*									
Water-strider	Rhagovelia collaris Burm.					*									*
Mussel	Quadrula rubiginosa Lea.					*					*	*			
Mussel	Lampsilis luteola Lam.					*					*	*			
Mussel	Lampsilis ventricosa Bar.					*					*	*			
Mussel	Lampsilis ligamentina Lam.					*					*	*			
Mussel	Unio gibbosus Bar.					*					*	*			
Snail	Campeloma integrum DeK.					*					*	*		*	*
Damsel-fly nymph	Enallagma sp.					*						*			
Bloodworms	Chironomus sp.					*									
Leech	Placobdella rugosa Ver.					*						*			
Snail	Planorbis bicarinatus Say.					*							*		*
Caddis-worm	Limnophilidae					*							*		*
May-fly nymph	Chirotenetes siccus Walsh.					*							*		*
Water-boatmen	Corixa sp.					*							*		*
Snail	Amnicola limosa Say.					*									*
Snail	Amnicola cincinnatiensis Lea.					*									*
Dragon-fly nymph	Aeschnidae.					*									
Back-swimmer	Notonecta sp.					*								*	*
Snail	Physa integra Hald.					*									
Burrowing May-fly nymph	Hexagenia sp.					*									
Snail	Campeloma subsolidum Ant.					*									*

TABLE XIX

ANIMALS OF SPRINGS AND SPRING BROOKS

The meaning of the letters in the column headed "Location" is as follows: Cs = Cary spring; Gs = Gaugars spring; Zs = Zion spring; Sb = Suman spring brook; Cb = Cary spring brook.

Common Name	Scientific Name	Location		
Amphipod............	*Gammarus fasciatus* Say........	Cs,	Gs,	Zs
Planarian.............	*Planaria dorotocephala*.........	Cs		
Planarian.............	*Dendrocoelum* sp..............	Cs		
Dragon-fly nymph......	*Aeschna* sp...................	Cs	Zs	
Midge larva...........	*Tanypus* sp..................			Sb
Black-fly larva.........	*Simulium* sp.................			Sb, Cb
Caddis-worm	*Hydropsyche* sp..............			Cb
Midge larva...........	*Chironomus* sp..............			Cb
Fly larva.............	*Dixa* sp....................			Cb
May-fly nymph........	*Heptagenia*.................			Cb
Bivalve..............	*Musculium*.................			Cb
Amphipod............	*Eucrangonyx gracilis* Smith.....			Cb
Crayfish	*Cambarus propinquus* Gir.......			Cb
Snail................	*Physa gyrina* Say.............			Cb
Damsel-fly nymph......	*Calopteryx maculata* Beauv.....			Cb
Parnid...............	*Elmis fastiditus* Lec...........			Cb

TABLE XXI

THE DISTRIBUTION OF FISH (NOMENCLATURE AFTER 79) IN HICKORY CREEK (AND ITS WEST BRANCH) IN THE SUMMER OF 1909

Those starred were in the pool nearest the source. I, the first mile of the stream, measured from the fish pool nearest the source, toward the mouth; II, the third and fourth miles; III, at the head of erosion, five miles from the pool nearest the source; IV, six miles from the pool nearest the source; V, nine miles from same; stream much larger with good riffles and one weedy cove.

Common Name	Scientific Name	I	II	III	IV	V
Horned dace*.........	*Semotilus atromaculatus..*	*	*	*	*	*
Golden shiner*........	*Abramis crysoleucas......*	*	*	*	*	
Johnny darter*........	*Boleosoma nigrum.......*	*	*	*	*	
Stone-roller*..........	*Campostoma anomalum...*	*	*	*		*
Straw-colored minnow*...	*Notropis blennius........*	*	*		*	*
Blue-spotted sunfish*....	*Lepomis cyanellus.......*	*	*	*		
Blunt-nosed minnow.....	*Pimephales notatus.......*	*	*	*	*	*
Common sucker*.......	*Catostomus commersonii...*	*	*			*
Mud minnow..........	*Umbra limi.............*	*				
Top minnow..........	*Fundulus notatus........*	*	*			
Red-bellied dace.......	*Chrosomus erythrogaster...*	*	*			
Chub-sucker..........	*Erimyzon sucetta........*	*		*		
Black bullhead........	*Ameiurus melas........*	*				*
Blackfin	*Notropis umbratilis......*		*	*		
River chub............	*Hybopsis Kentuckiensis..*		*	*	*	
Fan-tailed darter.......	*Etheostoma flabellare.....*			*	*	*
Rainbow darter........	*Etheostoma coeruleum....*	*		*	*	*
Least darter...........	*Microperca punctulata....*				*	
Sucker-mouthed minnow.	*Phenacobius mirabilis....*				*	
Cayuga minnow.......	*Notropis cayuga........*				*	*
Rock bass.............	*Ambloplites rupestris.....*				*	*
Common shiner........	*Notropis cornutus.......*					*
Rosy-faced minnow.....	*Notropis rubrifrons......*					*
Banded darter.........	*Etheostoma zonale......*					*
Bluegill	*Lepomis pallidus.......*					*
Long-eared sunfish.....	*Lepomis megalotis.......*					*
Stonecat	*Noturus flavus.........*					*
Yellow perch..........	*Perca flavescens.........*					*
Small-mouthed black bass	*Micropterus dolomieu.....*					*
Hogsucker............	*Catostomus nigricans.....*					*
Common red-horse......	*Moxostoma aureolum.....*					*

TABLE XXII

The Fish (Nomenclature after 79) of Thorn Creek, Collection Made at the Headwaters in 1908 and 1909 and at Other Points in 1909 and 1910

A = the first fish pool; B = four miles downstream; C = ten miles downstream.

Common Name	Scientific Name	A	B	C
Horned dace............	*Semotilus atromaculatus*........	*	*	*
Blunt-nosed minnow.......	*Pimephales notatus*............	*	*	*
Blue-spotted sunfish	*Lepomis cyanellus*.............	*	*	*
Stone-roller.............	*Campostoma anomalum*........	*	*	*
	Notropis umbratilis...........		*	?
Banded darter...........	*Etheostoma zonale*		*	?
Common shiner..........	*Notropis cornutus*.............		*	?
Striped-top minnow.......	*Fundulus dispar*.............		*	?
Black-sided darter........	*Hadropterus aspro*............		*	*
Johnny darter...........	*Boleosoma nigrum*............		*	*
Mud minnow............	*Umbra limi*..................		*	*
Cayuga minnow..........	*Notropis cayuga*..............		*	*
Golden shiner...........	*Abramis crysoleucas*..........		*	*
Large-mouthed black bass..	*Micropterus salmoides*.........			*
Small-mouthed black bass..	*Micropterus dolomieu*.........			*
Bluegill	*Lepomis pallidus*.............			*
Crappie.................	*Pomoxis sparoides*............			*
Pirate perch.............	*Aphredoderus sayanus*.........			*
Yellow perch............	*Perca flavescens*.............			*
Carp...................	*Cyprinus carpio*..............			*
Black bullhead	*Ameiurus melas*..............			*
Common sucker.........	*Catostomus commersonii*.......			*
Short-headed red-horse.....	*Moxostoma breviceps*..........			*
Pike...................	*Esox lucius*.................			*

TABLE XXIII

ANIMALS OF THE DESPLAINES, CHICAGO, AND DUPAGE RIVERS

The meaning of the letters in the column headed "Location" is as follows:
L=Libertyville (still–silt); W=Wheeling (mud–gravel); D=DuPage; R=Riverside
(swift–stones). Libertyville is the farthest upstream, and the other situations follow
in the order named. C=Chicago River at Edgebrook, which is added without
regard to longitudinal order.

Common Name	Scientific Name	Location		
Crayfish	*Cambarus virilis* Hag.	L		C
Snail	*Lymnaea humilis modicella* Say	L		
Dragon-fly nymph	*Basiaeschna janata* Say	L		
Snail	*Ancylus tardus* Say	L		
Snail	*Ancylus rivularis* Say			C
Crayfish	*Cambarus propinquus* Gir.	L	D	W
Crayfish	*Cambarus diogenes* Gir.			C
Mussel	*Anodonta grandis* Say		D	W
Mussel	*Anodontoides ferussacianus* Lea			C
Mussel	*Quadrula undulata* Bar.		D	W
Mussel	*Lampsilis luteola* Lam.		D	W
Bivalve	*Musculium truncatum* Lins.		D	W
Snail	*Goniobasis livescens* Mke.		D	W
Mussel	*Alasmidonta calceola* Lea.		D	
Snail	*Amnicola limosa* Say		D	C
Snail	*Planorbis bicarinatus* Say		D	
Snail	*Physa gyrina* Say			C
Mussel	*Lampsilis ellipsiformis* Con.		D	C
Snail	*Pleurocera subulare intensum* Ant.	R	D	
Snail	*Pleurocera elevatum* Say	R		C
Stone-flies	*Perla* sp.	R		C
Dobson	*Corydalis cornuta* Linn.	R		
Amphipod	*Hyalella knickerbockeri* Bate.			
Caddis-worm	*Hydropsyche*	R		C
Isopod	*Asellus communis* Say			
Damsel-fly nymph	*Argia* sp.	R		
Dytiscid	*Hydroporus vittatus* Lec.			
Newt	*Diemictylus viridescens* Raf.	R		
Polyzoan	*Plumatella* sp.			C
Parnid	*Elmis*			C
Caddis-worm	*Helicopsyche* sp.			C
Leech	*Haemopis grandis* Verrill			C
Caddis-worm	*Phryganeidae*			C
Sialid	*Sialis* sp.			C
Sphaerid	*Sphaerium stamineum* Con.		D	C
Burrowing dragon-fly nymph	*Gomphus exilis* Selys			C

TABLE XXIV

MUSSELS OF THE CALUMET–DEEP RIVER. ARRANGED IN ORDER OF LONGITUDINAL
SUCCESSION BEGINNING WITH THE UPPER PARTS OF THE
RIVER AT AINSWORTH

The letters indicate place of collection. A=Ainsworth; G=East Gary;
M=south of Miller, in the Little Calumet; and C=Clark, in the Grand Calumet.

Common Name	Scientific Name	Location			
Mussel.............	*Symphynota costata* Raf......	A			
Mussel.............	*Lampsilis ventricosa* Bar......	A			
Mussel.............	*Quadrula undulata* Bar.......	A	G	M	
Mussel.............	*Lampsilis luteola* Lam.......	A	G	M	
Mussel.............	*Symphynota complanata* Bar...	A	G	M	
Mussel.............	*Unio gibbosus* Bar...........			M	
Mussel.............	*Quadrula rubiginosa* Lea......			M	
Mussel.............	*Anodonta grandis* Say........			M	C

TABLE XXV

ANIMALS FROM A SLUGGISH PORTION OF FOX RIVER

The meaning of the letters in the column headed "Location" is as follows: Gm = gravel in mid river in eight feet of water; G = gravel near shore; S = sand; M = mud or silt; V = vegetation.

Common Name	Scientific Name	Location				
Snail	*Goniobasis livescens* Mke..	Gm	G	S		
Mussel	*Anodonta grandis* Say....	Gm	G	S		
Mussel	*Lampsilis ligamentina* Lam	Gm				
Mussel	*Quadrula undulata* Bar...	Gm			M	
Red midge larva...	*Chironomus*	Gm				
Green midge larva..	*Chironomus*	Gm				
Caddis-worm	*Hydropsyche*	Gm				
Polyzoan	*Plumatella*	Gm				
Dragon-fly nymph..	*Macromia taeniolata* Ram.		G	S		V
Crayfish	*Cambarus propinquus* Gir.		G	S		
Snail	*Campeloma integrum* DeK.		G	S		
Snail	*Pleurocera elevatum* Say...		G	S		
Mussel	*Unio gibbosus* Bar		G	S		
Mussel	*Quadrula rubiginosa* Lea..		G			
Mussel	*Lampsilis luteola* Lam....				M	V
May-fly nymph....	*Hexagenia*				M	
Fly larva	*Stratiomyia* sp				M	V
May-fly nymph....	*Callibaetis* sp					V
Amphipod	*Hyalella knickerbockeri* Ba.					V
May-fly nymph....	*Caenis* sp					V
Beetle	*Donacia*					V
Sialid larva	*Chauliodes* sp					V
Snail	*Physa integra* Hald					V
Water-boatman....	*Corixa* sp					V
Water scorpion....	*Ranatra fusca* Beau					V
Amphipod	*Gammarus fasciatus* Say..					V
Bug	*Zaitha fluminea* Say					V
Parnid	*Elmis 4-notatus* Say					V
Creeping bug	*Pelocoris femoratus* Pal Beauv					V
Back-swimmer	*Notonecta variabilis* Fieb..					V
Dragon-fly nymph..	*Ischnura verticalis* Say....					V
Leech	*Glossiphonia fusca* Castle.					V
Top minnow	*Fundulus diaphanus menona* J. and C					V
Snail	*Planorbis bicarinatus* Say.					

CHAPTER VII

ANIMAL COMMUNITIES OF SMALL LAKES

I. Introduction

Lakes are difficult to classify on the basis of animal relations. This is because size, shape, exposure to wind, depth, and age are all important in determining conditions that affect animals. A classification into *coastal lakes* and *morainic lakes* will serve our purposes best, because, other things being equal, it represents age and depth (near Chicago).

Morainic lakes are depressions in the moraine due to irregularities of deposition, which stand below ground-water level. They are of various sizes. We shall apply the term *lake* only to those bodies of water that are large enough to produce an area of at least a few square rods of sandy shore, which supports gilled snails, mussels, etc. The principal lakes included in our area are shown on the map facing p. 52. The largest of these are the Fox, Pistakee, Maria, and Grass lakes in northern Illinois; Hudson, Cedar, Stone, and Flint lakes in Indiana; and Paw Paw and Pipestone lakes in Michigan. The only coastal lakes of any size are Wolf Lake and Calumet Lake. These are located in the old Lake Chicago plain.

1. CONDITIONS IN LAKES

Depth is important in determining the conditions at the bottom, but is of little importance to the other parts of the lake. Little is known of the depths of our lakes. Exposure to wind is of importance in affecting the waves and circulation of the water (see p. 61), both of which are important to animals. A lake well protected by high hills will be likely to be less affected by wind than others. Shape is also a factor. Long lakes whose long axes are parallel with the direction of the prevailing winds are more strikingly affected by the wind than those with the long axis at right angles to the wind.

Waves are never large on small lakes, but are usually effective in determining the kind of bottom by controlling erosion and deposition. The general circulation of all our lakes has not been studied. On account of their small size it is probable that the deeper ones at least have an incomplete circulation like that indicated in Fig. 11, p. 61. Those that get warmed throughout in summer probably have a complete circulation. The dissolved content of the waters of lakes is usually

124

similar to that of the large lakes and rivers. Oxygen is usually abundant in the surface waters, but is often wanting in the bottoms of lakes (74) with incomplete summer circulation. Muck bottoms in deep water or in bays have little or no dissolved oxygen. Dissolved nitrogen is important, but has been little studied. In the open water light and pressure are governed by the same factors as in the large lakes (see pp. 62–64). The bottom in small lakes varies with exposure to waves. Where the waves are eroding, the bottom is stony or sandy; where depositing, it contains silt and humus. There are often deposits of marl, which is a calcium carbonate deposit, frequently reaching a depth of 18 feet in the Indiana lakes. It frequently reaches to the surface of the water, but when it does so is often covered by muck. Muck bottom is common in the deeper water and in bays. The vegetation in such lakes is very much like that in base-level streams. The vegetation of the shores of rivers like Fox River is duplicated in these lakes, and in fact, small lakes are strictly comparable to sluggish rivers in many respects. We have patches of vegetation, patches of sand and gravel bottom, but also much bottom which has more organic matter than river silt. The principal difference is that currents in the lakes vary with the wind, and in sluggish streams are mainly in one direction.

II. COMMUNITIES OF SMALL LAKES

(Stations 30, 30*a*, 31; Table XXVI)

These are divided into the limnetic formation, the formations of sandy and stony shores, the formations of muck bottom in shallow water, the formations of the vegetation, and the formations of deep water (anaerobic).

I. THE LIMNETIC FORMATION (104)

(List II)

The limnetic formation of the smaller lakes is very similar to that of the larger lakes. It is made up of the same groups, but with the addition of a few pelagic insects such as the phantom larva (*Corethra* sp.). The species of crustaceans, rotifers, and protozoa are different. The characters of the formation are similar to those of Lake Michigan (p. 75).

2. SHALLOW WATER FORMATIONS

a) *Terrigenous bottom formation* (105).—Vegetation sparse or absent —water 0–3 meters. Crawling over the sandy bottom are usually found caddis-worms (*Goera* sp. or *Molanna* sp.) (Figs. 70, 71). These forms belong to different families, but have similar cases and similar habits. This is a good example of what is meant by *mores*. The *forms* are very different, but their *mores* are similar. The Johnny darter, the straw-colored minnow (Fig. 72), and the blunt-nosed minnow are usually found (105) in the shallowest water. The Johnny darter, the blunt-nosed minnow, the miller's thumb, and probably other minnows breed in these situations (105, 106). Crayfish are common here (in Wolf Lake, *Cambarus virilis*).

Snails (such as *Pleurocera subulare* [Fig. 73], and sometimes *Goniobasis livescens*) are common on the shoals, crawling over the bottom which is always covered with diatoms, desmids, etc. These algae serve as food

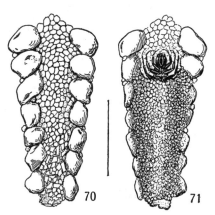

for the mussels. Miss Nichols found 16 species of algae on the shell of a specimen of *Pleurocera* taken from a Wolf Lake shoal. In the deeper waters (3 ft.) we find the same crayfishes and the same snails fewer in number than in the shallower parts of the shoals. Associated with them are the mussels (especially *Lampsilis luteola*, *Anodonta marginata* and *grandis*). Such sandy and gravelly bottomed shoals in 1–3 ft. of water are especially important to the food fishes. There are many first-class food fishes in all such lakes. Of those in Wolf Lake seven breed

FIG. 70.—The case of a caddis-worm (*Molanna* sp.), sandy bottom (Fox Lake, Ill.) (original).

FIG. 71.—The same from below.

in these shallows. There are the large-mouthed black bass (Fig. 74), the bluegill, the pumpkinseed, the green sunfish, the perch (Fig. 75), the speckled catfish, and the crappie. Nearly all in making their nests scrape the bottom clear of all débris; the males guard the nests. The number of food fishes in a lake is related to the area of such shoals, which are accordingly of great economic importance and should be protected from destruction by the encroachment of vegetation and accumulation of débris. Associated with the fish are occasional musk turtles (*Aromochelys odorata*). Shoals are invaded by bulrushes and bare bottom may exist between them. Here the viviparous snail (*Vivipara contectoides*) (Fig. 76) sometimes occurs.

72

REPRESENTATIVES OF THE BARE SAND COMMUNITY

FIG. 72.—Straw-colored minnow (*Notropis blennius*) (from Forbes and Richardson).

FIG. 73.—Snail (*Pleurocera subulare*) crawling over sandy bottom; slightly enlarged (photographed in aquarium).

FIG. 74.—Large-mouthed black bass (*Micropterus salmoides*), juvenile; natural size (original).

Characters of the formation: The formation is distinctly dependent upon a clean bottom of sand or coarser materials, and is made up of creeping forms and those using the bottom as a breeding-place.

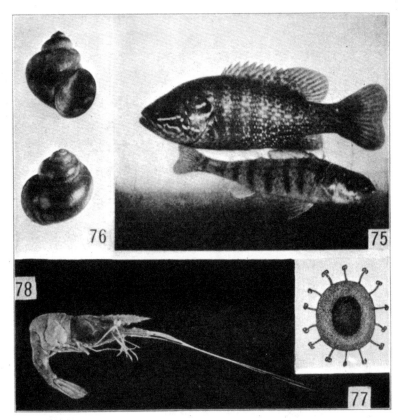

REPRESENTATIVE ANIMALS OF THE SUBMERGED VEGETATION

FIG. 75.—Upper fish, the green sunfish (*Lepomis cyanellus*); lower fish, the yellow perch (*Perca flavescens*); both juvenile; slightly reduced (original).

FIG. 76.—A viviparous snail (*Vivipara contectoides*); natural size.

FIG. 77.—A winter body or statoblast, of the gelatin-secreting polyzoan (*Pectinatella magnifica*); 10 times natural size (original).

FIG. 78.—A shrimp (*Palaemonetes paludosus*); twice natural size (original).

b) *Submerged vegetation association of the open waters.*—A lake of the coastal type is separated rapidly from the larger body of water in connection with which it is formed, or a morainic lake, when the ice retreats, is left with the greater part of its shallow water of the type which we have described. Vegetation is present from the first in the form of floating microscopic plants, and the dead bodies of these and of the animals present are swept into the depressions and protected situations where the waves do not drag on the bottom. Here vegetation grows in the greatest luxuriance and causes the production of more plant débris, which adds to that already in the protected situations. We then have, after a time, a covering of the bottom by the humus and conditions unfavorable for most bottom animals. The animals of the bare bottom shoals are no longer present in numbers. Small, apparently stunted forms of *Lampsilis luteola* are found for a time, but are soon driven out by the increase of humus and vegetation. The early vegetation is made up of scattered aquatic plants, such as *Myriophyllum* and *Elodea*, and in the shallower water usually bulrushes.

One of the most distinctive and characteristic forms of such lakes is a transparent true shrimp (*Palaemonetes paludosus*), about 2 inches long (Fig. 78), which is a close relative of some of the edible marine shrimps. In spring they are found carrying numbers of green eggs attached to the appendages of their abdomens. Another common animal in these situations is the large polyzoan (*Pectinatella magnifica*). This is a colonial form which reproduces by budding in several directions. It also secretes a clear and transparent jelly. As the number of animals increases the amount of jelly increases on all sides and the animals are arranged on the outside of the more or less spherical mass of jelly; the necessary increase in surface for the growth of the colony is supplied through additional secretion by each new animal added. Some of these masses of jelly reach a size of 6 inches in diameter. They are often attached about a stalk of *Myriophyllum* as a center. In the autumn they form bodies known as statoblasts (Fig. 77), which are disk-shaped, the center containing living cells and the rim being filled with air-bubbles. The rim of the disk is supplied with hooks which catch onto objects. Probably they must be frozen before they will grow into new colonies for they do so only in the spring.

Other characteristic animals of this open-water vegetation are shelled protozoa (Fig. 79), water-mites (Fig. 80), and ostracods (Fig. 81). On the stems of the water plants, such as bulrushes and pickerel weed, are the snails (*Ancylus*) which belong to the lunged group, but are said to take water into the lung and thus do not need to come to the surface for air. Occasional snails, leeches, and midge larvae occur. Water-mites fasten their eggs to the bases of the aquatic plants. Among the leaves of the divided leaved plants the midge larvae, damsel-fly nymphs, and May-fly nymphs (*Callibaetis* sp.) are usually numerous. All these are important as fish food. This area is the feeding-place for a number of fishes. Those feeding in the vegetation are the subfishes, basses and perches, most of which breed on the barren shoals. With them are also the carp, the chub-sucker, the warmouth bass, the brook silverside (*Labidesthes sicculus*), and the buffalo fish (84). This part of the lake is also the favorite haunt of the turtles (107), such as the soft shell (*Aspidonectes spinifer*), and in the parts with some bare bottom the musk

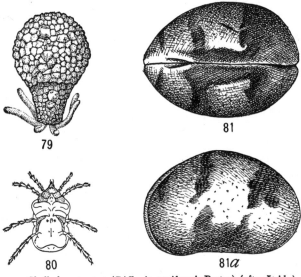

FIG. 79.—Shelled protozoan (*Difflugia pyriformis* Perty.) (after Leidy).

FIG. 80.—A red mite (*Limnochares aquaticus*); 6 times natural size (after Wolcott).

FIG. 81.—Dorsal view of an ostracod (*Cypridopsis vidua*); 80 times natural size (after Brady).

FIG. 81a.—The same seen from the side.

turtle (*Aromochelys odorata*), and the geographic turtle (*Graptemys geographicus*). The mud puppy (*Necturus maculosus*) is also found in such situations (*fide* Mr. Hildebrand). The muskrat (*Fiber zibethicus*) builds its nest (Fig. 82) in the shallow water adjoining these situations. The musk turtle frequently deposits its eggs on the nest in early summer (105). We have found them in these situations in the month of June. Various aquatic birds feed here (108). This formation may be characterized as belonging to the aquatic vegetation, but practically all the species are relatively independent of the atmosphere and of the bottom.

c) Emerging vegetation association of bays.—Such situations as are occupied by this association are found in bays and protected situations in the larger lakes and represent a stage which is last in the history of a lake. Water-lilies, water buttercups, and *Myriophyllum* are the principal plants. Filamentous algae are usually very abundant. Logs, sticks, and pieces of wood are not uncommon.

On the under side of logs, we find such forms as the polyzoan (*Plumatella*) and sponges (*Spongilla* sp.). On the under side of the water-lily pads are usually numbers of *Hydra* together with great numbers of

Fɪɢ. 82.—A muskrat's nest adjoining the lake border among the bulrushes on sandy bottom.

shelled protozoans and rotifers, especially sessile forms. Snails also are common here (*Segmentina armigera, Planorbis parvus, Physa gyrina* and *integra, Planorbis campanulatus*, and some species of *Lymnaea*).

A large number of species of aquatic insects cling in the vegetation with the abdomen near the surface of the water and secure air through various anatomical arrangements which conduct it to the spiracles; the most noteworthy of these are the water scorpion (*Ranatra*), the electric-light bugs (*Benacus* and *Belostoma*), the predaceous diving beetles (*Dytiscidae*) (99c), the water scavengers (*Hydrophilidae*), and the water-boatmen (*Corixa*). There are also a number of aquatic insects that are not dependent upon the atmospheric air in their young stages. They require, however, some object which reaches above the surface of the water when they emerge from the larval skin.   The prominent members of this group are the dragon-fly nymphs (*Anax junius* and *Ischnura verticalis*).

There are a few insects that are relatively independent of vegetation as a means of attachment.   The back-swimmers are an example.   They float or swim in the water among the vegetation.   The commonest of these are those belonging to the genera *Plea*, *Notonecta*, and *Buenoa*. There are a few fish that have a similar habit.   The top minnow (*Fundulus dispar*), which feeds at the surface, is an example.   It invades the pools near shore and devours mosquito larvae.   The young of such fishes as the basses and the sunfishes are sometimes taken in these situations.

In the mud of the bottom there are but few animals.   Some of these· are the same species as those found in the bottom in the region of open water and will be discussed later.   There are, however, forms that live only on the rhizomes of the water-lily.   Certain of the leaf-feeding beetles (*Chrysomelidae*, *Donacia*) (109) are aquatic in the young stages. The female eats a hole in the leaves of the water-lilies and reaches through with her ovipositor and deposits the eggs in a semicircle which has the hole as its center.   When these eggs hatch the larvae crawl to the rhizomes.   They are not provided with gills and do not come to the surface for air.   They have a pair of spines adjoining the spiracles. These spines are thrust into the plant and the spiracles which open at their bases come into contact with the holes;   the gas in the plant and the gas in the air tube of the insect's body interchange, and the animal is thus supplied with oxygen.   When the larva is ready to pupate it spins a cocoon in some unknown way under water, but when it is completed it is filled with gas, not water, and surrounds the body of the animal. The animal then eats a hole, connecting the cocoon with the air spaces of the plant.   It then pupates and is supplied with oxygen by the plant during the entire pupal period.

The common painted turtle (*Chrysemys marginata*) and the snapping turtle are common in such small bays.   They come out upon the logs and bask in the sun.   The pied billed grebe builds its floating nest, and many other aquatic birds feed in such situations (108).

Characters of the vegetation formation: This formation is of the old-pond type which will be especially discussed in the following chapter. There are two characters, one or the other of which is possessed by nearly all the animals.   They depend upon the atmospheric air or must have the support of the vegetation, or both.   The majority of the animals of this formation stick their eggs either in or on vegetation.   Such formations are quite similar in many respects to the formations of the vegetation in sluggish rivers but resist lack of oxygen and stagnant water much better.

d) The anaerobic formation.—This is the bottom and deep-water formation. We have already stated that the circulation of water (see Fig. 10, p. 61) is not known for any of the lakes discussed. Old lakes like those about Chicago are usually covered with humus on the bottom. In this humus and probably just above it there is little or no oxygen. Analyses of the bottom water from ponds with humus-covered bottoms showed that it contained no oxygen. The open water of the lakes with the incomplete circulation in summer is without sufficient oxygen to support life, below the level of circulation (Fig. 11, p. 61). There are, however, numbers of animals that pass the summer under these conditions (110, 111). These are protozoa belonging to eleven genera, worms belonging to two genera, one rotifer, one ostracod, and the small bivalve (*Pisidium idahoense*). Dr. Juday kept these animals in jars without oxygen and observed their activities. The rotifer was always active. The ostracod showed little activity, and the bivalve kept its valve closed, showing no activity whatever.

There are occasional midge larvae in the mud of such bottoms, but they are rare. Some of these have haemaglobin in their blood and are supposed to be able to use oxygen when it is present in the minutest quantities. In the open oxygenless water there are phantom larvae (*Corethra*) which are able to carry a supply of oxygen with them from the surface.

III. Succession in Lakes

The general tendency of succession in lakes has been indicated. The first formation is the bare-bottom type, which is locally transformed to the vegetation of open-water type. This usually begins in the protected situations first; the bays are ecologically oldest. These bays pass rapidly from the third open-lake type to the bay conditions. When such a stage has been reached the situations that have a less degree of protection from waves have reached the second stage and we have lakes as we find most of the larger ones about Chicago. They contain, at various points, the three formations which we have discussed. The lake is reduced in size by filling near its shores and the lowering of its outlet. The older stages are continuously encroaching on the younger. The area of barren shoal is constantly becoming less as the lake fills and the outlet, if it has one, is lowered. Around the shores the development of prairie or forest is usually well begun and one or the other of these types of land vegetation finally displaces the lake.

1. THE INFLUENCE OF SIZE AND DEPTH

Size and depth have a marked influence on the rate of succession. If the lake is large, like Lake Michigan, its waves beat upon the shores with such force as to prevent the development of vegetation or the establishment of any of the formations just discussed. Smaller lakes have proportionally less efficient wave-action, and situations which would not be protected to any marked degree in a lake like Lake Michigan are relatively free from effective wave-action. The formations succeed one another rapidly where wave-action is slight. The various parts of the shore of a small kettle-hole with a regular shore-line would pass through all these stages at nearly the same rate. Depth is an important factor also because the various formations cannot succeed over the deep water until the deeper parts are filled (or drained), which often requires long periods. The rate of succession in lakes is then directly proportional to their size and depth. The small lakes pass through all the stages more quickly than the larger lakes. Those considered here have for the most part, at present, become dominated by the late stages. The lakes of the inland type which are large enough to maintain all the formations discussed are among the most complex of all our habitats.

2. INFLUENCE OF MATERIAL AND MODE OF ORIGIN

At the very beginning the kind of material in which a lake is situated is important but as time goes on it becomes less and less important. If the lake is in clay, at the outset there are no sandy areas, but the action of the waves soon removes the finer material and leaves sand (the finer materials being deposited on the bottom of the lake). Young lakes in rock are probably very different from those in clay, but even here sandy shores are soon formed and occupied by the same animals as sandy shores of different origin.

The distinction between lakes and ponds is a purely artificial one. The ponds have the same communities at the outset as the lakes, but the changes proceed so rapidly that very young ponds are rare. All lakes and ponds tend to become ecologically similar, regardless of mode of origin and kind of material.

LIST II

The following *Entomostraca* have been taken from Wolf Lake: * indicates the species is found in Fox Lake; † in Butler's Lake; ‡ in the series of ponds at the head of Lake Michigan: Copepods: ‡*† *Cyclops serrulatus* Fischer; *†‡ *C. albidus* Jurine; ‡ *C. viridis brevispinosus* Herrick. Cladocerans: *Acroperus harpae* Baird; ‡ *Scapholeberis mucronata* Muel.; ‡ *Pleuroxus denticulatus* Birge; *Diaphanosoma brachyurum* Liev.; ‡ *Chydorus sphaericus* Muel.; *Polyphemus pediculus* Linn; *Macrothrix rosea* Jurine; ‡ *Ceriodaphnia reticulata* Jurine; ‡ *Simocephalus serrulatus* Koch; *Bosmina obtusirostris* Sars. Ostracods: *Potamocypris smaragdina* Vav.

TABLE XXVI

ANIMALS FROM SMALL LAKES

Meaning of letters occurring in the columns is as follows: "Habitat" column: S = bottom of sand; SH = bottom of sand and humus; B = bulrush vegetation; VO = vegetation of open water; VB = vegetation of bays; in "Lake Where Recorded" column: F = Fox Lake; W = Wolf Lake; G = Lake George; B = Butler's Lake.

Common Name	Scientific Name	Habitat from Which Collected	Lake Where Recorded
Caddis-worm	Goera sp	S	W
Caddis-worm	Molanna sp	S	F
Caddis-worm	Polycentropidae	S	
Snail	Pleurocera subulare Lea	S	W G
Snail	Goniobasis livescens Mke	S	W G
Crayfish	Cambarus virilis Hag	S	W
Turtle (musk)	Aromochelys odorata Lat	S	W
Geographic turtle	Graptemys geographicus LeS	S	W
Straw-colored minnow	Notropis blennius Gir	S	W
Johnny darter	Boleosoma nigrum Raf	S	W,F
Mussel	Lampsilis luteola Lam	S SH B	W,F G
Planarian	Planaria maculata Leidy	S,SH	
Mussel	Anodonta grandis Say	S	W.F
Mussel	Anodonta marginata Say	S	W
Polyzoan	Plumatella polymorpha Kraep	S,SH	
Leech	Placobdella parasitica Say	S,SH	
Brook silverside	Labidesthes sicculus Cope	S,SH,B	W
Snail	Ancylus fuscus Adams	B VO	
Snail	Segmentina armigera Say	B VO	W,F
Midges	Chironomus sp	B VO	W,F
Amphipod	Hyalella knickerbockeri Bate	B VO	
May-fly nymph	Callibaetis sp	B VO	W,F
Dragon-fly nymph	Ischnura verticalis Say	VO,VB	W B
Polyzoan	Pectinatella magnifica Leidy	VO,VB	W
Shrimp	Palaemonetes paludosus Gib	VO,VB	W
Cricket-frog	Acris gryllus Lec	VO,VB	W
Top minnow	Fundulus dispar Ag	VO,VB	W
Snail	Physa gyrina Say	VB	W
Snail	Planorbis campanulatus Say	VB	F
Snail	Planorbis parvus Say	VB	B
Damsel-fly nymph	Enallagma sp	VB	W
Dragon-fly nymph	Tetragoneuria cynosura Say	VB	F
Dragon-fly nymph	Anax junius Dru	VB	B
Back-swimmer	Buenoa platycnemis Fieb	VB	W
Back-swimmer	Notonecta variabilis Fieb	VB	F
Back-swimmer	Plea striola Fab	VB	F
Back-swimmer	Notonecta undulata Say	VB	F
Leech	Macrobdella decora Say	VB	W
May-fly	Ephemerella excrucians Walsh	VB	W
Isopod	Mancasellus danielsi Rich	VB	F
Bug	Zaitha fluminea Say	VB	B
Beetle	Coptotomus interrogatus Fab	VB	W B
Beetle	Donacia sp	VB	W

CHAPTER VIII

ANIMAL COMMUNITIES OF PONDS

I. Introduction

Ponds are fascinating to all, and do not lack interest from the scientific point of view. They are of especial interest to those familiar with the laboratory study of zoölogy. The common animals of the laboratory are pond animals, because pond animals are forms that will live in stagnant water. The common aquarium fishes are all pond fishes, as the brook forms die quickly if they are not supplied with running water. The frog, so much studied, is a pond form. The conditions in ponds are different from those in lakes and streams, because currents are not strong nor particularly important. The water doubtless piles up at one side or end of a pond during strong winds, and a complete circulation is effected, but this is not important. All of the conditions of lakes are duplicated in ponds, but on a smaller scale. One of the chief differences between ponds and lakes is the vegetation. Ponds are usually very largely captured by vegetation which is very much like that in the bays of lakes. Succession of plants in ponds is similar to that in lakes; the age of a pond is therefore a matter of first importance. The bottom materials are of most importance at the beginning (6, 112). The bottom materials in the ponds of the Chicago area are rock, clay, and sand. Rock-bottomed ponds have been but little studied, though there are a number of ponds in abandoned quarries of different ages which would make a good series for investigation. Clay bottom occurs in the moraine area. Nearly all the natural clay-bottomed ponds have reached a stage at which the bottom is not important, but one could no doubt find a good series if he were to make a special study. Sand-bottomed ponds are the commonest of all, and for the purpose of studying the effect of age upon ponds, a series of sandy-bottomed ponds, which differed chiefly in the matter of age, was selected.

II. Area of Special Study

The ponds that have been made the subject of special study lie in the sand area at the south end of Lake Michigan, within the corporate limits of the city of Gary, Ind. They may be reached from the stations known as Pine, Clark Junction, and Buffington (Fig. 84). The locality

136

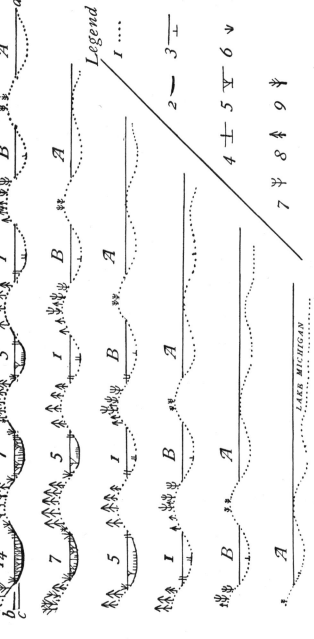

FIG. 83.—Diagram showing the relation of the horizontal and vertical series of pond stages. The horizontal series at the top represents the present ponds with the intermediate ones omitted and hypothetical stages *A* and *B* added. The line *a–b* represents the ground-water level; the line *a–c*, indicated by occasional dashes, the level of the lake. When read from bottom to top the entire figure represents the history of the area showing the addition of ponds near the lake shore and the aging of the others.

The left-hand vertical series, *A*, *B*, 1, 5, 7, 14, represents the history of the present Pond 14, and is used as the basis for the discussion of succession. The legend is as follows: (1) sandy bottom; (2) humus; (3) aquatic plants which do not reach the surface; (4) bulrushes; (5) aquatic plants which reach to the surface; (6) shrubs; (7) cottonwoods; (8) pines; (9) oaks.

is characterized by a series of ridges running parallel with the shores of the lake. Their average width is about 30 meters (100 ft.), and they are separated by ponds somewhat narrower. Most of the ponds are several miles long and vary in depth, during the spring high water, from a few inches to 4 or 5 ft. Originally there were probably a number of outlets to the ponds, either connecting them with the lake or with the Calumet River. This river flows across the long ponds at a small angle. The ponds and ridges were formed under water, and the river has cut its way across them with the falling of the lake level. The building of sewers associated with the growth of the Northern Indiana towns has drained a number of the ponds, and roads and railroads have isolated parts of others.

1. ORIGIN OF THE PONDS (62)

The waters of the lake appear to have fallen gradually from the 12-ft. level referred to on p. 47. There are at present usually two or three depressions along the shore of the lake under the water. The present submerged depressions and ridges appear to be strictly comparable to those found on the plain of Lake Chicago, and the ones with which we have to deal probably belong to a series formed by the continuous recession of the lake level (Fig. 83). This gives us a series of ponds differing principally in age, the oldest being farthest from, and the youngest nearest to, the lake.

As has been stated, the ponds have been partly drained, so that we have been obliged to study isolated portions. The younger members of the series (1st, 5th, 7th, and 14th, as counted from the lake) show the greatest differences and have, accordingly, been studied in detail. The arrangement of these ponds is shown in Fig. 83. In addition to the ponds named, the 13th, the 52d, the 93d, and the 95th have been studied, but with less care.

2. PHYSICAL CHARACTERISTICS (112)

The main facts of the topography of the isolated portions studied are shown in Table XXVIa.

TABLE XXVIa

Pond	Area in Sq. M.	Depth in Meters	Average Depth	Slope 3–7°	Slope 20°
1..........	3,500	0.6	0.3	Much	Little
5..........	3,500	0.9	0.5	Less	Much
7..........	25,000	0.9	0.5	Very little	Much
14..........	10,000	0.667	0.4	Very little	All
30..........	50,000	0.5	0.2	Very little	All
52..........	630	0.4	0.1	Very little	All

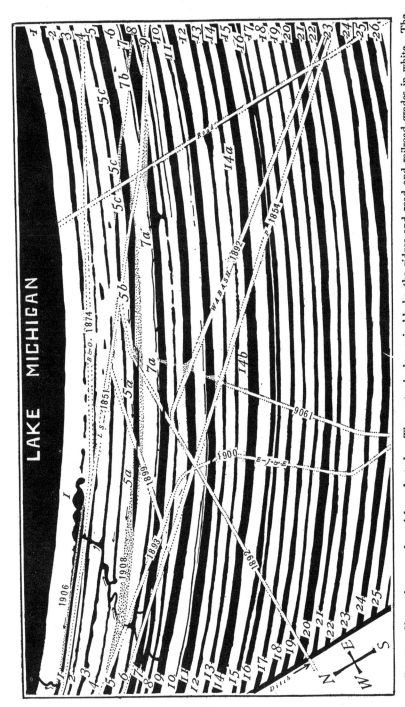

FIG. 84.—Shows the area of special pond study. The water is shown in black; the ridges and road and railroad grades in white. The numbers 1, 5, 7, 14, etc., refer to the numbers of the ponds and ridges counted from the lake shore. The dates associated with the indicated railroad grades refer to the date of building of the grade, and probable damming of the long ponds.

A decrease in depth, due to the accumulation of humus and the lowering of the ground-water level, is to be noted in the older ponds. The series is, then, an ecological age-series, and throughout our discussion we refer to earlier and later phases of the various associations concerned.

III. Communities of Ponds

1. the pelagic formation

We have in the ponds a pelagic formation. Though it is limited in number of species, many of which breed on the bottom, it is similar to that of larger lakes. We have found little difference in the pelagic species inhabiting younger and older permanent ponds. *Diaptomus reighardi* has not been taken from ponds filled with the vegetation which reaches the surface. Other species are about the same in the different permanent ponds. The pelagic formation is poorly developed.

2. pioneer formation (terrigenous bottom)

(Ponds, 1, 5, 7) (113) (Stations 9 and 32; Tables XXVII and XXXIV)

The youngest ponds of the Chicago area are near Waukegan. The outer end of the Dead River receives the force of the winter waves from the lake and the bottom is bare, with a few scattered aquatic plants. Here animals are few. We have taken only a few invertebrates. The fish present probably get their food from the older parts farther back from the lake. The fish are: the pike (*Esox lucius*) which prefers clear, clean, cool water (79); the red-horse (*Moxostoma aureolum*) which dies in the aquarium if the water is the least bit impure, and which also succumbs to any impurities in its natural environment (79); *Notropis cayuga*, which prefers clear waters; the common shiner (*Notropis cornutus*) which breeds on bare bottom (105), and the white crappie (*Pomoxis annularis*) which lives in streams. On the bottom at such a period one is likely to find the larvae of caddis-flies (*Goera* sp.), snails, mussels, etc., but we have found none in the Dead River.

Vegetation quickly captures parts of such a pond. Chara is the first plant to cover parts of the bottom. After this has happened, the pioneer formation may still continue. In Pond 1 of the series of special study (Fig. 85) we have a considerable area of bare sand, and the forms present are the caddis-worm (*Goera* sp.) and the mussels (*Anodonta marginata* and *grandis*, and *Lampsilis luteola*). These are preyed upon by muskrats (Fig. 86). There are a number of fish that belong to this formation because of their breeding relations. The large-mouthed black bass, the bluegill, the pumpkin-seed, and the speckled bullhead all make nests on the sand, the male fish guarding the nests and driving off other fish that approach. These species are the same as those of the bare-bottom formations of a lake. In their feeding the fish belong in part to another formation in the pond, namely, that of the chara.

Character of the formation: The formation may be designated as the bare-bottom formation, the forms present being those that are dependent

Fig. 85.—Shows Pond 1 at the extreme low water of the drought of 1908. In the spring the old boat is usually covered with water. In the foreground a large area of bare sand bottom is shown; to the right a few rushes and sedges. The absence of shrubs near the water's edge should be noted.

upon bare bottom in their most important activities—the fish in breeding, the caddis-worms in making their cases, the mussels in their general activities. It is necessary for the mussels to be on bare bottom in order to maintain themselves in an upright position.

Tendencies in the formation: This formation is similar to that of the bare bottom of lakes. The vegetation comes in, as has been indicated in the protected situations, and the bare bottom disappears, its place being taken by the chara. The chara gives rise to humus. upon which chara will grow for a long time, so the bottom becomes a humus- and chara-covered bottom.

3. THE SUBMERGED VEGETATION ASSOCIATION
(Ponds 1, 5, and 7; Stations 32, 33, and 34)

The *Chara* community is entirely different from that of the bare bottom, and differs also from that of other vegetations. *Chara* is highly siliceous. It is probably eaten only accidentally by animals or at least forms no important part of their food. It should be considered simply as a covering for the bottom and a resting- and living-place for animals. Some fish culturists (113) have said that it is very rich in life. This may be true under certain artificial pond conditions; but the chara ponds are poorer than any others of our series. *Chara* differs from some other plants in not reaching to the surface of the water. Many aquatic insects that carry air beneath the surface must cling to objects which reach the surface when obtaining a fresh supply, and others must crawl to the surface on some object in order to emerge from the nymphal skin (96). Associated with chara are often growths of bulrushes near the sides of the ponds and on the sterile bottom. In the sparse chara the most characteristic animal forms are *Anodonta grandis footiana* (Fig. 86), and the musk turtle (*Aromochelys odorata*), which is abundant on these bottoms but is not found elsewhere. There are often nests of a few un-identified fishes that clear off the bottom in building. The burrowing dragon-fly nymph (Fig. 87) lives on the bottom among sparse chara, in the presence of but little oxygen. It lies half buried in the mud, with its abdomen protruding a little at the end. The mud minnow (*Umbra limi*) (Fig. 88), the golden shiner (*Abramis crysoleucas*) (Fig. 88), the chub-sucker (*Erimyzon sucetta*), bullheads, the little pickerel (*Esox vermicula-tus*), the tadpole cat (*Schilbeodes gyrinus*), and occasionally the warmouth bass (*Chaenobryttus gulosus*) spend their time in the denser chara. The shiner and mud minnow place their eggs on the chara or other plants.

Among the most abundant forms in the association are the midge larvae (*Chironomus*); these (Figs. 89, 90, 91) are present sticking to the vegetation in their small silken cases in great numbers (81). They are important articles in the food of the fishes. Aquatic insects are not numerous except for the midge larvae and a little May-fly. Others are occasional horseflies (Fig. 92), damsel-fly nymphs, May-fly nymphs (*Siphlurus* sp.), and occasional dragon-fly nymphs (*Tramea, Anax, Leucorhinia*). There are also a number of dytiscid beetles, many of which are common in all shallow waters, even rain pools, because of their powers of flight.

Ecologically one of the most interesting insects is a caddis-worm (*Leptoceridae*), which creeps over the *Chara* and submerged wood. It (Fig. 93) has a case made of the minutest sand grains and pieces of humus, such as are stirred up by the waves and which are to be found

REPRESENTATIVES OF A YOUNG POND COMMUNITY

FIG. 86.—The shell of a mussel (*Anodonta grandis footiana*) that has been broken open by a muskrat; slightly enlarged.

FIG. 87.—The burrowing dragon-fly nymph (*Gomphus spicatus*), with the mask extended.

FIG. 88.—Some fishes of the pond. The dark fish which rests near the bottom is the mud minnow (*Umbra limi*). The fish swimming about is the golden shiner (*Abramis crysoleucas*); 1/5 natural size.

among the chara. This species is the successor of the bottom species (*Goera*). It belongs to a different group and has structural characters which distinguish it from *Goera*, but which probably have no relation to its habitat or habits. On the other hand, the *mores* as indicated by case-building is also different but is related to the environment. The crustaceans constitute an important element in this association. The smaller amphipod (*Hyalella knickerbockeri*) is abundant among the chara. The crayfish (*Cambarus immunis*) occurs here sparingly. In ponds there is an important element of small crustaceans that belong to the vegetation and the bottom; this element is composed chiefly of

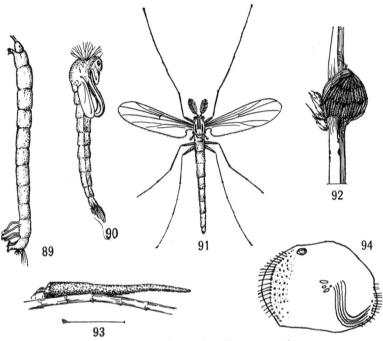

REPRESENTATIVES OF THE SUBMERGED VEGETATION ASSOCIATION

FIGS. 89, 90, 91.—Larva of a midge (89), pupa of the same (90), the adult. Midges are inhabitants of the chara-covered bottom; enlarged about 4 times (after Johannsen, *Bull. N.Y. State Museum*).

FIG. 92.—The eggs of the common large black horsefly on the tip of the bulrush stalk.

FIG. 93.—The chara-inhabiting caddis-worm (*Leptocerinae*); enlarged as indicated.

FIG. 94.—Ostracod (*Notodromas monacha* Müll.); 30 times natural size (after Sharp).

Ostracoda (Fig. 94), which are small bivalved forms resembling the bivalved *Mollusca*. They form food for fishes to a small degree.

Especially abundant just under the chara are the red water-mites (*Limnochares aquaticus*) (Fig. 80, p. 130). One sees numbers of these when he stirs the bottom. Creeping over the plants are the small snails (*Amnicola limosa*) (Fig. 100, p. 146). These respire by means of gills. Other snails are also occasionally present. *Physa* and *Lymnaea*, etc., are always small or juvenile. We have never taken an adult specimen of these from the young ponds and in all only a few specimens have been taken. These animals get into the ponds that are formed by the removal of sand. We are not at all sure but that the few forms found in Pond 1 are the result of such entrance, rather than the regular establishment of the species.

Among the bulrushes are a few aquatic insects that belong to the vegetation that comes above the surface. One of the most characteristic forms is the neuropterous larva (*Chauliodes rastricornis*) (Figs. 110, 111, p. 150), which is a marsh form and will drown in water.

Characters of the association: This association differs from the preceding and from the others generally in being distinctly aquatic and also essentially independent of the bare bottom and of the surface. The animals of this association are, however, strictly dependent upon the vegetation for nesting-places, shelter, etc. The mud minnow has been studied experimentally and shows avoidance of direct light.

Tendencies in the association: This association, like all the others, is destined not to last; changes are taking place all the time. The chara is filling the pond at the rate of one inch a year (58) and is making a fine soil for roots of other plants. As soon as the dense chara stage has existed for a time we find other plants, such as *Myriophyllum*, *Potamogeton*, and water-lilies. As soon as these have become established we have the commencement of the next association. These plants usually appear in spots, and in many cases the zones are much less important than in the lakes because of the small areas of the plants. We can, however, recognize a zone of water-lilies, and zones or patches of other plants.

Just as we noted that the formations of the bare-bottom type existed in the small ponds with the *Chara*, we see also that the surface-reaching vegetation occurs with the *Chara* association and often all three occur together. Pond 5 contains a poorly developed phase of all three, the bare bottom being of minor importance. Pond 7 contains the chara association and the surface-reaching association. Ponds 14 and 30 are the best expressions of the surface-reaching type, and Pond 52 is the last stage of it. This will be discussed more fully, and we will pass directly to the association of the vegetation which reaches the surface.

4. THE ASSOCIATION OF EMERGING VEGETATION

(Stations 34–37, 39; Ponds 5, 7, and 14) (Fig. 101) (30 and 52)

With the incoming of the water-lilies and the fine-leafed plants, we have the inauguration of a new state of affairs. Among the new animals

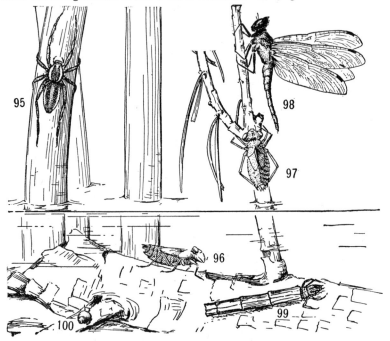

REPRESENTATIVES OF THE DENSE BULRUSH ASSOCIATION (POND 5)

(All about natural size)

FIG. 95.—The common diving spider (*Dolomedes sexpunctatus*). The individual from which this drawing was made was taken with a nymph of the dragonfly shown, in its jaws.

FIGS. 96, 97, 98.—Various stages of a dragon-fly (*Leucorhinia intacta*): 96, nymph; 97, about to shed its outer covering; 98, the adult. (Modified from Needham.)

FIG. 99.—The larva of a caddis-worm (*Phryganeidae*), which makes its case from bits of grass blades, etc.

FIG. 100.—Small gill-breathing snail (*Amnicola limosa*).

that come in, the bivalved mollusks deserve special mention. The *Unionidae* must have bare bottom for their activities; they are too large and heavy to climb on such small vegetation, and the development of such a habit has not taken place. They disappear with the sparse

Chara. Their place is taken by other bivalves, viz., the *Sphaeridae*, such as *Musculium partumeium*, which lives in the humus of the bottom, and *Musculium secure* and *truncatum*, which live in the vegetation and are able to climb on the vegetation and on the side of aquarium jars.

In the early phases, shrubs and young trees have begun to grow by the sides of the ponds and these from time to time fall into the water, thus forming a resting-place for many forms that are not found in the other situations. Diving spiders (Fig. 95) are common on the bulrushes which

FIG. 101.—Showing Pond 14 at moderate low water. In contrast with Pond 1 we see that it is choked with emerging vegetation and the margin occupied by shrubs and bulrushes, etc.

are here growing on a bottom of humus outside leaf-bearing plants (Fig. 101), inside the shrubs. These spiders dive for the immature aquatic insects which are here at their maximum. We find numerous damsel-fly nymphs and dragon-fly nymphs, both the creeping form (*Leucorhinia intacta*) (Figs. 96, 97, 98) and the climbing form. The burrowing dragon-fly nymph has gone, or is present in small numbers only, and there are but few May-fly nymphs. Those that persist creep about on submerged sticks in company with *Amnicola* and are especially likely to occur in the earlier phases of this community. With these occur the caddis-worms (*Phryganeidae: Neuronia*) (Fig. 99), which are also abundant in the later stages of dense vegetation. This worm's case is somewhat similar in form to that of *Leptoceridae*, being a circular tube, but it is made of pieces of grass blades or other pieces of plant fragments instead of sand grains. The pieces are fastened together with silk. The worm is found creeping among the vegetation, drawing its case after it. *Amnicola* (Fig. 100), the river-dwelling snail, is common, especially on twigs and logs. In the mature stage represented by Pond 14 (Fig. 101) the common newt (Fig. 102) probably reaches its maximum abundance. The snails which are at best advantage in these ponds are the lung breathers. They can here come to the surface for air, and food is abundant, as the surfaces of the plants are covered with algae and these form the food of the snails. Those snails which come to the surface for air are common. *Planorbis campanulatus* (Fig. 103) is characteristic of the mature stage and *Lymnaea reflexa* (Fig. 104) in the older stages. The individuals in this case are larger than those of the temporary marshes (cf. Figs. 104 and 125, pp. 149, 175). *Planorbis parvus* (Fig. 105) is commonest in the earliest phases and *Planorbis hirsutus* (Fig. 106) in the later. Diving beetles (Fig. 107), which are common throughout, are most numerous in the denser vegetation. The soldier-fly larvae (Fig. 108) are often common in the dense filamentous algae of the mature phases of the association; here the number of all dipterous larvae is greater than at any other point. Midge larvae occur in great numbers, having their cases among the algae. Horseflies (Fig. 92), also *Tanypus, Ceratopogon*, and some mosquitoes are present. Specific identification, however, is not possible, and whether or not the species differ in modes of life or reactions from those inhabiting the earlier stages in the pond series has not been determined.

Adult aquatic insects have increased with the increase in vegetation, in a remarkable fashion. The prominent forms are the larger bugs, such as the electric-light bugs (*Zaitha fluminea* and *Belostoma americana* Leidy, with *Benacus griseus* Say). The water-boatmen are also common. The species of these are not well known, and we cannot say whether or not they are the same in the older and younger ponds. Back-swimmers are also abundant (*Notonecta variabilis* and *undulata*, *Buenoa platycnemis*, and the small form, *Plea striola*, occur here). They are few in number or absent from the younger ponds.

Some animals particularly abundant in the older stage are the common leech (*Placobdella parasitica*) (Fig. 109), the larvae of a netted-winged insect (*Chauliodes rastricornis*) (Figs. 110, 111), the large flat snail (*Planorbis trivolvis*) (Fig. 112), and the amphipod (*Eucrangonyx gracilis*) (Fig. 113). All these occur in the senescent stage, where in dry years the pond goes almost dry. The vertebrates of the mature and later stages are not numerous. The fish are limited to mud- and muck-preferring species, the black bullhead (*Ameiurus melas*) and the mud minnow (*Umbra limi*) (106). The grass pickerel and the dogfish are found in such vegetation-choked ponds.

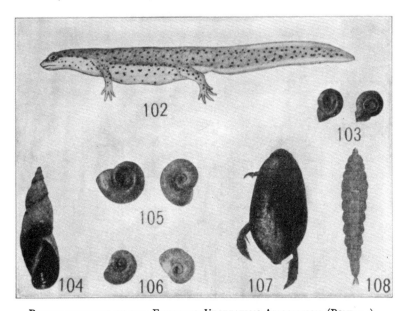

REPRESENTATIVES OF THE EMERGING VEGETATION ASSOCIATION (POND 14)

FIG. 102.—The common newt (*Diemictylus viridescens*); natural size (after Hay).

FIG. 103.—A flat pond snail (*Planorbis campanulatus*); natural size.

FIG. 104.—The common pond snail (*Lymnaea reflexa*); natural size.

FIG. 105.—Small flat snail (*Planorbis parvus*); 3 times natural size.

FIG. 106.—A snail (*Planorbis hirsutus*); 3 times natural size.

FIG. 107.—A predaceous diving beetle (*Cybister fimbriolatus* Say); natural size.

FIG. 108.—A soldier-fly larva—unidentified; twice natural size.

The amphibia are the frogs which occur in all stages of the association, and the common salamander (*Amblystoma tigrinum*), which burrows in the soft mud where it remains during the greater part of the year. It comes out in spring (February or March) and deposits eggs in the pond, where the young are found later. Of the turtles the common painted turtle (*Chrysemys marginata*) is abundant, basking on the fallen trees. The geographic turtle and the snapping turtle are found also in the younger phases. Garter-snakes pick up their food along the ponds (Fig. 114), while muskrats, occasional minks, and various aquatic birds (108) feed in the ponds.

SENESCENT POND INHABITANTS

FIG. 109.—A leech with young attached to the ventral side (*Placobdella parasitica*); natural size.

FIG. 110.—The larva of a netted-winged insect (*Chauliodes rastricornis*).

FIG. 111.—Pupa of the same (slightly enlarged).

FIG. 112.—A snail (*Planorbis trivolvis*); natural size.

FIG. 113.—Common amphipod (*Eucrangonyx gracilis*); twice natural size.

FIG. 114.—Pond 58 in a dry season, showing dead fish (mud minnows) both on bottom and out of water and in the water. A garter-snake (*Thamnophis* sp.) feeding on the fish.

Consocies of logs.—This is the chief place to find the sponge and the polyzoa. Their numbers vary from year to year but they are usually present. With them are often found leeches, especially *Macrobdella decora*, which is a brilliant red-and-green form. The only character-istic insect is the dytiscid beetle (*Agabus semipunctatus* Kirby) (99c), a slender reddish-brown form. The other forms found here are inci-dental in the vegetation. Hollow logs are probably used for breeding-places by the fishes, such as the bullheads (105), while the eggs of *Physa* and of water-mites, and some of the aquatic insects, are also placed here. The mammals of these ponds are the muskrat, which occurs in all the stages, and the mink, which is now rare.

Tendencies of the association: This association is unstable. Its fate is heralded by the incoming of different amphibious plants at the sides. This is the form *Proserpinaca*, with the divided leaves above water and the entire ones below. This is often associated with *Equisetum* and plants that have the growth form of grasses. Following these are the shrubs, such as the buttonbush (6). Before these have captured the entire pond it becomes dry during the dry season and the end of the aquatic community is come. The formation which follows is the tempo-rary pond, swamp, or marsh type.

Characters of the formation: The formation composed of the two associations mentioned may be characterized as made up of forms which require but little oxygen, and no bare bottom. The reproduction is one of two types: either the young are carried or the eggs are attached to plants. Some of those carrying the young are the *Sphaeridae*, the amphipods, and the isopods. Those sticking the eggs onto or into the vegetation are the snails (all), the *Dytiscidae*, all the species recorded, the *Hydrophilidae*, the *Notonectidae*, the *Belostomidae*, the *Ranatras*, the caddis-flies, the *Donacias*, and in fact most of the forms of the formation.

IV. SUCCESSION

The first formation to take possession of a pond when it is first separated from a lake like Lake Michigan is the bare-bottom formation; chara soon makes its appearance in the deeper parts and we have the beginning of the chara association. The chara association so acts upon the bottom by covering it with humus and vegetation that it renders the continued existence of the bare-bottom formation impossible (6, 112, 114, 114a). At the same time it prepares a way for the vegetation which reaches to and above the surface. This, in turn, fills the pond still further, and the strictly marsh vegetation takes possession. The history of the true pond is then at an end and the story of the marsh begins. Our series of 95 ponds illustrates the series of stages. The vegetation which comes to the surface of the water and the later marsh and swamp vegetation encroach from the sides toward the center. *Entomostraca* do not ordinarily show so clear a succession of species as do other groups and our collections are very incomplete. The following have been noted: Cladocerans: *Ceriodaphnia reticulata* Jurine, *C. pulchella* Sars, and *C. quadrangula* Muel. from Ponds 52 to 93. Copepods: *Cyclops albidus* Jurine appears more common throughout the series and *C. viridis* Jurine is common in the older ones. *Diaptomus reighardi* Marsh is in the younger ponds and its place is taken by *D. leptopus* Forbes beginning with Pond 30. Of the ostracods, *Cypria exsculpta* Fisch. is common throughout the series. *Cypridopsis vidua* Müll. is common in the semi-temporary ponds.

I. FATE OF THE PONDS

In the late stages the pond dries during extreme droughts and passes rapidly from the stage at which it dries occasionally during a dry season to the stage when it dries every season. It is then known as a marsh or swamp, or often vernal marsh or swamp, or summer dry pond. At such a stage it is a land habitat in summer and a water habitat in spring. As the pond bottom is built up higher by the accumulation of peat, and the surrounding ground-water level is lowered by the forces of erosion, the question of what is to become of the pond brings us to a question of great importance in connection with climatic formations. It will become whatever the surrounding climatic formation may be. If it is forest, directly or indirectly, the pond becomes forest, and if it is steppe the pond becomes steppe, while if prairie or savanna the pond becomes savanna.

We have already noticed that the area of study is on the border of the forest and prairie (steppe formations). A pond in the area of study may therefore become prairie or forest. Ponds with sloping sides usually become prairie, and those with steep abrupt banks or shores turn into forest. There is no marked difference between the animal life of the two. Collections made in a series of three prairie ponds which are situated near Wolf Lake, Ind., and which in ecological age may be compared with Ponds 1, 7, and 14 of the Lake Michigan series, are almost parallel with the collections from the Lake Michigan ponds. The differences to be noted are that the snail *Planorbis trivolvis*, which usually occurs in old ponds only, is found in the earliest pond of the prairie pond series, while the snail *Vivipara contectoides* and the shrimp *Palaemonetes paludosus*, which usually occur only in streams and small lakes, also occur in the prairie pond series. The presence of the latter two may be explained, however, by the fact that the ponds were once connected with Wolf Lake.

In the pond formation proper, the fate of the pond early becomes evident along the margin. This will be discussed in connection with swamps and marshes. The discussion of the areas properly called marshes and swamps is the most complex of all our discussions, and will be taken up in the chapter on swamps, marshes, and temporary ponds. Tables XXVII–XXXIII show animals recorded from the series of ponds at the head of Lake Michigan (Stations 32–37).

TABLE XXVII
SPONGES

Name	Pond Numbers							
	1	5	7	14	30	52	93	95
Meyenia(?) *crateriformis* Pot....	*	*						
Meyenia fluviatilis Auct........			*					
Heteromeyenia argyrosperma Pot.			*					
Spongilla fragilis Leidy........			*				*	

TABLE XXVIII
LEECHES

Name	Pond Numbers							
	1	5ᶜ	7ᵃ	14	30	52	93	95
Glossiphonia fusca Castle........	*							
Dina fervida Verrill.............	*	*	*	*				
Erpobdella punctata Leidy.......	*	*	*		*	*	*	
Macrobdella decora Say..........		*	*	*				
Haemopis grandis Verrill.........		*	*	*				
Placobdella parasitica Say........		*	*					
Placobdella rugosa Verrill........		*	*			*	*	
Glossiphonia heteroclita Linn......						*	?	*
Haemopis marmoratis Say........						*		*

TABLE XXIX
Sphaeridae AND *Unionidae*

Name	Pond Numbers							
	1	5ᶜ	7ᵃ	14ᵇ	30	52	93	95
Unionidae—								
Lampsilis luteola Lam.........	*							
Anodonta grandis Say..........	*	*						
Anodonta marginata Say........	*	*	*					
Anodonta grandis footiana Lea...		*	*					
Sphaeridae—								
Musculium truncatum Lins......	*	*			*	*	*	
Musculium secure Prime........		*	?	*			*	
Musculium partumeium Say.....		*	?					

TABLE XXX

SNAILS

Name	Pond Numbers							
	1	5c	7a	14b	30	52	93	95
Amnicola—								
Amnicola limosa Say	*	*	*	*				
Amnicola cincinnatiensis Lea	*	*	*	*				
Amnicola limosa parva Lea			*	*				
Physa—								
Physa gyrina Say	F	F	C	C	C			
Physa heterostropha? Say				*				
Lymnaeidae—								
Lymnaea reflexa exilis Lea	*							
Planorbis bicarinatus Say	F	F						
Lymnaea humilis modicella Say	*	*	*					
Lymnaea obrussa Say	*	*	*					
Planorbis parvus Say	*		*	*				
Planorbis campanulatus Say	*	*	*	*				
Planorbis hirsutus Gld	*	*	*	*	*	*		
Planorbis exacuous Say						*	*	
Lymnaea reflexa Say	F	F	C	A	A	*	*	
Planorbis deflectus Say			*					
Planorbis trivolvis Say				C	A	*	*	
Segmentina armigera Say				*	?	*	*	?

TABLE XXXI

Crustacea

Name	Pond Numbers							
	1	5c	7a	14b	30	52	93	95
Hyalella knickerbockeri Bate	C	C	C	F	F			
Eucrangonyx gracilis Smith		F	C	A	A	A	A	F
Mancasellus danielsi Rich					*	*	*	
Asellus communis Say					*	*	*	*
Cambarus immunis Hagen	F	F	C	C	C	*	*	*
Cambarus blandingi acutus Girard			F	F	?	*		

TABLE XXXII

AQUATIC INSECT LARVAE AND NYMPHS

Name	Pond Numbers							
	1	5c	7a	14b	30	52	93	95
May-flies—								
Siphlurus sp	*	*	*	*		*	*	
Caenis sp	*	*	*	*	*			
Callibaetis sp					*	*		
Neuroptera—								
Chauliodes rastricornis Ram	*	*	*	*	*	*	*	*
Damsel-flies—								
Lestes sp	*							
Enallagma sp	*	?	*	*			*	*
Ischnura verticalis Say		*	?	*	*			
Dragon-flies—								
Tramea lacerata Hagen	*							
Celithemis eponina Drury	*							
Libellula pulchella Drury	*	*						
Gomphus spicatus Selys	*	*	*					
Leucorhinia intacta Hagen	*	*	*	*				
Anax junius Drury	*	*	*	*	*	*	*	
Sympetrum rubicundulum Say		*						
Sympetrum sp			*	?	*			
Pachydiplax longipennis Burm						*		
Epiaeschna heros Fab								*
Caddis-worms—								
Goera sp	*							
Leptocerinae sp	C	F						
Neuronia sp			˙	C	A	*	*	*
Diptera larvae—								
Chironomid larvae	*	*	*	*	*	*	*	☼
Stratiomyid larvae	*	*	*	*	*			
Tanypus sp		*	?	*		*	*	☼
Tipulid larvae				*	?			
Ceratopogon sp					*			
Hemiptera—								
Ranatra kirkaldyi Buen	*	*						
Corixa sp	*	*	*					
Ranatra fusca P.B	*	*	*	?	?			
Zaitha fluminea Say		*	*	*	*	*	☼	
Notonecta undulata Say		*	*	☼	*			
Buenoa platycnemis Fieb		F	C	?				
Notonecta variabilis Fieb								
Plea striola Fieb				*	*			
Water-striders—								
Gerris rufoscutellatus Lat	*							
Gerris marginatus Say	*	?	*					
Mesovelia bisignata Uhl	*		*					

TABLE XXXIII

DISTRIBUTION OF FISH: PONDS ARRANGED ACCORDING TO ECOLOGICAL AGE
For meaning of numbers and letters see Fig. 84, p. 139.

Common Name	Scientific Name	Ponds							
		1	5b 5c	7a 7b	5a	14a	14b	56 58	52
Large-mouthed black bass..	*Micropterus salmoides*..	*							
Bluegill	*Lepomis pallidus*......	*							
Blue-spotted sunfish.......	*Lepomis cyanellus*.....	*						*	
Pumpkin-seed...........	*Eupomotis gibbosus*....	*							
Warmouth bass..........	*Chaenobryttus gulosus* ..	*							
Yellow perch.............	*Perca flavescens*.......	*	*						
Chub-sucker.............	*Erimyzon sucetta*......	*	*	*					
Spotted bullhead.........	*Ameiurus nebulosus*....	*	*	*					
Pickerel.................	*Esox vermiculatus*......	*	*	*	*	*		*	
Mud minnow............	*Umbra limi*...........	*	*	*	*	*		*	
Golden shiner...........	*Abramis crysoleucas*....		*	*	*	*			
Yellow bullhead..........	*Ameiurus natalis*......		*						
Black bullhead...........	*Ameiurus melas*.......		*	*	*	*		*?	
Dogfish.................	*Amia calva (juvenile)*..								*

TABLE XXXIV

HIGHER VERTEBRATES

Name	Pond Numbers				
	1	5c	7a	14b	30 52
Aromochelys odorata Lat.................	*	*			
Rana pipiens Sch.......................	*	*	*	*	*
Chrysemys marginata Ag.................	*	*	*	*	*
Graptemys geographicus LeS..............		*			
Diemictylus viridescens Raf..............		*	*	*	*
Fiber zibethicus Linn...................	?	?	?	?	?

CHAPTER IX

CONDITIONS OF EXISTENCE OF LAND ANIMALS

I. INTRODUCTION

Man being a land animal, it is natural that he should be more familiar with the conditions of existence of land animals than with those of aquatic forms. The reader will recognize that the primary divisions into which land animals may be divided are (*a*) those living exposed to the atmosphere on the surface of the soil and of plants and animals, and (*b*) those out of direct contact with the atmosphere, in the soil, in wood, and in the tissues of living plants and animals. The solid substances in and upon which animals live are called *materials for abode* (55, 115) and, aside from soil, materials are just as varied as are the living and decaying bodies of plants and animals. For this reason, an adequate discussion of such materials for abode would require a separate treatise. Since the laws governing the physical conditions surrounding animals living hidden away, for example in the bodies of living and dead organisms, are little known, we will pass directly to a discussion of the conditions of existence of animals living in soil and exposed to atmosphere.

II. SOIL (116)

Because of its importance in agriculture, the relation of plants to soils has been much studied. The laws governing plants in their relation to soils apply in the main to soil-inhabiting animals, all the various properties of soils being of some importance in this connection.

1. TEXTURE

The texture of soils is of importance to animals because of the varying difficulty with which they may burrow into it, and the ease with which their burrows are maintained when once dug. Particular animals prefer soils of a particular texture, some preferring rock, some sand, etc.

2. WATER

Most subterranean animals are submerged in water during rains. The amount of water which they encounter in the soil at other times is determined to a large extent by their relation to the water table (57), and by the character of the soil. The water-holding power of different soils is different. It increases with the decrease in size of the soil particles and with the addition of humus which takes up water by imbibition. The amount of water in the soil is usually expressed in terms of per cent of weight, but a soil with 8 per cent of moisture may not give up water to an organism as readily as another soil with only 2 per cent. It is necessary therefore to determine the capacity of a soil to retain or give up moisture. This has been determined for a number of soils (117, 118), in terms of what is called the moisture equivalent. The moisture equivalent of a soil is the percentage of water which it can retain in opposition to a centrifugal force 1,000 times that of gravity. The maintenance of turgor in plants is believed to be a purely physical matter. If the roots of a plant are in a mass of soil, the plant gradually reduces the water content until the permanent wilting occurs. The *wilting coefficient* of a soil is the moisture content (in percentage of dry weight) at the time when the leaves of the plant growing in the soils first undergo a permanent reduction in moisture content, as a result of a deficiency of moisture supply. The *moisture equivalent* of a soil is 1.84 times the *wilting coefficient for wheat*, used as a standard plant. Fuller (119) states that the wilting coefficient of dune sand is about 0.75 per cent, while the usual moisture content of the cottonwood dune sand is two or three times this amount. For the clay soil of the oak-hickory forest, according to McNutt and Fuller (119*a*), the coefficient is about 8 per cent. These standards of soil moisture indicate the amount of water available to animals through direct contact with the soil or available for evaporation into the air of cavities which they construct for themselves beneath the surface of the soil. A soil gives water to or takes water from the body of a subterranean animal in proportion to the availability of water in the soil in question. The amount of available water increases with depth (119)

3. TEMPERATURE

Transeau found that the temperature of bog soil and bog water is below that of other soils and waters. This has, however, not been observed for different dry soils. The differences between soil on the beach at Sawyer, Mich., August 19, 1911, at 3:00 P.M. and in the beech woods near at hand was as follows: Air 20° C., upper half-inch of beach sand 38°–39° C., sandy soil of beech woods 19°–20° C., a difference of 19° C. The upper half-inch of bare sand goes as high as 47° C. on the hottest days of summer, while the soil in the beech woods is probably always a little cooler than the air at the time of the air maximum. Dune sand temperature on the hottest summer days at about 3:00 P.M. has been found to be as follows:

TABLE XXXV

SHOWING VARIATION OF SAND TEMPERATURE WITH DEPTH AND MOISTURE CONTENT
AIR 36° C.

	Dry Sand	Moist Sand
1.25 cm. below surface....................	47° C.	32° C.
3–4 cm. below surface....................	38° C.	31° C.
8–9 cm. below surface....................	35° C.	29° C.
10–11 cm. below surface....................	33° C.	.....
12–13 cm. below surface....................	32° C.	27° C.
17–18 cm. below surface....................	30° C.	.....

It will be noted from the table that temperature decreases with depth and with increasing moisture.

4. PLANTS AND ANIMALS

Cowles (120) mentions the importance of soil bacteria which increase with the increase of the humus, and the development of substances toxic to the plants producing them (121, 114a). Little is known of the effect of animals upon the soils in which they live but if excretory products ever accumulate in any quantity, they probably have a detrimental effect, especially upon the animals which produce them (114). On the other hand, many burrowing animals bury organic material and bring mineral soil to the surface. The digger wasps add much to the sand by burying many insects for their young. Earthworms contribute to soil formation (30). Cowles states further on the authority of Transeau (122) that humus accumulation alters soil aeration. It follows that the atmosphere available to subterranean animals differs in different soils.

III. ATMOSPHERE

Animals living fully exposed to the atmosphere are usually those most dependent upon the various physical factors of the air, viz., light, temperature, pressure, humidity, currents, electrical conditions, etc.

1. LIGHT

Animals are either positive or negative to the actinic rays of the spectrum (45, 123). Considerable work has been done by plant ecologists, on the measurement of light with photographic papers, but its bearing on plant problems is questioned by some because the nonactinic portion of the spectrum is most important in the process of photosynthesis. It appears that these measurements are of much greater significance for animals than for plants. Zon and Graves (124) have brought together the literature and discussed the methods of study (see especially several papers by Wiesner). The light in which animals live varies from that of the strongest sunlight of mid-day to the darkest recess of soil, etc. Many animals show diurnal migration due to changes in light.

2. TEMPERATURE

The temperature of the air varies with light (insolation). Cloudy summer days are about 4° cooler than sunny days. Cloudy winter days are warmer (125, p. 136) than sunny ones. The temperature of the lowest strata of air on sunny days varies in some inverse ratio with the distance from the soil, vegetation, etc. The temperature immediately above bare soil may be very high in summer (see Table XXXV).

3. PRESSURE

According to experimental work by Cohnheim and others (126, 127), man is sensitive to variations in atmospheric pressure. Many other animals, such as rabbits, dogs, etc., are probably also sensitive. Bird movements are often correlated with variation in atmospheric pressure. In all cases the pressure, as meteorologically recorded, represents a variation in humidity, etc., and relations to pressure alone have been but little studied.

4. HUMIDITY

Atmospheric humidity (128) is very important to animals and determines the sensible temperature and rate of evaporation to a large degree (see under "Evaporation," below).

5. COMPOSITION OF THE ATMOSPHERE

(Table II, p. 59)

The amount of carbon dioxide varies (125) in different localities but is usually greatest near the ground where decomposition is taking place. Animals living among decaying organic substances probably live in the presence of much more carbon dioxide than animals upon vegetation. Carbon dioxide is probably important to animals because of its effect upon respiratory activity. Carbon dioxide is believed by some physiologists to be a necessary stimulus to the brain to cause all respiratory movements. It is further held by some that mountain sickness (associated with high altitude) is due to decreased carbon dioxide pressure.

6. CURRENTS

Currents of wind are important in scattering animals and in affecting the rate of evaporation from their bodies. Some animals take up definite positions with reference to wind (anemotaxis) (128*a*), as for example some flies hover in the air in one position with the head toward the wind. Some animals, such as the land salamanders, frogs, toads, millipedes, spiders, and insects turn away from currents of air because of increased evaporation.

7. ATMOSPHERIC ELECTRICITY (125)

The effect of atmospheric electricity upon organisms is little known. It varies with variations in other conditions of the atmosphere. It will probably be found to be important in the life of animals.

IV. COMBINATIONS OR COMPLEXES OF FACTORS

As has already been pointed out (55), the animal environment is a combination of moisture, temperature, light, pressure, materials for abode and food, all of which factors taken together constitute a complex of interdependences. These various factors are so dependent upon one another that any change in one usually affects several others. This property of environmental complexes is what makes ecology one of the most complex of sciences, and experimentation in which the environment is kept normal except for one factor, an ideal rarely realized in practice, even under the best conditions.

The efforts of ecologists, geographers, and climatologists have long been directed toward the finding of a method of measuring the environment which shall include a number of the most important environmental factors. De Candolle undertook to base the efficiency of a climate, for supporting plants, upon the mean daily temperatures above 6° C., this temperature being taken as the starting-point of plant activity. Merriam has followed this lead and calculated total temperatures for many places in North America and made maps and zones based upon such totals. This system, however, has been rejected by botanists and plant ecologists on account of much evidence, both experimental and observational, which is quite out of accord with this view. The scheme has not been generally accepted by zoölogists outside of the United States Biological Survey. There is practically no evidence of an experimental sort for the application of such a scheme to animals. Relative humidity has been suggested as an important index (128) but does not properly express the influence of atmospheric humidity upon the animal body (125, p. 53). The saturation deficit has also been suggested but does not take temperature into account.

I. EVAPORATION

"The total effect of air temperature, pressure, relative humidity, and average wind velocity upon a free water surface in the shade or in the sun is expressed by the amount of water evaporated" (125, p. 72). Since temperature in the season without frost is directly due to the sun's rays, light is in part included. In our latitude, clouds in summer slightly decrease the air temperature (125, p. 72). In winter, however, the temperature of cloudy days is higher. The strongest light is usually associated with the greatest evaporation. Yapp (129) found that the rate of evaporation was directly correlated with temperature and illumination, but most closely correlated with relative humidity. From the standpoint of including many factors, the evaporating power of the air is by far the most inclusive and is therefore by far the best index of physical conditions surrounding animals wholly or partly exposed to the atmosphere. It is not, however, to be expected that it will hold good for all the factors under all climatic conditions, and for this reason, records of light, temperature, pressure, carbon dioxide, etc., should be made.

The data are usually obtained by using a porous cup atmometer. Evaporation from the atmometer is more nearly like that from an organism than is evaporation from any other device; it was devised by Livingston (130). "It consists of a hollow cup of porous clay 12.5 cm. high, with an internal diameter of 2.5 cm. and a thickness of wall of about 3 mm. It is filled with pure water and connected by means of glass tubing to a reservoir usually consisting of a wide-mouthed glass bottle of one-half liter capacity. The water, passing through the porous walls, evaporates from the surface, the loss being constantly replaced from the supply within the reservoir. Readings are made by refilling the reservoir from a graduated burette to a certain mark scratched upon its neck. For convenience in handling, a portion of the base of the cup is coated with some impervious substance and, before being used in the field, the instrument is standardized by comparing its loss of water with that from a free water surface of 45 sq. cm. exposed under uniform conditions. As a further check against error this standardization is repeated at intervals of six to eight weeks throughout the season" (Fuller, 131). In Fuller's work, the bottles were arranged so that the evaporating surface of the instrument was 20–25 cm. above the surface of the soil.

a) Effect of evaporation upon animals.—In the case of man some observations have been made. According to Pettenkoffer and Voit (*fide* 125), an adult man eliminates 900 gms. of water from his skin and lungs daily. Of this amount 60 per cent or 540 gms. come from the skin alone and changes in relative humidity of only 1 per cent cause perceptible changes in the amount of evaporation from the skin. If evaporation from the skin and lungs is diminished, the amount of urine is increased, as in many cases are also the secretions of the intestines. Sudden changes in humidity make themselves felt in sudden increased or decreased blood pressure. The less dilute blood of dry climates operates as a stimulant and increases the functions of the nervous system. The consequences are excitement and sleeplessness (125, pp. 56–57).

Little has been done on the physiological effect of evaporation or desiccation upon cold-blooded animals. Various writers have found a loss of water associated with hibernation. Greeley (132) obtained the same results with desiccation as with freezing (132; 133; 51, pp. 182–88). The reactions of animals to evaporation gradients have been studied by the writer (134). A high rate of evaporation is advantageous to some animals and decidedly detrimental to others. Animals inhabiting dense woods turn back when they encounter air with a high evaporating power. This is true of frogs, salamanders, insects, and millipedes. The frogs and salamanders die in an hour or more in an atmosphere of high evaporation power but centipedes and ground beetles and other heavily armored animals do not die for many hours or even days though they react negatively to the dry air when they encounter it in a gradient. Animals from hot, dry, sand areas usually select air of high evaporating power and die in air of high evaporating power only after very long exposure. The results of a long series of experiments may be summarized as follows: (1) the animals studied react to air of a given high rate of evaporation whether the evaporation is due to moisture, temperature, or rate of movement; (2) the sign and degree of reaction to the given rate of evaporation are in accord with the comparative rates of evaporation in the habitats from which the animals were collected; (3) the animals of a given habitat are in general agreement in the matter of sign and degree of reaction; the minor differences which occur are related to vertical conditions (see below) and kind of integument; (4) there is a rough agreement between survival time in air of high evaporating power, and kind of integument, but no agreement between survival time and habitat when a number of members of a community are taken together. The relation of warm-blooded animals to rate of evaporation has been sufficiently studied so that, when it is taken with the work on cold-blooded animals, we are warranted in concluding that the evaporating power of the air is probably the best *index* of *environmental conditions* of land animals.

b) *Evaporation in different habitats.*—The evaporating power of the air varies in different situations (Fig. 115). There are great differences between open prairies and closed forests. Shimek (135) found that the evaporation in the undisturbed groves in Eastern Iowa during July and August was very much less than that in the prairies adjoining. From the free surfaces of pans set in the ground so that the water which they contained was level with the surface of the soil, the evaporation of the groves was about 27 per cent of that of the prairie; with cup evaporimeters about 37 per cent, and with Piche evaporimeters

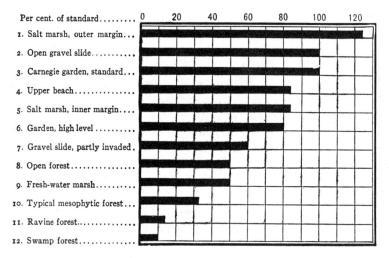

FIG. 115.—Showing the comparative evaporation rates in the ground stratum of several animal habitats on Long Island during July and August (after Transeau, courtesy of the *Botanical Gazette*).

about 47 per cent. This is about the same as the difference on Long Island between the inner side of Transeau's salt marsh dominated by grasslike plants and his mesophytic forest. Sherff (135) found the evaporation in a marsh forest to be a little less than that in the beech-maple and from 1.8 to 2.6 times as great as in the lowest stratum of an open marsh.

c) *Vertical differences in evaporating power and other conditions.*—The evaporating power of the air is usually greater at the higher levels of a habitat.

TABLE XXXVI

EVAPORATION FROM POROUS CUP EVAPORIMETERS IN DIFFERENT STRATA OF A
SUMMER DRY MARSH, CAMBRIDGESHIRE, ENGLAND, DURING THREE
PERIODS BETWEEN JULY 9 AND SEPTEMBER 8, 1907

(Yapp, 129, p. 299)

HEIGHT ABOVE GROUND	RATIO OF EVAPORA-TION	TEMPERATURE			
		Mean Max.	Mean Min.	Mean	
5 ft. 6 in. to 4 ft. 6 in..	100.00	22.1	6.6	16.5	Well above vegetation
2 ft. 2 in...........	32.8	23.0	....	....	A little above the mid-height
5 in...............	6.6	18.0	7.1	14.1	
Soil...............	.....	12.7	11.2	11.8	

Table XXXVI shows marked differences in the rate of evaporation
and considerable differences in temperature at the different levels, both
due largely to vegetation. Differences in light are also to be expected.
Sherff (136, p. 420) has found conditions similar to the above by a two
months' study of evaporation on Skokie Marsh near Chicago. The
evaporation there was three times as great at a height of 1.95 m. as at
the surface of the soil in among the plants of Phragmites. Mr. Harvey
has secured similar (unpublished) results on the prairie at Chicago
Lawn, Chicago, also Mr. Fuller, in the beech woods.

Division into strata: Plant and animal habitats are commonly
divided into strata as shown below.

Plant (12) after Warming

1. No such stratum recognized.

2. Ground stratum made up of algae,
mosses, immediately above the surface of the ground.

3. Field stratum; grasses and herbs.

4. Shrub stratum; formed of shrubs
taller than the herbaceous vegetation.

5. Tree stratum.

Animal

1. Sub-aqueous stratum made up of
animals requiring water during
their active reproductive stages.

1a. Subterranean stratum made up
of animals or stages in the life
histories of animals which inhabit
the ground, especially during the
breeding season.

2. Ground stratum made up of animals or early stages in the life histories of animals, as 1.

3. Field stratum; the inhabitants of
the herbaceous vegetation on land.

4. Shrub stratum; inhabitants of
shrubs.

5. Tree stratum; inhabitants of trees.

V. Quantity of Life on Land (137)

The quantity of life on land has been but little studied. While it is evident that some habitats have more animals than others, we have no exact data. As a rule the number of species is small in pioneer situations. While the number of individuals in some one or two species may be large, the grand total is probably not so large as in later stages. In forest development it appears from naturalistic observations that the number of both species and total number of individuals increases with age up to the oak-hickory stage, the maximum being in the oak-hickory stage. The beech and maple forest is qualitatively and quantitatively poor in animals. Felt (137) records pest species on the trees of the white-oak, red-oak, hickory forest as follows: Oak in general, 157; red oak, 12; white oak, 31; hickory, 30; wild cherry, 38; hazel, 33; total 401. He records pests on trees of beech and maple forest as follows: beech, 92; sugar maple, 19; pawpaw, 5; total, 116.

1. FOOD SUPPLY

The food supply of land animals is in part dependent upon soil. All the chief principles governing the elementary food substance of plants and animals in water are given on pp. 65–68. Since all these processes are dependent upon water (as a solvent) and since soils at all times contain some water (116), the reader will easily apply most of the principles there stated to the soil problem. There is probably no kind of organic matter found that is not food for some animals. Some require plants or their juices, some decayed fruits, some wood, some living animals, and some carrion. Each stage of its decay, a dead plant or animal is food for *some animal.*

Certain animals, usually plant-eaters, reproduce very rapidly and are preyed upon by many other animals. Mice, aphids, grasshoppers are examples (26). These form small centers about which many of the activities of a community rotate. The centers are indicated by the convergence of lines in Diagram 6.

2. EQUILIBRIUM

The balance in land communities is probably less perfect than in aquatic communities even under strictly primeval conditions. This is due to the fact that there are many small (feeding) groups of organisms centering around each of several rapidly reproducing groups such as aphids, mice, and grasshoppers. It is accordingly probably possible for a land community to be out of adjustment in some particular corner without the maladjustment being felt as far as in an aquatic community of corresponding magnitude.

To illustrate the character of land communities in the matter of food supply and equilibrium we have chosen a number of prairie animals and constructed them into an arbitrary community. This community is graphically represented in Diagram 6. The arrows point from the animal eaten to the animal doing the devouring, many such relations being shown on the basis of actual published records.

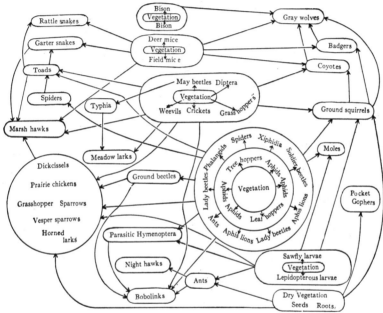

DIAGRAM 6.—Showing the food relations of land animals. Circles and ellipses inclose groups of organisms which are commonly eaten by the same animals, and groups eating similar food. Arrows point from the animals eaten to those doing the eating. For explanation see text.

From the diagram we note that wolves destroy the bison. If for any reason the wolves increased, they would destroy so many bison that the bison would decrease because wolves were abundant. The greater destruction of mice in summer by the numerous wolves would cause a decrease of mice. Finally, wolves would decrease because of lack of big game in winter and mice in summer. This would give the bison and mice an opportunity to recover their former number and the whole chain of changes would be duplicated and a general equilibrium be

re-established.   The decrease of mice just noted might, however, cause
the coyote to eat more ground squirrels and thus cause an increase of
insects because of the removal of the ground squirrel as a check upon
their numbers.   The numerous checks upon the numbers of insects
would tend to prevent their increasing greatly, but would no doubt
affect the greater part of the community.   The reader will be able to

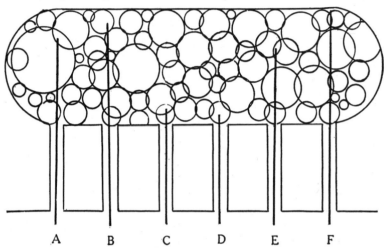

A   B   C   D   E   F

DIAGRAM 7.—Representing the food relations of the animals of a land community.
The circles represent life histories which come into contact or overlap at the point
where one species feeds upon another.   The vertical shafts represent the animals
which feed upon the vegetation (herbivora and phytophaga).   The extent to which
the shaft penetrates the community indicates its importance as food of the forms
whose life histories are represented.   The letters refer to the vertical lines (shafts)
above them.   These lines (shafts) represent the various central groups of Diagram 6
and other comparable groups as follows: *A*, large herbivores such as the bison;
B, the mice, rats, squirrels, and rabbits; *C*, vegetation-eating birds; *D*, boring insects
secured by the woodpeckers; *E*, the large plant-eating insects; *F*, the small soft-
bodied insects such as aphids, scales, etc.   The animals represented by the shafts are,
figuratively speaking, the propellors which keep the life histories shown above them,
turning.

trace out many such possible fluctuations and equilibrations.   The
number of possibilities is great even in an arbitrary community, though
much greater in an actual one.

Diagram 7 is a graphic representation of the relations of life histories in land communities to elementary food substances.   The number
of plant-feeders which serve to lock the inorganic substances to the main
part of the community is far greater than in aquatic communities.

CHAPTER X

ANIMAL COMMUNITIES OF THE TENSION LINES BETWEEN LAND AND WATER

I. INTRODUCTION

Margins of bodies of water, swamps and marshes, and temporary ponds are on the border-line between land and water. Swamps and marshes are areas occupied by plants whose stems, leaves, and blossoms are in the air and whose roots are in the water or very moist soil, throughout the year. Areas covered by grasslike plants are commonly called marshes, while those covered by trees are called swamps. Swamps and marshes usually contain water the year round and are commonly either directly connected with some permanent body of water or are fed by springs. Others are dry in summer, and possess an active aquatic fauna only in spring and after heavy rains. Our area, being in a region of glaciation, represents a portion of one of the great marsh areas of the world. Geologically speaking, however, these features represent the positions of lakes and serve to show us the fate of our small lakes and ponds. Classification of these communities is difficult, but they may be divided into temporary and permanent swamps and marshes and into margins of lakes, ponds, and rivers.

II. COMMUNITIES

1. PERMANENT WATER, SWAMP, AND MARSH COMMUNITIES

a) Lake-margin marsh sub-formation (senescent pond, or emerging vegetation pond association) (Stations 30, 30*a*, 31).—About the margins of lakes and ponds there is often a girdle of bulrushes and cattails (Fig. 116) which has a characteristic animal community. The sub-aquatic stratum is made up of pond animals and has been considered already in chap. viii. There are a few characteristic animals which live chiefly above the water. The diving spider (*Dolomedes sexpunctatus*) (Fig. 95) crawls about on the marsh vegetation and dives beneath the water for prey. The long slender spider (*Tetragnatha laboriosa*) is common among the bulrushes (138). At the base of the rushes and sometimes crawling near the top is the snail (*Succinea retusa*) (91). Common frogs (*Rana pipiens* and *clamata* Lat.) (Fig. 116) and the cricket-frog (*Acris gryllus*) hop about in the water (139).

169

There are also a number of insects which live upon the vegetation and never go into the water. These are the blue and yellow moth (*Scepsis fulvicollis*), which is most characteristic, flies which breed in the water, such as horseflies (*Tabanidae*) (140), *Tetanocera*, etc., also midges, mosquitoes, dragon-flies, damsel-flies, May-flies, etc. These are associated with grasshoppers, such as *Stenobothrus*, *Xiphidium*, and various

PERMANENT WATER MARSH AND ITS INHABITANTS

FIG. 116.—General view of an open bulrush marsh at Wolf Lake.

FIG. 117.—Similar but closer view of a marsh at Nippersink Lake, showing the yellow-headed blackbird (*Xanthocephalus xanthocephalus* Bonap.) perched on the bulrushes. Photo by T. C. Stephens.

bugs and beetles which belong to drier places but which alight on the vegetation above the water. These will be discussed in connection with low prairie communities.

The birds deserve especial attention (108, 141). The pied billed grebe, the black tern, and coot are especially aquatic. The grebe builds a nest from decayed floating rushes; its bottom is usually wet and the eggs commonly lie in moisture. The black tern builds a nest of weeds and trash similarly situated; the coot is less aquatic. The yellow-head blackbird (Fig. 117), mallard, pintail, American bittern, the least bittern (Fig. 118), the Florida gallinule (Fig. 119), the long-billed marsh wren, and sometimes the Virginia, sora, and king rails, and the red-winged blackbird nest in such situations. These birds build nests, either woven from grasses or in the form of crude piles of dead vegetation, each species having its characteristic method.

The muskrat breeds here and builds a nest from bulrushes (Fig. 82,

PERMANENT WATER MARSH AND ITS INHABITANTS

FIG. 118.—Nest of the least bittern (*Ardetta exilis* Gmel.) in a marsh at Nipper-sink Lake. Photo by T. C. Stephens.

FIG. 119.—Nest of the Florida gallinule (*Gallinula galeata* Licht) in a marsh at Nippersink Lake. Photo by T. C. Stephens.

p. 131). The mink likewise is found in this kind of situation (22, 142, 143). The grassy outer edges of such ponds are the favorite breeding-places of frogs (*Rana clamata*) which stick their eggs to grass. Points about such lakes, especially where there are shrubs and willows, are the favorite haunts of the bullfrog (*Rana catesbeiana* Shaw) (Fig. 117).

b) *Spring-fed marsh sub-formations* (Figs. 120–22) (Stations 10, 51).—These are very similar to the marshes which adjoin bodies of water, but the water of such marshes, however, gets very warm in summer, while the spring-fed marsh water is usually cool. It is the subaquatic stratum which differs most.

One of our best examples of spring-fed marsh is at Cary, Ill. (Fig. 120). This contains watercress, which is usually associated with springs. The most characteristic animals are the flatworms or planarians. *Planaria dorotocephala* (Fig. 121) is common on the under sides of leaves, etc.;

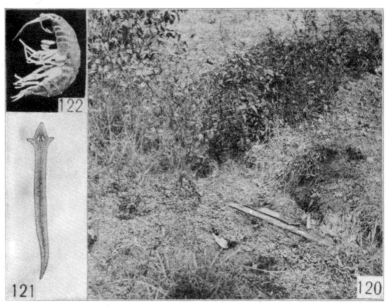

SPRING MARSH AND INHABITANTS

FIG. 120.—A spring marsh at Cary, Ill.

FIG. 121.—*Planaria dorotocephala;* 1½ times natural size (original).

FIG. 122.—The brook amphipod (*Gammarus fasciatus*); twice natural size.

if one puts a piece of meat into the water it will be covered with worms within a short time. The worms follow the diffusing meat juices, often passing through the direct sunlight, which they usually avoid. When they reach the piece of meat they crawl to the under side.

Associated with this planarian is *Dendrocoelum* (144), a larger, light-colored species, which does not come to the meat but is found with the former, under boards, chips, leaves of plants, etc. The brook amphipod (*Gammarus fasciatus*) (Fig. 122) occurs here also. The animals of the vegetation above the water including the birds are about the same as in the preceding sub-formations.

Permanent and temporary swamps are covered with trees. The most important permanent swamps are the tamarack swamps. The aquatic phase of these will be discussed in connection with the tamarack swamp itself (p. 193). Temporary swamps will be discussed under the head of temporary forest ponds.

2. TEMPORARY POND OR TEMPORARY SWAMP AND MARSH FORMATIONS

The situations known as temporary ponds, temporary marshes or swamps, or summer dry ponds, are common about Chicago and usually contain water in early spring, drying before the first of June. At some points at the south end of Lake Michigan much sand has been removed for commercial purposes and frequently the workmen remove it to points below the ground-water level of the spring months and accordingly make temporary ponds which have pure white sand bottoms. A few of these have been studied, one when it was one year old, another when about twelve years old. These were compared with ponds of the horizontal series which are much older.

a) Bare-bottom association.—Twelve-months-old pond association (Station 40; Table XXXVII): In April, 1910, we found this full of filamentous algae, and containing rotifers, copepods, and ostracods, the eggs of all of which will probably withstand drying and may have blown into the pond during the preceding dry seasons. There was a single full-grown snail (*Physa gyrina*), a small individual (probably *Physa heterostropha*), and a small long snail, *Lymnaea* (probably *exigua*). These snails may have been carried into the pond, from other ponds a few rods away, on the feet of turtles or frogs.

Twelve-year-old pond association (Station 40; Table XXXVII): As such a pond as we have just described grows older, the algae continue and the reed (*Juncus balticus*) comes in, together with some sedgelike plants. In such ponds the number of species is usually greater than at an earlier period.

In addition to the species found in the twelve-months pond, we obtained water-beetles, which are, however, not particularly significant because they may occur in rain pools. *Cladocera*, the flat snails (*Planorbis* sp.), and the nymphs of damsel-flies and dragon-flies are also found. The difference between this pond and the preceding one is not great. Indeed, it is only when the bottom of the pond becomes covered with sedges that we find marked differences in the ponds of different ages.

b) *Vegetation choked temporary pond association* (Stations 41, 42, 43; Table XXXVII).—Sedges soon take possession of the bottom of such a pond as we have been discussing. Just how long a time is required is not known, though the pond which we are about to discuss is probably several hundred years old. Here we find nearly all the groups mentioned as occurring in the younger ponds, but also certain ecological types which are characteristic of sedge-bottomed ponds. Most notable is the small green, flat, cigar-shaped worm (*Vortex viridis*) which usually occurs in numbers, and a small brown species of *Mesostoma* similar in form but brown in color. With them are often small larvae of dytiscid beetles (species unknown), caddis-worms (*Phryganeidae*) with cases made from pieces of grass (their relation to those in permanent ponds is not known), and the snail (*Lymnaea modicella*).

As such a pond grows older the sedge becomes thicker and other plants make their appearance. What is known as low prairie develops. At such a stage the small ponds like those we have been describing usually become partially filled and so never contain the best development of the older temporary pond community. We accordingly turn to the later history of the ponds discussed in the preceding chapters, which represent the best development of the temporary pond communities.

In a forest climate when ponds are filled and drained they are occupied by forest. In the steppe climate they are occupied by steppe or prairie. In the forest border area, where our studies have been carried on, some ponds when filled are occupied by prairies, others by forest. Dr. Cowles is of the opinion that ponds with gently sloping sides and bottoms become covered with prairie, while those with steep slopes become covered with forest (Fig. 123).

As ponds, such as we have discussed in the preceding chapter, become ecologically old, they dry in dry seasons. They usually become occupied by cattails, equisetum, or other grasslike plants. The red-winged blackbird (Fig. 124) occasionally nests in them. At such a stage the isopods (*Asellus communis* and *Mancasellus danielsi*), amphipods (*Eucrangonyx*) (Fig. 113), and snails (*Lymnaea reflexa*) (Fig. 125; compare with Fig. 104, p. 149) are common. The fringe-legged mosquito (145) and the common marsh mosquito (Fig. 126) breed in such situations while the crayfishes and various of the old-pond species continue.

When such a stage is reached, it is only a step to the typical temporary pond. If the ground-water level is lowered, as is the case in many

SEMI-TEMPORARY POND OR MARSH AND INHABITANTS

FIG. 123.—General view of Pond 93, which is occupied by *Sagittaria* and grass-like plants.

FIG. 124.—Side view of a red-winged blackbird's nest.   Photo by T. C. Stephens.

FIG. 124a.—The same from above.

FIG. 125.—A temporary pond form of the snail (*Lymnaea reflexa*): natural size.

of the ponds south of Lake Michigan, such ponds usually become grassy in the middle and often typical temporary prairie ponds. Here we find the green flatworm (*Vortex*), vernal planarians (*Planaria velata*), great

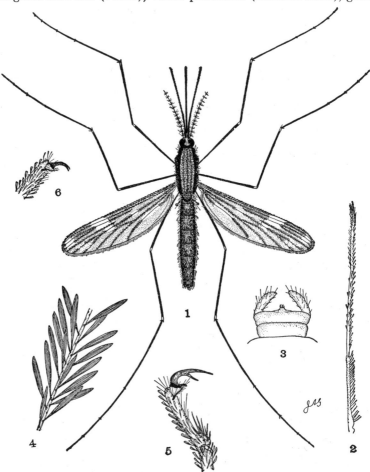

Fig. 126.—The common marsh mosquito (*Anopheles punctipennis* Say); much enlarged (from Williston after Smith). The details are such as to enable one to recognize this species of mosquito: (1) adult female; (2) her palpus; (3) her genitalia; (4) part of a wing-vein showing scales; (5) anterior, and (6) middle claws of the male.

numbers of *Entomostraca*, belonging to all orders. Of the last there are many very large cladocerans, the copepods (146) (*Cyclops viridis americanus*)(Fig. 127), the red copepod (*Diaptomus stagnalis*) (Fig. 128), the ostracod (*Cyprois marginata*)(147) (Fig. 129), and the fairy shrimp (*Eubranchipus*) (148) (Fig. 130), all of which are characteristic of temporary ponds. Red mites (Fig. 131) are also common (149).

Professor Child (unpublished) has noted that the distribution each spring of *Eubranchipus* and of other temporary pond species is modified

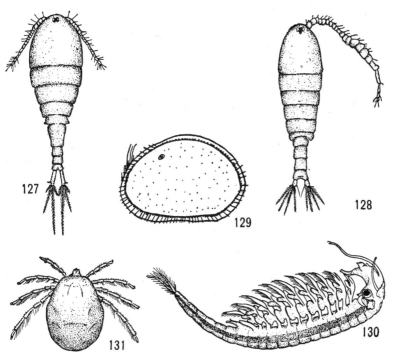

TEMPORARY GRASSY POND ANIMALS

FIG. 127.—A temporary pond copepod (*Cyclops viridis americanus* Marsh); 35 times natural size (after Herrick and Turner).

FIG. 128.—The red copepod (*Diaptomus stagnalis*) from temporary pond; 12 times natural size, left antenna omitted (after Herrick and Turner).

FIG. 129—The temporary pond ostracod (*Cyprois marginata*); 35 times natural size (after Sharp).

FIG. 130.—The fairy shrimp (*Eubranchipus*); 3 times natural size.

FIG. 131.—The red mite (*Hydrachna* sp.); 10 times natural size.

by the rainfall of the preceding season. When the rainfall of the preceding season has been great, the temporary pond species are found only in the smallest and highest (above ground-water) ponds such as would develop in the place of one of the small ones with sandy bottom.   Follow-ing dry seasons the temporary pond species are found in ponds which do not usually dry in summer, but which were dry the preceding summer. It has been shown that the eggs of *Eubranchipus* must be dried and

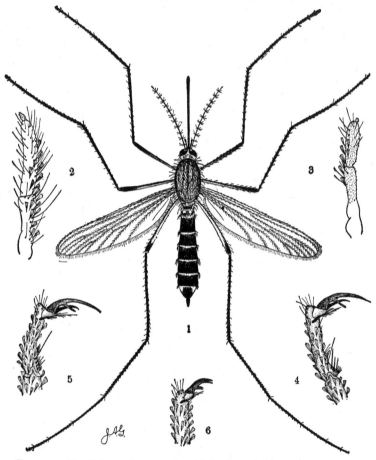

Fig. 132.—The little smoky mosquito (*Aedes fusca* O. S.); much enlarged (from Williston after Smith): (1) adult female; (2) her palpus; (3) palpus of the male; (4) anterior; (5) middle, and (6) posterior claws of the male.

frozen before they will hatch.   The relation of their distribution, follow-ing the seasons of different rainfall, suggests that some definite degree of drying must be attained to insure hatching as well as that the eggs are probably blown about by wind.   One autumn, about 1900, there was early freezing and cold weather followed by warm weather of a very springlike character in December. Professor Child observed that the *Eubranchipus hatched during this period of warm weather.* Cold weather came on soon after and most of those that had hatched died before reaching sexual maturity, and for several years after the species was very scarce in the vicinity of Chicago. *Eubranchipus* is found only in grassy ponds, possibly because the forested ponds do not dry sufficiently in summer. We have found it on one occasion in woods, but this was

THE BARE SAND WATER MARGIN AND INHABITANTS

FIG. 133.—Margin of Lake Michigan at Buffington.
FIG. 134.—The beach tiger-beetle (*Cicindela hirticollis*); 1½ times natural size.
FIG. 135.—The beach ground beetle (*Bembidium carinula*); 1½ times natural size.

in flood-plain pools following an early spring flood and might have been due to the washing-in of eggs or young.

c) Forest temporary pond sub-formation (association) (Station 50; Table XXXVII).—These are characterized by the absence of both *Diaptomus* and *Eubranchipus*. The *Entomostraca* are chiefly ostracods, such as *Cyprois marginata*, which occurs in grassy ponds. *Vortex*, mosquito larvae, the little bivalve (*Musculium*), small earthworms (*Lumbriculus*), and the larvae of a beetle (*Dascyllidae*) are also very common.

The amphipods and sowbugs of the earlier stages are still present. This is the breeding-place of such mosquitoes as the little smoky mosquito (*Aedes fuscus*) (Fig. 132, p. 178) (145, 99c).

3. COMMUNITIES OF MARGINS OF BODIES OF WATER

There is always a narrow area along the margins of bodies of water which is difficult to classify as water or as land. The association of this area is the one with which we now have to deal.

Along the margins of young ponds and lakes is an area which is characterized by being made up of wet sand or mud which is sub-merged at high water and moist at other times.

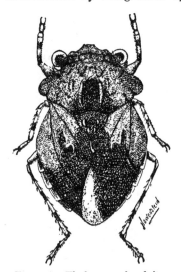

a) *Association of the terrigenous margins of large lakes* (Fig. 133) (Stations 57, 58; Table XXXVIII).—Here we find the springtails the simplest insects, the shore bugs (150), *Saldidae*, especially *Salda humilis* Say, a large number of tiger-beetles (151) (*Cicindela hirticollis*) (Fig. 134) (*C. cuprascens*), together with numerous small flies.

The ground beetle (*Bembidium carinula*) (Fig. 135) and numerous scavengers are common because the beach is often strewn with dead animals which have floated ashore. The relations of the drift to other communities will be discussed in the chapter on dry forests. The spotted sandpiper feeds here, and with the piping plover often breeds not far from the water's edge. Under conditions of rapid recession of the lake such a margin is separated from the wave-action. It is then rapidly transformed into the next association.

FIG. 136.—The bare pond and river-margin toadbug (*Gelastocoris oculatus*); greatly enlarged (after Lugger).

b) *Association of the terrigenous margins of ponds and small lakes* (Stations 30, 40; Table XXXIX).—This association differs from that of the large lake in that the scavengers are absent and the animals much less active, not moving about so rapidly. Here we find springtails, *Saldidae* of another species, and the toadbug (*Gelastocoris oculatus*) (150) (Fig. 136), which is colored like the ground and is found hopping about close to the water. The tiger-beetle of the Lake Michigan shore is displaced by that of another (*Cicindela repanda*) which is less active. With these is the hooded grouse locust (*Paratettix cucullatus*) (Fig. 137) (40, p. 419). The small semiaquatic snail (*Lymnaea modicella*) is frequently present in numbers.

The nests of the spotted sandpiper (108, 141) and the yellowlegs are found here, and the birds no doubt feed upon the invertebrates present on the margins of the ponds and of the shallow water.

c) *Association of sedge margins of ponds and small lakes* (Stations 32–34; Tables XL, XLI).—As time goes on, the sandy margin is captured by sedges which are scattered at first, so that the animals just discussed continue for a time among them (Figs. 138, 139). Finally, however, the ground becomes sodded over with sedges and a low prairie animal community comes in, and the bare ground animals disappear. In the case of ponds which are to develop into forest this stage is found only along the young ones. The sedges are soon displaced by shrubs and the sedge communities give way to shrub.

d) *Associations of shrub margins of ponds and small lakes* (Fig. 140) (Stations 34, 37, 44; Tables XLI, XLII).—Mr. Allee has verified my observations to the effect that the aquatic part of this formation is almost entirely barren; however, in summer we get the short-winged and armed grouse locust (*Tettigidea armata* Morse, and *parvipennis* Harr.) (40) and the slimy salamander (*Plethodon glutinosus*) (152) (Fig. 141). Of the birds associated with the water we have here the wood-duck and the green heron.

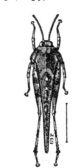

Fig. 137.—Hooded grouse locust (*Paratettix cucullatus*) (after Lugger).

4. MARGINS OF RIVERS
(Station 29)

Here the sandy margin is similar to that of the ponds and lake. Along the Fox River we find the mole cricket (40) which burrows into the sand. Mud margins are rather barren except for occasional beetles. The margins of rivers which are grassy or marshy are like those of ponds and lakes. The margins of the Calumet and lower Deep rivers are covered with marsh plants and saturated with water in spring. They are the nesting-places of the long-billed marsh wren (Figs. 142, 143) and many other marsh birds (108, 153).

a

b   138   c

FIG. 138.—Prairie-like stage of a pond margin. Habitat of *Cicindela tranquebarica* in the pine zone of the ridges at the south end of Lake Michigan. The dark portion in the foreground is the shadow of a tree. At the left is the cattail zone of the depression; between *a* and *b*, the sedge zone; between *b* and *c* the zone of high-depression plants. The white blossoms here are those of *Parnassia caroliniana;* their distribution, September, 1906, corresponds approximately to the distribution of the larvae of *C. tranquebarica*, which arose from eggs laid in May and June, 1905. The portion to the right and above *c* represents the higher portion of the ridge and the habitat of *C. scutellaris.* Reprinted from the *Journal of Morphology.*

FIG. 139.—The upper part of the burrow of *C. tranquebarica*, pupal cell shown by dotted line; ⅓ natural size. Reprinted from the *Journal of Morphology.*

139

III. General Discussion

The areas which we have been discussing in this chapter are the tension lines between the land and the water. It is in such areas that ecologists have learned most about succession and about the tendencies

Fig. 140.—Pond 95, showing the death of the pond by the growth of buttonbush.

Fig. 141.—The shiny salamander (*Plethodon glutinosus*); about twice natural size (after Fowler).

and processes in animal formations and associations. In this chapter we first considered the marshes which border the lakes and ponds about Chicago. Dr. Cowles and others have pointed out that lakes and ponds are filled by organic débris and that bulrushes invade from the shore and "capture" the ponds and lakes. As the bulrushes and other plants invade, the girdle of marsh which is the nesting site of the birds mentioned moves farther and farther toward the center of the pond or lake, the former positions being occupied by shrubs, such as buttonbush or willow, or in some cases by prairie. Such a situation is in unstable equilibrium.

Turning to the margins of ponds, lakes, and rivers, we note that at the beginning we often have the bare sand. This is first occupied

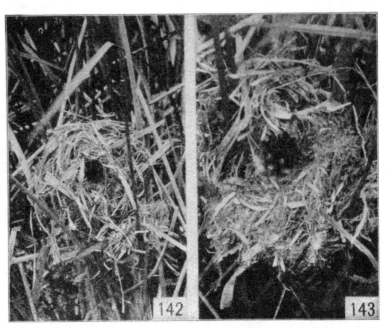

Fig. 142.—The long-billed marsh wren's nest. The nest unopened.
Fig. 143.—The nest torn open showing the eggs.

by reeds and sedges, and finally by shrubs. It is this reed and sedge group, or the buttonbush, that invades the swamp as it fills with bulrushes and cattails. We note accordingly that the vegetation which appears on the shore invades the pond as it fills. The last stage of a pond is either a buttonbush swamp or a low prairie which we shall discuss in later chapters.

TABLE XXXVII

TEMPORARY PONDS OF DIFFERENT AGES

The numbers standing at the heads of the columns where months or years are not indicated refer to the number of the pond in question when counted from the lake. J.P. is a pond south of Jackson Park.

COMMON NAME	SCIENTIFIC NAME	UNDIFFEREN-TIATED		PRAIRIE			FOR-EST
		12 Mo.	12 Yr.	4	55	J.P.	94
Rotifer		*					
Copepod		*	*				
Cladoceran			*	*			
Ostracod		*	*				
Snail	*Physa gyrina* Say	*					
Snail	*Lymnaea obrussa exigua* Lea	*					
Ground beetle	*Bembidium* sp	*					
Scavenger beetle	*Aphodius fimetarius* Linn		*				
Small water-bug	*Zaitha fluminea* Say	*					
Scavenger beetle	*Hydrophilidae*			*	*		
Water-strider	*Gerridae*			*	*		
Water-mite	*Hydrachna* sp			*	*	*	
Flat snail	*Planorbis*			*			
Dragon-fly nymph	*Enallagma*			*	*		
Toad-shaped bug	*Gelastocoris oculatus* Fabr			*			
Green flatworm	*Vortex viridis* M. Sch.			*	*	*	
Brown flatworm	*Mesostoma* sp			*	*	*	*
Predaceous beetle	*Dytiscidae*			*	*	*	
Springtail	*Collembola*			*	*	*	
Caddis-worm	*Neuronia?* sp			*	*	*	
Pond snail	*Lymnaea reflexa* Say			*	*	*	
Red copepod	*Diaptomus stagnalis* For				*	*	
Fairy shrimp	*Eubranchipus serratus* Forbes				*	*	
Vernal planarian	*Planaria velata* Str				*	*	
Amphipod	*Eucrangonyx gracilis* Sm				*	*	*
Isopod	*Asellus communis* Say				*	*	*
Mosquito larva	*Culicidae*				*	*	*
Bivalve	*Musculium secure* Prime				*	*	*
Ostracod	*Cypris fuscata* Jurine					*	
Ostracod	*Cyprois marginata* Strauss					*	*
Beetle larva	*Dascyllidae*						*
Annelid worm	*Lumbriculus inconstans* Smith						*

TABLE XXXVIII

ANIMALS FREQUENTING THE MOIST MARGIN OF LAKE MICHIGAN

(Stations 57, 58)

Common Name	Scientific Name	Month
Flesh-fly	*Sarcophaga* sp	4–9
Flesh-fly	*Chrysomyia macellaria* Fab	4–9
Flesh-fly	*Cynomyia cadaverina* Des	4–9
Hister beetle	*Saprinus patruelis* Lec	4–9
Ground beetle	*Bembidium carinula* Chd	4–9
Tiger-beetle	*Cicindela hirticollis* Say	6–8
Tiger-beetle	*Cicindela cuprascens* Lec	7–8

TABLE XXXIX

ANIMALS RESIDENT ON THE MARGIN OF A TWELVE-YEAR-OLD ARTIFICIAL POND AND
OF WOLF LAKE (SANDY)

(Stations 30, 40)

Common Name	Scientific Name	Month
Ground beetle	*Bembidium variegatum* Say	....
Snail	*Lymnaea humilis modicella* Say	4–9
Toad-shaped bug	*Gelastocoris oculatus* Fabr	4–9
Tiger-beetle	*Cicindela repanda* Dej	4–9
Hooded grouse locust	*Paratettix cucullatus* Burm	4–9

TABLE XL

ANIMALS RECORDED FROM SEDGE-COVERED POND MARGINS

(Stations 32, 33)

Common Name	Scientific Name	Month
Snail...............	*Lymnaea humilis modicella* Say.......	4-9
Snail...............	*Succinea retusa* Lea................	4-9
Snail...............	*Succinea avara* Say................	4-9
May-fly............	....	4-9
Toad...............	*Bufo lentiginosus* Sh...............	4-9
Diving spider........	*Dolomedes sexpunctatus* Htz.........	4-9
Spider..............	*Pirata insularis* Em...............	....
Walking-stick........	*Diapheromera femorata* Say..........	8-9
Grasshopper nymph.......	*Acrididae*......................	6-10
Damsel-bug..........	*Reduviolus ferus* Linn..............	8-9
Centipede...........	*Lithobius* sp....................	4-9
Spider..............	*Chiracanthium inclusa* Htz..........	8-9
Long-bodied spider........	*Tetragnatha laboriosa* Htz...........	8-9
Spider..............	*Tibellus duttoni* Htz................	8-9
Spider..............	*Eucta caudata* Em................	5
Orb-weaving spider........	*Epeira foliata* Koch................	5-9
Short-tongue bee..........	*Augochlora confusa* Rob.............	7-8
Ant................	*Ponera coarctata* La'r.............	9
Root beetle..........	*Diabrotica 12-punctata* Oliv..........	8
Root beetle..........	*Diabrotica vittata* Fab.............	8
Spider..............	*Dictyna sublata* Htz................	8
Ambush-bug..........	*Phymata fasciata* Gray.............	8
Negro-bug...........	*Thyreocoris unicolor* P.B............	8
Bug................	*Philaronia bilineata* Say	9
Red-legged grasshopper.....	*Melanoplus femur-rubrum* DeG........	8-9
Stinkbug............	*Cosmopepla carnifex* Fab............	....
Ant................	*Formica fusca* var. *subsericea* Say......	....
Midge..............	*Chironomidae*....................	....

TABLE XLI

ANIMALS RECORDED FROM THE MARGIN OF POND 8 (MIXED SEDGES AND SHRUBS)
BY MR. B. F. ISELY

(Station 34)

Common Name	Scientific Name	Month
Lygaeid..................	*Cymus angustatus* Stal..............	7–8
Apple-leaf hopper..........	*Empoasca mali* LeB...............	....
Lacebug	*Physatochila plexa* Say.............	7–8
Jassid...................	*Draeculacephala mollipes* Say........	7–8
Dusky plant-bug..........	*Adelphocoris rapidus* Say...........	7–8
Cranberry lygaeid.........	*Ischnodemus falicus* Say.............	7–8
Beetle...................	*Nodonota tristis* Oliv...............	7–8
Ground beetle............	*Anomoglossus pusillus* Say..........	7–8
Ant-like flower-beetle......	*Stereopalpus mellyi* Laf.............	7–8
Leaf-beetle...............	*Chalepus hornii* Sm................	7–8
Mordellid................	*Mordellistena aspersa* Mel..........	7–8
Case-bearer...............	*Pachybrachys abdominalis* Say.......	7–8
Strawberry beetle.........	*Typophorus canellus sellatus* Horn.....	6–9
Lampyrid.................	*Lucidota punctata* Lec..............	7
Lampyrid.................	*Pyractomena borealis* Rand..........	7
Lampyrid.................	*Lucidota atra* Fab.................	7
Metallic wood-borer	*Pachyscelus laevigatus* Say..........	7
Dascyllid................	*Ptilodactyla serricollis* Say..........	7
Maia or buck moth.......	*Hemileuca maia* Dru...............	7
Short-winged brown grass-hopper.................	*Stenobothrus curtipennis* Harr........	7–8
Slender meadow grasshopper.	*Xiphidium fasciatum* DeG...........	7–8
Short-winged meadow gr'hop.	*Xiphidium brevipenne* Scud..........	7–8
Fly......................	*Tetanocera umbrarum* Lin...........	7–8
Fly......................	*Tetanocera plumosa* Loew...........	7–8
Fly......................	*Tetanocera combinata* Loew..........	7–8
Fly......................	*Tetanocera saratogensis* Fitch........	7–8
Horsefly.................	*Chrysops aestuans* V.W..............	7–8
Horsefly.................	*Chrysops callidus* O.S..............	7–8
Syrphus fly..............	*Mesogramma marginata* Say.........	7–8

TABLE XLII

ANIMALS RECORDED FROM THE WILLOW AND BUTTONBUSH. MARGINS OF PONDS
52 AND 93. RECORDS BY ALLEE ARE INDICATED

(Stations 37, 44)

Common Name	Scientific Name	Month
Plant-bug................	*Adelphocoris rapidus* Say (young).....	7–8
Stinkbug.................	*Cosmopepla carnifex* Fabr...........	7–8
Fulgorid.................	*Amphiscepa bivittata* Say...........	7–8
Jassid...................	*Parabolocratus viridis* Uhler.........	8
Smeared dagger-moth.......	*Acronycta oblinita* S and A (Allee)....	8
Ant.....................	*Aphaenogaster treatae* Forel (Allee).....	8
Fly.....................	*Tetanocera* sp. (Allee).............	8
Spider..................	*Epeira trivittata* Key. (Allee).........	8
Leaf-beetle...............	*Calligrapha multipunctata* var. *bigsbyana* Kirby (Allee)................	....
Leaf-beetle...............	*Typophorus canellus aterrimus* Oliv....	....

CHAPTER XI

ANIMAL COMMUNITIES OF SWAMP AND FLOOD-PLAIN FORESTS

I. Introduction

Swamp forests are those which arise in the areas formerly occupied by ponds and lakes and which grow in water or very wet soil. About Chicago the many coastal and morainic lakes of earlier periods have been filled by organic detritus and more or less completely occupied by trees. Often the trees have grown upon floating bogs such as sometimes occur about lakes, though sometimes they have sprung up on solid ground and compact organic detritus.

II. Swamp Forest Formations and Associations

We shall consider these forests genetically: the marsh which often appears first, the shrub stage which follows, and finally the forest.

1. THE ELM-ASH SWAMP FOREST COMMUNITIES

a) The marsh association (Station 52; Table XLI).—One of the best examples of this community is at the north end of Wolf Lake, Ind. The youngest part is occupied by bulrushes and *Hibiscus*, and covered in the spring by about a foot of water which teems with small crustaceans, mosquito larvae, and red water-mites. *Lymnaea reflex*, usually about half the size of the specimens (100) of permanent ponds, and the small bivalve (*Musculium*) are present. As the season advances the water dries up and the eggs of the crustaceans and adult mollusks live through the dry season on the bottom of the pool. Above the water on the *Hibiscus* are the small *Succinea retusa* (91, 100), which belong to the forest edge and low prairie.

b) Shrub association (forest edge sub-formation) (Station 52; Table LXIII).—Surrounding the central pool which we have described is usually a girdle of buttonbush. Here we recognize several strata. The subterranean stratum has few inhabitants. We have recorded none. The ground stratum is not inhabited by many animals. The wood-cock and the northern yellowthroat (108, 153) probably occasionally nest here on the ground, possibly also the common shrew (*Sorex personatus* St. Hil.) (142). There is no distinct field stratum, as the

189

thickness of the shrubs prevents the growth of herbaceous vegetation. The shrub stratum is the chief habitat.

The buttonbush is remarkably free from plant-feeding animals. Occasionally some of the willow-eaters, such as the larva of the smeared dagger-moth, are found on it, but never in any numbers. This stratum is the resting-place of many of the insects whose early stages inhabit water. When the plants are in blossom, it is visited by many flower-frequenting insects, such as the bumblebee (40).

Mr. Visher has recorded a number of nesting birds in this girdle. The wood-duck usually makes its nest here in some hollow tree and lines it with feathers; where stumps or rotten trunks are found the prothonotary warbler sometimes nests; Traill's flycatcher (*Empidonax trailli* Aud.) places its nest well up in the branches and leaves of the bushes.

c) Forest formation (Stations 52 and 53; Table L).—As time goes on and the marsh fills with organic detritus, the buttonbush which is continually encroaching upon it comes to occupy a position farther in, while its former location is taken by the ash, which is the next girdle outside.

The ash is succeeded by the American elm and the basswood. These are frequently considerably mixed with the ash, but the two girdles can be distinguished in the Wolf Lake Forest. For convenience we shall treat these two girdles (associations) together under the head of the wet forest formation.

The subaqueous and subterranean-ground strata: The subterranean portion is inhabited by earthworms. On the higher parts there are doubtless other subterranean forms. Where the roots of the grapevine are in the drier soil, the vines are infested with the aphid (*Phylloxera*) which makes galls on both roots and leaves. The depressions of these forests are filled with water in spring and support temporary pond animals such as we have discussed on p. 179.

In the Wolf Lake woods we noted in the spring of 1910 that the small red spiders (*Trombidium* sp.) were numerous. Centipedes, crane-fly larvae, and ground beetles occur under the leaves. Hancock (40, p. 419) states that the obscure and Indiana grouse locusts (*Tettix obscura* Han. and *Neotettix hancocki* Bl.) are found in such forests. Under pieces of rotten wood are sometimes found specimens of the small snail (*Zonitoides arboreus*), which is first to appear in forests developing from the buttonbush swamp stages. On highest ground we get two other snails (*Polygyra monodon* and *Pyramidula striatella* Ant.) (91, 100). In the fallen logs we find a considerable number of borers (*Parandra brunnea* Fabr and others) and under the loosened bark are centipedes (*Lithobius*), milli-pedes (*Polydesmus*), and beetle larvae (*Pyrochroidae*) which are flattened. While we have no actual records of mammals in such situations, doubtless the varying hare (*Lepus americanus* Erx.) which frequents marshy woods with thickets such as the buttonbush, the common shrew, which nests under logs, and the mink (*Mustela vison lutreocephala* Harlan), all have been visitors if not residents in such situations in the past. The wood-duck, the woodcock, and prothonotary warbler often nest in such woods.

Field stratum: The field stratum is inhabited by small flies, such as crane-flies, midges, and mosquitoes (*Chironomidae* and *Culicidae*), occasional spiders, such as *Theridium frondeum*, parasitic hymenoptera (*Pimpla inquisitor* Say, *Ichneumon mendax* Cress.) and the scorpion-fly (*Panorpa*), which breeds in the ground.

Shrub stratum: The shrubs consist chiefly of buttonbushes and low-hanging grapevines. The vines frequently have conspicuous insect galls. One called the grapevine apple gall, because of its shape, is due to the larva of a small fly (*Cecidomyia vitis-pomum* W and R) (137); another which is a pointed tube on the leaf is due to *Cecidomyia viticola*. Wartlike galls on the under side of the leaves are due to *Phylloxera vastratrix* Pl. (Fig. 277, p. 273) (150), an aphid which, when introduced into France, threatened to destroy the vine industry. These occur only on the vines on high ground where the roots, upon which a part of the life of the aphids is spent, are out of water. Several grape insects, including the fulgorid bug (*Ormenis pruinosa* Say), have been taken.

Tree stratum: The tree stratum of this girdle has not been studied, because the study of the tree stratum in general is difficult. The white ash is, however, attacked by many insects. Felt (137) and Packard (154) record a number. One of the most difficult groups to secure is the "borers" of the solid wood or sapwood of trunk and twigs. These are chiefly beetle larvae, especially the *Cerambycidae* (155), or long-horned beetles. The larvae of these are legless and only slightly larger at the anterior end. Another prominent family is the metallic wood-borers or *Buprestidae* (155). The larvae of these are also legless and may be distinguished from the preceding by a broad, flattened enlargement just behind the head. The ecology of these two families alone is a subject for a work the size of this volume (see 137 and 154). The four-marked borer (*Eburia quadrigeminata* Say) is said to occur on the ash throughout Indiana (156). The elm and basswood likewise have many borers, some in common with the ash.

The insects feeding on the leaves are numerous on all the trees. The following are common to the three trees mentioned (137): the caterpillars of the hickory tussock-moth, the American dagger-moth, the forest tent caterpillar, the white-marked tussock-moth; each has a preference for one of the trees. The larvae of several other common moths occur on two of the trees, a few are confined to one. Beetle and sawfly larvae also attack the leaves. Each tree has its characteristic gall insects and galls; for example, on the elm, the coxcomb gall (*Colopha ulmicola* Fitch), on the ash, the midrib gall (*Cecidomyia verrucicola* O.S.). These are believed to be confined to particular tree species.

According to Wood (21) such forests are the chief haunts of the gray squirrel. The green heron is especially likely to nest on the low trees of such a forest if they are near water.

2. OTHER TYPES OF SWAMP FOREST COMMUNITIES

The swamp forest formation is well developed in the Skokie marsh area. We have visited these woods at a point west of Dempster Street, Evanston. This was originally characterized by trees very much larger than those at Wolf Lake. The soil at Wolf Lake is sand, while that at Evanston is clay, which is probably more favorable for trees. However, the most important cause of the greater luxuriance is greater age. The subterranean stratum has not been studied.

The ground stratum: Here we find, in addition to those species of the temporary ponds at Wolf Lake, a snail (*Aplexa hypnorum* Linn.) which is characteristic of very transient ponds (100).

On November 27, 1903, the condition of the animals of this stratum was noteworthy. In the lower moister parts of the wood we found the mollusks, especially *Pyramidula alternata*, in groups under logs. One of these groups contained 12 individuals. Under another log was a group of about 50 ground beetles (*Platynus* sp.). Under one small piece of bark were found three ground beetles, three rove-beetles, one slug, and two snails. Under another, one tetrigid or grouse locust, several ground beetles, and a rove-beetle. Under the bark of a log on the above date we found the hibernating parasitic hymenoptera (*Ichneumon extrematatus* Cress., *galenus* Cress. and *mendax*), also a queen white-faced hornet (*Vespa maculata*), which with its colony builds a large spherical nest in a tree in summer.

Most noticeable of all was a group of several hundred small blue chrysomelid beetles (*Haltica ignita* Illig.). They were under the leaves at the base of a tree down the sides of which individuals of the same species were moving. Such groupings are common among hibernating insects and are believed to keep the temperature a little higher. Baker (100) has studied the wet forest near Shermerville, Ill. In his Stations 7-17 the forests are ecologically older. (For birds and mollusks present, see 100, p. 468.)

3. THE TAMARACK SWAMP COMMUNITIES

(Stations 54, 54a; Tables XLIII–XLV)

Tamarack swamps develop about deep lakes. Floating plant débris supports first water-lilies and later bulrushes and cattails. Upon these grow shrubs, such as the leather-leaf and the willows; these make conditions suitable for the poison sumac and young tamaracks. The semi-aquatic plants are thus succeeded by the shrubs and finally by the tamarack.

The aquatic communities have not been studied in a typical tamarack lake, but there is no reason to suppose that they differ in any important way from the aquatic communities of other old bodies of water.

a) Floating bog and forest edge association (Tables XLIII, XLIV).— The floating bog of cattails and bulrushes is usually full of low places in which water is present the year round. Here we find the typical animals of semi-temporary ponds, as described on p. 150. The various frogs of the marsh probably breed here. Another aquatic habitat of some interest is the water-holding leaves of the pitcher-plant (158). The pitcher-plant mosquito (*Wyeomyia smithii*) is known to breed in the leaves of pitcher-plants only. These are accompanied by the larvae of midges and large numbers of dead insects which crawl into the pitchers and cannot get out on account of the presence of many hairs which project inward along the wall of the entrance.

The surface of the bog is frequented by marsh spiders, insects, and frogs, only a few of which belong especially to pre-tamarack bogs. The inhabitants of the vegetation (field stratum) are like those on the vegetation over other marshes, belonging chiefly to low prairies. The edge of the tamarack woods (Fig. 144) is a characteristic forest margin. Except for the presence of some of the tamarack leaf-feeders, such as the larch sawfly larva and measuring-worm, it possesses few species different from the margins of other marshes (Fig. 145).

b) Tamarack forest formation (Table XLV).—Pools: The pools within the forest proper contain old-pond animals together with some mosquito larvae (such as those of *Culex canadensis*) which are characteristic of pools in all moist and mesophytic forests (see 99c).

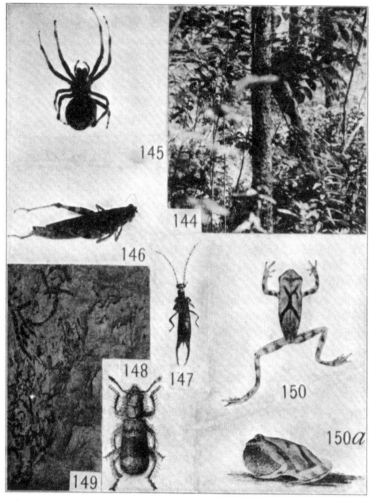

REPRESENTATIVES OF THE TAMARACK SWAMP COMMUNITY

Fig. 144.—View in the dense vegetation of the tamarack swamp.

Fig. 145.—Female orb-weaver (*Epeira gigas*); about natural size.

Fig. 146.—The brindled locust (*Melanoplus punctulatus*); natural size.

Fig. 147.—An earwig (*Apterygida aculeata*); natural size.

Fig. 148.—An engraver beetle destroyer (*Cleridae, Thanasimus dubius*); 3 times natural size (from Blatchley after Wolcott).

Fig. 149.—The bark of the tamarack, showing the work of the engraver beetle (*Polygraphus rufipennis*).

Figs. 150, 150a.—Pickering's tree-frog (*Hyla pickeringii*); about two-thirds natural size (after Fowler).

Ground stratum: On the sphagnum, which sometimes occurs in the pools, various insects and spiders occur, including, according to Hancock (40), two species of sphagnum crickets. On the higher ground numbers of typical moist forest animals occur sparingly. Frogs are often numerous. The common frogs (*Rana pipiens* and *clamata*) and the marsh tree-frog (*Chorophilus nigritus*) occur in summer. The wood-frog and Pickering's tree-frog (*Rana sylvatica* and *Hyla pickeringii*, Fig. 150) are regular residents; probably both breed in the pools (139) between the hummocks. Farther north the hermit thrush nests on the hummocks amid the dense undergrowth. This is also the typical haunt of the varying hare (*Lepus americanus* Erx.) (83, 142, 143), which is white in winter and brown in summer; it is common in tamarack swamps farther north. The lynx (p. 15) was probably once common near Chicago and is most likely to have frequented these swamps. Adams (83, 42) records its tracks on the hummocks of the tamarack swamps on Isle Royale in Lake Superior. Judged by its tracks it wanders far. It feeds largely on hares, the numbers of which fluctuate (inversely) with the numbers of lynx. The otter (*Lutra canadensis* Schr.) and Cooper's lemming mouse might be added as probable former residents (143, 21).

Field stratum: This is confined to hummocks supporting herbaceous plants. Insects, spiders (159), etc., are common; some characteristic species occur.

Tree stratum: The brindled grasshopper (*Melanoplus punctulatus*) (Fig. 146) has been found on the low branches of the tamarack and deposits its eggs on the bark of the trunk or on stumps. Several other insects have been recorded as common on the tamarack, among which are a sawfly, an earwig (Fig. 147), a lappet moth, and a woolly aphid, but we have not taken all of them. (See 137, II, 838, and I, Plate 18.) The tamarack is infested by bark beetles. In the swamp at Mineral Springs, Ind., we found one (*Polygraphus rufipennis*) (137), sometimes also *Dendroctonus simplex* Lec, common under the bark of partially dead trees (Fig. 149). The larvae of the clerid beetle (*Thanasimus dubius*) (Fig. 148) (137) occur with the bark beetles and feed upon them. The adult of the clerid (137) appears in spring, having wintered over as adult or in the late larval or pupal stage. It goes about on the bark of trees, seizing the bark beetles and later laying eggs at the openings of their galleries. The larvae invade the galleries and feed upon the eggs and larvae of the bark beetles. Felt states that two other bark beetles attack the tamarack (160). In this marsh the bark beetles have killed a number of trees. In summer the area of dead ones may be seen a mile away.

Farther north the blackburnian warbler nests here. The tree stratum of primeval conditions usually included the pine marten (*Martes americana* Tur.). It lives in trees in dark coniferous forests. Merriam (142) says that it nests in a hollow tree or log, rarely on the ground. It preys upon partridges, rabbits, squirrels, chipmunks, mice, shrews, birds' eggs, young birds, and frogs and toads. It disappears when civilized man settles the country. The marten's close relative, the fisher (*Martes pennanti* Erx.), is said to be the wildest of all wild animals. It is somewhat similar (21, 22, 162) to the marten in habits.

c) *The pine-birch transition girdle* (Station 54; Table XLVI).—This succeeds the tamaracks and contains a few old trees of this species. The pools are all dry in summer, though they may contain water in spring. The subterranean stratum has not been investigated.

The ground stratum includes the frogs of the tamarack formation (*Hyla pickeringii*). Insects, spiders, centipedes, and snails, which belong chiefly to mesophytic forest, are more numerous than in the tamarack stage. Nesting of the ruffed grouse likewise indicates that the swamp stage is past. The field and shrub strata likewise include more of the mesophytic forest animals than the true tamarack stage.

The tree stratum has not been studied. The trees are white pine, yellow birch, and an occasional maple. Felt (137) records no insect common to these two trees. There are several common to the white pine and tamarack (larch lappet, engraver beetle, etc.). Pines have many borers and few leaf-feeders. Each borer usually prefers a certain part, as the trunk, limbs, or growing shoots; some, as the white-pine weevil (*Pissodes strobi* Peck) (161), attack young pines. Felt records about 25 injurious insects common to birches and maples in general and one or two which occur only on yellow birch. The great crested flycatcher nests in holes in dead limbs; the wood pewee nests on horizontal limbs, and the red-eyed vireo builds a nest in trees from 5 to 40 ft. from the ground. Dead birches form suitable nesting-places for woodpeckers. The Canada porcupine (142) which we have noted in the ground stratum is a good climber and feeds largely in the trees, which it often girdles.

d) *The geographic relations of the animals.*—Most of the non-aquatic animals of the swamps are commonly said to belong to species common farther north where conifers dominate. However, our lists and the unpublished work of Messrs. Wolcott and Gerhard do not bear out this conclusion. Some of the species of these swamps doubtless formerly occurred among the hemlocks of Southern Michigan.

e) Fate of the formation.—Most of our tamarack swamps are in the regions which are commonly dominated principally by beech and maple. In the higher portions of the tamarack swamps are found several species characteristic of beech woods and other mesophytic woods. These are the wood-frog, the large slug, the snail (*Polygyra albolabris*) and the red-backed salamander (*Plethodon cinereus*) and the spider (*Castianeira cingulata*). These indicate that beech and maple are to follow.

4. FLOOD-PLAIN AND RAVINE FOREST COMMUNITIES

As we have noted on pp. 87–93 and 108–113, streams often develop by head-on erosion into uplands of rock or clay.

a) Streams developing in rock.—In case the upland is of rock, the beginning of the stream is a lower place in the slope of the rock through which water flows when it is raining. Vegetation is usually absent. If there are broken pieces of rock at the sides or in the course of the intermittent stream, some of the forms mentioned on p. 218 may be present. Until it becomes permanent or has cut itself a deep, straight-sided channel, it is inhabited by the animals which inhabit the bare rock of hills or hillsides. After the stream has cut such a channel, there are always small piles of fine soils which support nettles and other mesophytic plants similar to those of the old mesophytic flood-plain. Flood-plain animals appear early in the development of the stream.

b) Streams developing in clay.—Along the north shore we have an opportunity to study the vegetation of ravines of all ages corresponding to the aquatic stages described on pp. 87–93. The slightly lower places on the bluff side in which water runs only when it is raining are usually too steep to support plants and animals as regular residents, and have the same incidental forms as the steep bluff (p. 210). Later, when the sides of the gully become less steep, it is similar to if not identical with the second bluff stage (pp. 212–214); later, like the third (p. 215), and still later, like the young forest stage. There appears to be little or no difference between the bluff and the sides of young ravines. The outer ends of ravines as much as a mile and a half long are usually in the shrub stage and possess the shrub community. In favored situations the sapling forest, apparently identical in its animal associations with that of the bluff (p. 215), grows up. Up the stream, well back from the lake, a distance of a fourth of a mile, conditions become very different. A very mesophytic forest grows up. In this we have possibly some special features under primeval conditions, but in the ravines along the north shore where the forest is so much disturbed, ravines do not differ particularly from the rest of the forest, but animals of the forest collect in the ravines in dry seasons and apparently leave the ravines in the wet seasons.

We have noted that the animal species living at the headwaters of a stream may move inland as the headwaters move inland. This is true of aquatic species. In the case before us none of the species of the young stream are at the headwaters of the older streams because the headwaters of the older streams are in the forest of the upland while the young streams are in the unforested and exposed bluff of the lake.

c) Flood-plain communities.—In streams not more than a mile long we get suggestions of a small flood-plain near the mouth. Here we find ragweeds and other pioneer plants with their full quota of animals, such as the plant-bug (*Lygus pratensis*) and other common insects of rank pioneer vegetation; willows with their quota of cecropia caterpillars, viceroy larvae, willow-beetles, etc., are found here as elsewhere. The flood-plains of such small streams are hardly typical because the streams are cutting downward so rapidly. They doubtless possess many special features of interest which are subjects for detailed and special investigation.

Flood-plain forest is best developed among such streams as the DesPlaines River and Hickory Creek. As the stream meanders from side to side of its valley, it presents points of deposition and erosion. The points of deposition are best for the study of the development of flood-plain forest.

Girdle of bare sand or gravel (Station 66): On the wet portions of the sandy margins one finds the ground beetles (*Bembidium*) (156), sometimes toadbugs (p. 180), and more rarely the mole cricket. On the higher and drier portions we have taken the Carolina locust (*Dissosteira carolina*) (40) and the two-lined locust (*Melanoplus bivittatus*) (40) hopping over the ground.

Girdle of ragweed and helianthus (sub-formation) (Stations 66, 71a; Table XLVII): Here (in September) we found several species of spiders, the meadow grasshopper, long-legged flies, the leaf-hoppers, and the common plant-bug. This girdle is later displaced by willows.

Willow girdle (sub-formation) (Stations 66, 71a; Table XLVII): When herbaceous plants have grown for a few years they become mixed with willows which are inhabited by animals common in low forest margins. Here (in September) continues the same meadow grasshopper, the same plant-bug of the earlier stage. Two different spiders are recorded (*Pisaurina* and *Epeira*). From willows along other streams we have taken the larvae of the viceroy butterfly (163) and the larvae of the cecropia moth (*Samia cecropia* Linn.). Doubtless forest-edge birds nest here also.

The belt which succeeds the willow is commonly found farther from the water and has not been so much studied. It is commonly made up of larger willows, river maples, young elms, young ashes, and small hawthorns. These are usually much tangled with weeds such as nettles, and masses of flood trash and vines. General collecting in such a situation along Thorn Creek (August) secured for us the large green stink-bug (*Nezara hilaris*), the spiny assassin-bug (*Acholla multispinosa*), and the broad-winged fulgorid (*Amphiscepa bivittata*). On the maples are frequently larvae of *Symmerista* (Fig. 151). On a small hawthorn were a number of larvae of the handmaid moth (*Datana*). At this stage the trees and shrubs become the nesting-places of the yellow warbler and American goldfinch, which are probably our most characteristic early flood-plain birds.

In the wet ground of the flood-plain, especially in any small depressions made by overflows, the crayfish (*Cambarus diogenes*) is the characteristic resident. Under driftwood and on the plants of the water margin is the slug (*Agriolimax campestris*), and often also the snails (*Succinea retusa* and *avara*).

Such situations are also the chief haunts of the beaver, which cuts away the saplings to make its dams. The otter (*Lutra canadensis*) is particularly fond of stream margins. It feeds upon crayfishes, fishes, frogs, etc. It has particular powers of traveling long distances and a curious habit of sliding down mudbanks and snowbanks for sport (142). In winter it progresses on ice by repeatedly running a distance and then sliding as far as the momentum will carry the body. The nest is nearly always in the stream bank, with the entrance below water. The skunk, the mink, and the raccoon are also fond of the stream-margin thicket, the latter picking up fish or crayfish if they can be had at night. This animal is said to wet its food before devouring it; hence the "waschbär" of the Germans. The skunk likewise devours almost anything that is to be had at the water's edge.

d) *Flood-plain forest association* (Station 68; Table XLVIII).—As times passes the river cuts lower, the forest develops, and we have a dense forest of elm, hawthorn, ash, and basswood, with sometimes walnut and butternut, these being partially displaced on the higher ground by the oaks. This we may regard as the typical flood-plain forest.

Subterranean-ground stratum: The nymphs of the seventeen-year cicada and the two-year cicada together with earthworms are always numerous. The latter comes out on the ground under a log and ascends under the bark of dead trees during wet weather.

On the ground one finds slugs (*Agriolimax campestris*). Under leaves, logs, and bark are snails (*Circinaria concava, Polygyra profunda,*

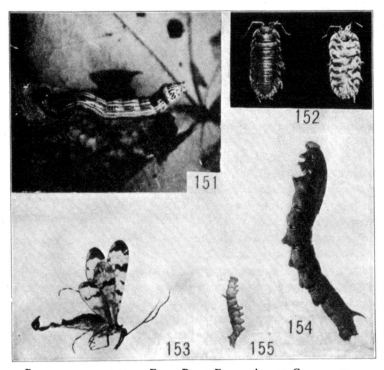

REPRESENTATIVES OF THE FLOOD-PLAIN FOREST ANIMAL COMMUNITIES

FIG. 151.—A caterpillar (*Symmerista albifrons*) on the leaf of the soft maple; natural size.

FIG. 152.—The common land sowbug (*Porcellio rathkei*); twice natural size.

FIG. 153.—The scorpion fly (*Panorpa venosa*); much enlarged.

FIG. 154.—A sphinx caterpillar from Virginia creeper; natural size.

FIG. 155.—The unicorn larva from dogwood; enlarged.

Pyramidula alternata, and *Polygyra clausa,* and rarely *thyroides*). Land sowbugs are common (Fig. 152). Of the centipedes we note the long ground form (*Geophilus* sp.) and sometimes the large millipede (*Spiro-*

bolus marginatus). The white-footed wood-mouse (*Peromyscus leucopus noveboracensis* Fisch.) nests usually under a stump or a log though sometimes slightly under ground or in hollow trees (21). The short-tailed shrew (*Blarina brevicauda* Say) and the common shrew (*Sorex personatus* St. Hil.) are common residents.

In the earlier days (22) the ground stratum was occupied by the larger mammals. The black bear doubtless found the delicate herbaceous plants desirable at certain times of the year. The Virginia deer occurred here commonly, and the bison and elk invaded the flood-plain forest in going to the rivers to drink. The timber wolf and the common fox, both of which formerly frequented all parts of Illinois, were no doubt also to be found.

Under fallen logs we find all the animals that are found on the forest floor, and some others also. When a tree first falls to the ground, if it be still solid or living, the animals which attack it are the same as those which attack it when it is standing. If the tree be an oak or a basswood, one of the first of these is the weevil (*Eupsalis minuta*) (Fig. 156) (155), which burrows into the wood. Later the larvae of some of the long-horned beetles are found working under or in the inner

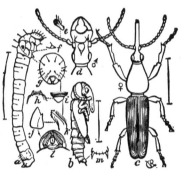

Fig. 156.—An oak borer (*Eupsalis minuta* Drury): *a*, larva; *b*, pupa; *c*, adult female; *d*, head of adult male; details of parts are indicated (after Riley).

layers of the bark. These are followed by the *Tenebrionidae* and the *Buprestidae* larvae, or flat-headed borers (137). All these tend to let the water between the trunk and bark, which meanwhile has been loosening with every rain, then drying, freezing, and thawing, until it soon becomes quite loose. The space between bark and log is loosely filled with the castings of the many animals that have worked over the outer wood and bark, and with wood and bark that have decayed without the aid of these animals. At such a time the space between bark and log becomes the abode of the flattened larvae of *Pyrochroidae*, centipedes, slugs, ground beetles, and nearly all of the small animals mentioned as belonging to the ground stratum proper. Fallen logs are also the nesting-places of the weasel (*Mustela noveboracensis*) (142, 143).

In the autumn we find many hibernating animals under the leaves of the floor of the flood-plain forest. Here we have found water-striders, the cutworms from the field stratum, the stinkbugs and leaf-bugs from the river margin, and large white-faced hornets (*Vespa maculata*).

Field stratum: In early summer the forms of the field stratum are most in evidence. There are scorpion-flies (Fig. 153, p. 200), the males of which have curious clasping organs at the posterior end of the abdomen. *Bittacus*, the long-legged insect, closely related to the former, flies about among the nettles; it has the curious habit of seizing flies with its hindermost pair of legs and holding them while they are being devoured. In their breeding both of these insects belong to the ground stratum.

The harvestmen, or daddy-longlegs, are always in evidence, crawling over the nettles (*Liobunum dorsatum* and *ventricosum* being most common). Several spiders (*Leucauge hortorum* and *Theridium frondeum*) occur. Numerous bugs including *Reduviolus annulatus*, syrphus flies (*Syrphus americanus*), and aphids, with the various enemies which occur with them, are common here. After rains we find many animals of the ground stratum on the nettles and the trunks of trees. We have noted the slugs (*Agriolimax campestris*) and the snails (*Polygyra profunda* and *thyroides*) here.

Shrub stratum: The shrub stratum is well developed. The dogwood is one of the characteristic shrubs, and in early summer its leaves usually are covered with small bunches of foam which upon inspection are found to contain a small insect, the spittle insect (*Aphrophora*). The unicorn larva (Fig. 155) (163) feeds on the leaves of the dogwood, and the sphinx larva on the Virginia creeper (Fig. 154*a*).

Tree stratum: The tree stratum has not been especially studied. The trees above the level of the shrub stratum are inhabited by many borers, lepidopterous larvae feeding on the leaves, and many birds nesting in the branches. The raccoon is especially fond of nesting high in hollow trees of the flood-plain forest. The opossum, which was never abundant near Chicago, found a suitable place in the trees of the flood-plain with its wild grapes and tender herbs. Under natural conditions this is one of the chief haunts of the gray squirrel, now familiar in our parks (21). For birds frequenting the flood-plain, see Baker (100, pp. 476–78).

The most striking peculiarity of the flood-plain forest is its frequent inundation. In the spring of 1908 we found the flood-plain of the north branch of the Chicago River inundated at a time when the nettles were but a few inches high. On the small nettles we found the common small slug (*Agriolimax campestris*) and the snails (*Succinea avara* and *retusa*) in great abundance. Caught in a corner behind a tree in some driftwood we found a carpenter ant (*Camponotus*), some flood-plain cutworms, crane-fly larvae, and ground beetles. These had been swept into this position by the current. Wood (21) says that the white-footed mice and shrews climb the trees when the stream is in flood. As the number of animals does not seem to be decreased after floods, the animals of the lower strata of the flood-plain forest must be able to withstand submergence for days at a time. The fact that these floods come in spring and winter when the animals are inactive doubtless assists in preserving them because of the low ebb of their metabolic processes.

e) Succession in the flood-plain forest.—As the stream works over its flood-plain, it is constantly destroying the forest at some points and depositing new materials upon which a new series develops at other points. The depositing sides of the curves present the early forest stages. Back of these and higher above the stream are the older stages. Thus the horizontal series which we see when we pass from a depositing bank across the various terraces is a duplication of the vertical series at the oldest point or on the highest terrace.

The higher and drier parts of the plain left by the lowering of the river bed, and much of the flood-plain proper, are often well drained, rarely flooded, and when thus drained pass rapidly into the oak-hickory type. At such a stage the oak-hickory animal association is present and the characteristic flood-plain animals have disappeared.

f) Comparison with other moist forests.—There are a few species common to the marsh and flood-plain forests. This is true of several mammals and insects. One of the most characteristic of the insects is the scorpion-fly. Many of the others belong particularly to the trees common to the two, such as the ash, elm, basswood, etc.

TABLE XLIII

ANIMALS FROM THE OPEN PRE-TAMARACK BOG DOMINATED BY BULRUSHES,
SEDGES, AND SIMILAR PLANTS

Animals recorded from Tamarack Swamps; M from the swamp at Mineral Springs, Ind. (Station 54), and P from that near Pistakee Lake, Ill. (Station 54*a*). Numbers refer to month in which the specimens were taken. † indicates that the species has been taken from low prairie; * that it has been taken from high.

Common Name	Scientific Name	Habitat	Marsh and Month
Midge larva	*Chironomus* sp..............	Pitcher-plant	M 5
Mosquito larva.....	*Wyeomyia smithii* Cq	"	M 5
Caterpillar.........	*Noctuinae* larva.............	"	M 5
Ground beetle (dead)	*Amara polita* Lec............	"	M 5
*Orb-weaving spider (dead)...........	*Epeira foliata* Koch..........	"	M 5
* Jumping spider (dead)..........	*Phidippus podagrosus* Htz......	"	M 5
Snout-beetle (dead)..	*Listronotus callosus* Lec.......	"	M 5
Ant (dead)........	*Dolichoderus mariae* Forel......	"	M 5
Water-beetle.......	*Helophorus lineatus* Say.......	Pool	P 8
Flat snail..........	*Planorbis parvus* Say	"	P 8
Crayfish...........	*Cambarus diogenes* Gir........	"	M 5
Entomostraca..	....		...
Snail..............	*Succinea retusa* Lea...........	"	M 5
Snail......	*Circinaria concava* Say........	"	M 5
†Spider (*Pisauridae*) .	*Pisaurina undata* Htz.........	Surface ground	M 5
Running spider.....	*Pirata montana* Em...........	"	M 5
†Marsh rattlesnake..	*Sistrurus catenatus* Raf	"	M 5
Ground beetle......	*Platynus picipennis* Kirby......	"	M 5
Harvestman........	*Liobunum grande* Say.........	Vegetation	M 5
*Orb-weaving spider..	*Epeira prompta* Htz..........	"	M 5
†*Orb-weaving spider.	*Plectana stellata* Htz.........	"	M 5
†Garden spider.....	*Argiope trifasciata* For	"	P 8
†Dictynid	*Dictyna sublata* Htz..........	"	...
*Crab spider.......	*Runcinia aleatoria* Htz	"	P 8
†Crab spider.......	*Tibellus duttoni* Htz	"	P 8
Jumping spider.....	*Dendryphantes octavus* Htz.....	"	M 5
Jumping spider.....	*Thiodina puerpera* Htz........	"	M 5
† Orb-weaving spider.	*Eugnatha straminea* Em:......	"	M 5
Meadow grasshopper.	*Orchelimum glaberrimum* Burm..	"	P 8
Hoosier locust......	*Paroxya hoosieri* Bl...........	"	P 8
†*Grasshopper (*Acrididae*)...........	*Stenobothrus curtipennis* Harr....	"	P 8
†Toad-bug.........	*Pelogonus americanus* Uhl	"	M 5
Bug (*Cercopidae*)....	*Lepyronia quadrangularis* Say...	"	P 8
Bug (fulgorid)	*Pentagramma vittatifrons* Uhler..	"	P 8
Braconid............	....	"	P 8
Ant...............	*Formica fusca* Lin...........	"	P 8
Fly (*Sciomyzidae*)....	*Sepedon pusillus* Loew........	"	M 5
Snout-beetle.......	*Baris confinis* Lec...........	"	M 5
Snout-beetle	*Brachybamus electus* Germ......	"	M 5
*Lampyrid beetle....	*Telephorus lineola* Fab........	"	M 5
Snout-beetle	*Listronotus inaequalipennis* Boh.	"	M 5

TABLE XLIV

ANIMALS RECORDED FROM MARGIN OF TAMARACK FOREST—FOREST EDGE

Abbreviations as in Table XLIII

Common Name	Scientific Name	Habitat	Locality and Month
Spider.............	*Argiope trifasciata* For........	Ground	P 9
Snail..............	*Vitrea indentata* Say..........	"	M 5
Slug..............	*Agriolimax campestris* Binn.....	"	M 5
Camel cricket.......	*Ceuthophilus* sp..............	"	M 5
Millipede..........	*Polydesmus* sp...............	"	M 5
Firefly larva.......	*Lampyridae* sp...............	"	M 5
Tortoise beetle......	*Coptocycla bicolor* Fab..........	"	M 5
Ground beetle.......	*Pterostichus lucublandus* Say....	"	M 5
Tree-frog..........	*Hyla versicolor* Lec	"	M 5
Crab spider........	*Philodromus ornatus* Bks.......	"	M 5
Stinkbug	*Euschistus tristigmus* Say.......	Shrubs	M 5
Jumping spider......	*Dendryphantes militaris* Htz.....	"	M 9 P 8
Orb-weaving spider...	*Epeira trifolium* Htz...........	"	P 8
Orb-weaving spider...	*Epeira trivittata* Key	"	P 8
Katydid...........	*Amblycorypha* sp.............	"	P 8
Ant...............	*Formica fusca* Linn...........	"	M 5
Pickering's frog	*Hyla pickeringii* Hol...........	"	M 9
Beetle.............	*Languria gracilis* Newm........	"	M 5
Earwig............	*Apterygida aculeata* Scud.......	Young tamarack	M 9
Brindled locust......	*Melanoplus punctulatus* Scud....	"	M 9
Orb-weaving spider...	*Epeira gigas* Lea.............	"	M 9
Jumping spider......	*Dendryphantes militaris* Htz.....	"	M 9
Sawfly larva........	*Nematinae*	"	P 8
Caterpillar.........	*Geometridae*................	"	P 8
Fly...............	*Mesogramma marginata* Say	"	M 5
Spider.............	*Chiracanthium inclusa* Htz	"	P 8
Long-bodied spider...	*Tetragnatha grallator* Htz.......	"	P 8
Harvestman........	*Liobunum dorsatum* Say........	"	P 8
Jumping spider......	*Dendryphantes octavus* Htz......	"	P 8
Red mite..........	*Trombidium sericeum* Say	"	M 5
Ground beetle.......	*Platynus decens* Say..........	"	M 5

TABLE XLV

TAMARACK FOREST

For meaning of abbreviations see Table XLIII

Common Name	Scientific Name	Habitat	Locality and Month
Mosquito	*Culex canadensis* Theob	Pools	M 5 P 8
Amphipod	*Eucrangonyx gracilis* Sm	"	M 5 9
Sowbug	*Asellus communis* Say	"	M 5 9
Copepod	*Canthocamptus northumbricus* Br.	"	M 6 9
Copepod	*Cyclops viridis americanus* Mar.	"	M 9
Copepod	*Cyclops serrulatus* Fisch	"	M 9
Copepod	*Cyclops albidus* Jurine	"	M 9
Spider (lycosid)	*Pirata piratica* Cl	Sphagnum	M 9
Spider	*Mangora maculata* Key	"	M 9
Millipede	*Scytonotus granulatus* Say	Under log and bark	M 9
Millipede	*Polydesmus* sp	Under log and bark	M 9
Snail	*Zonitoides arboreus* Say	Under log	M 9 5
Slug	*Philomycus carolinensis* Bosc	"	M 9
Snail	*Circinaria concava* Say	"	M 9 5
Crane-fly larva	*Tipulidae*	....	M 9
Ground beetle	*Pterostichus coracinus* Newm	Under log	M 9
Swamp tree-frog	*Chorophilus nigritus* Lec	"	M 9
Wood-frog	*Rana sylvatica* Lec	"	M 5 9
Snail	*Polygyra multilineata* var. *algonquinensis* Nason	Ground under bark	P 8
Ground beetle	*Pterostichus adoxus* Say	Ground under bark	M 5
Ground beetle	*Pterostichus pennsylvanicus* Lec.	Ground under bark	M 9
Engraver beetle	*Polygraphus rufipennis* Kirby	Early decay	M 5 9
Engraver destroyer (larva)	*Thanasimus dubius* Fab	"	M 5 9
Borer (larva)	*Cerambycidae*	Decaying wood	M 5 9 P 8
Spider	*Epeira ocellata* Cl	"	P 8
Spider	*Habrocestum pulex* Htz	Under loose bark	P 8
Spider	*Theridium frondeum* Htz	Herbs	P 8
Leaf-hopper	*Gypona striata* Burm	"	P 8
Leaf-bug	*Poecilocapsus lineatus* Fab. ?	"	P 8
Tree-hopper	*Ceresa borealis* Fair	"	P 8
Caterpillar		"	P 8
Beetle (*Melandryidae*)	*Synchroa punctata* Newm	"	P 8
Tortoise beetle	*Coptocycla clavata* Fabr	"	P 8
Jumping spider	*Zygoballus bettini* Peck	Undergrowth	M 9
Orb-weaving spider	*Epeira gigas* Lea	"	M 5 9
Orb-weaving spider	*Epeira foliata* Koch	"	M 9
Long spider	*Tetragnatha grallator* Htz	"	M 9
Jumping spider	*Dendryphantes octavus* Htz	"	M 5
Jumping spider	*Dendryphantes militaris* Htz	"	M 5
Dictynid spider	*Dictyna foliacea* Htz	"	M 5
Leaf-hopper	*Cicadula variata* Fall	"	M 9
Bug	*Scaphoideus immistus* Say	"	M 9
Leaf-bug	*Lygus plagiatus* Uhler	"	...
Leaf-hopper	*Gypona octolineata* Say	"	...
Spider	*Pisaurina undata* Htz	"	...

TABLE XLV—*Continued*

Common Name	Scientific Name	Habitat	Locality and Month
Beetle (*Dermestidae*)..	*Cryptorhopalum haemorrhoidale* Lec......................	Undergrowth	M 9
Beetle (*Scarabaeidae*)	*Chalepus nervosa* Panz.........	"	M 5
Lady-beetle.........	*Psyllobora 20-maculata* Say.....	"	M 5
Beetle (*Dascyllidae*)..	*Cyphon variabilis* Thunb.......	"	M 5
Beetle (*Dascyllidae*)..	*Cyphon padi* Linn............	"	M 5
Beetle (*Melandryidae*)	*Allopoda lutea* Hald..........	"	M 5
Pummice-fly........	*Drosophila amoena* Loew......	"	M 5
Ant...............	*Formica fusca* Linn...........	Low shrubs	M 5
Dictynid spider.....	*Dictyna sublata* Htz..........	"	M 5
Spider.............	*Hypselistes florens* Cam.......	Shrubs	M 5

TABLE XLVI

ANIMALS OF THE BIRCH-MAPLE BELT, WHICH SUCCEEDS THE TAMARACK AT MINERAL SPRINGS

(Station 54)

Common Name	Scientific Name	Habitat	Month
Red mite..........	*Trombidium sericeum* Say	Ground	5
Snail..............	*Polygyra albolabris* Say.......	......	...
Ground beetle......	*Pterostichus adoxus* Say.......	"	5
Melandryid beetle ...	*Phloeotrya quadrimaculata* Say...	"	5
Rove beetle........	*Listotrophus cingulatus* Grav....	"	5
Clubionid spider.....	*Castianeira cingulata* Koch	"	5
Horntail...........	*Xiphydria maculata* Say.......	"	5
Wood-frog.........	*Rana sylvatica* Lec...........	"	5
Salamander........	*Plethodon cinereus* Gr.........	"	5
Salamander........	*Plethodon glutinosus* Gr........	"	5
Spider.............	*Agelena naevia* Wal..........	Herbs	9
Tree-frog..........	*Hyla pickeringii* Hol..........	"	...
Spider.............	*Phidippus audax* Htz.........	Shrubs	5
Spider.............	*Dendryphantes militaris* Htz....	"	5
Spider.............	*Misumessus oblongus* Keys......	"	5
Beetle.............	*Cyphon padi* Linn............	"	5
Beetle.............	*Photinus corruscus* Linn........	"	5
Ant...............	*Camponotus herculeanus ligniperdus* var. *noveboracensis* Fitch..	"	9
Leaf-beetle........	*Calligrapha multipunctata* var. *bigsbyana* Kirby	"	9

TABLE XLVII

ANIMALS OCCURRING IN THE RAGWEED AND WILLOW THICKET STAGES OF FLOOD-PLAIN FOREST DEVELOPMENT

(Stations 66, 67, 71a)

RAGWEED STAGE

Common Name	Scientific Name	Habitat	Month
Snail...............	*Succinea avara* Say............	....	...
Snail...............	*Succinea retusa* Lea...........	....	...
Meadow grasshopper.	*Xiphidium brevipenne* Scud.....	....	8–9
Tarnished plant-bug..	*Lygus pratensis* Linn..........	....	...
Spider.............	*Argiope trifasciata* For........	....	8–9
Long-bodied spider...	*Tetragnatha laboriosa* Htz.......	....	8–9
Meadow grasshopper.	*Orchelimum glaberrimum* Burm..	....	8–9

THICKET STAGE

	The Succineas above continue	Plants	
Snail...............	*Succinea ovalis* Say...........	Willow	...
Katydid...........	*Amblycorypha oblongifolia* DeG..	"	...
Grape scarabaeid....	*Pelidnota punctata* Lin........	Grape	...
Fulgorid bug.......	*Amphiscepa bivittata* Say......	Weeds and willow	...
Cercopid bug.......	*Lepyronia quadrangularis* Say...	Haw	...
Assassin-bug	*Acholla multispinosa* DeG......	"	...
Stinkbug...........	*Nezara hilaris* Say..........	Willow	...
Spider.............	*Pisaurina undata* Htz..........	....	...
Spider.............	*Epeira gigas* Lea.............	Willow	...
Bythoscopid bug.....	*Idiocerus snowi* G. and B	"	...
Sawfly larva.......	*Cimbex americana* Lea.........	"	...
Leaf-beetle........	*Crepidodera helxines* Lin	"	...

TABLE XLVIII

ANIMALS USUALLY COMMON ON HERBACEOUS VEGETATION (CHIEFLY NETTLES) OF THE RIVERSIDE FLOOD-PLAIN FOREST (OAK-ELM STAGE) IN JUNE AND JULY

Those starred occur in the corresponding stages of marsh forests.

Common Name	Scientific Name
*Dictynid spider....................	*Dictyna foliacea* Htz.
*Spider..........................	*Theridium frondeum* Htz.
Spider (*Epeiridae*).................	*Leucauge hortorum* Htz.
*Harvestman.....................	*Liobunum dorsatum* Say
*Harvestman.....................	*Liobunum ventricosum* Wood
False crane-fly....................	*Bittacus strigosus* Hag.
*Scorpion-fly.....................	*Panorpa venosa* Westw.
*Snail...........................	*Polygyra thyroides* Say (moist days)
Lampyrid beetle..................	*Podabrus rugulosus* Lec.
Long-horned beetle................	*Strangalia acuminata* Oliv.
Click-beetle......................	*Limonius interstitialis* Melsh.
Scarabaeid beetle.................	*Chalepus scapularis* Oliv.
Snout-beetle.....................	*Rhinoncus pyrrhopus* Lec.
Damsel-bug......................	*Reduviolus annulatus* Reut.
Capsid bug......................	*Plagiognathus fuscosus* Prov.
Fly (*Psilidae*)....................	*Loxocera pectoralis* Loew.

CHAPTER XII

ANIMAL COMMUNITIES OF DRY AND MESOPHYTIC FORESTS

I. INTRODUCTION

The forest communities discussed in the preceding chapters are those displacing aquatic communities. In a climate suitable for forests, trees spring up on high, well-drained surface materials of all kinds. Forest appears on rock, sand, clay, etc., first as shrubs or scattered trees, later as dense mesophytic forest. In the region about Chicago we have forest in all stages of development and on several kinds of material.

The bluffs of the lake and artificial exposures of clay along the drainage canal and the till uplands afford examples of development peculiar to this type of soil. The few outcrops of Niagara limestone and the quarries and rock dumps present scattered data on the history of forests on rock. The extensive sand areas afford examples of all stages of development peculiar to sand. From all these situations, we find forests leading toward some type related to climate, either the typical forest of the forest climate, or the forest of the savanna climate.

II. FOREST COMMUNITIES ON CLAY

(Fig. 157) (55)

The chief areas of more or less active erosion are along the west side of the lake, from Waukegan to Winnetka, and on the east side of the lake from South Haven to Benton Harbor. The old bluffs of the Tolleston and Calumet stages as represented north of Waukegan and at various other points offer valuable areas for comparison. There are also similar bluffs along many of our streams, some of those in Michigan being very old.

When the ice sheet receded entirely and left the outline of Lake Michigan much as it is now, doubtless the shore presented a more or less rounded profile. However, since that time waves have gradually changed the shore profile. By washing away the clay at the base of such a shore, a bluff has been developed (62).

1. STEEP BLUFF ASSOCIATION

(Station 56; Table XLIX)

a) *Ground stratum* (55) (Fig. 157).—In spring, when the frost goes out of the ground, leaving the clay somewhat loosened, the ground-water level is high, and gravitation overcomes the viscosity of the clay, and great masses, whose consistency is that of thick mud, slump down in the form of landslides. This process naturally decreases the angle of slope at the points where the slumping takes place. Slumping does not occur equally everywhere and the bank becomes very irregular. Under such conditions the only animals present are the *Collembola*. In summer the steep bank dries. No animals are present as actual residents. The bank serves only as a casual alighting-place for tiger-beetles, butterflies, bees, flies, and other insects. Few or no plants are present.

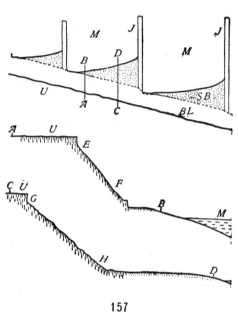

157

Fig. 157.—Upper figure is a diagram showing Lake Michigan bluff as seen from the zenith. *U*, level surface of upland; *BL*, bluff; *SB*, sandy beach; *M*, water of Lake Michigan; *J*, piers; toward the left is north; sand has lodged on the north side of the piers. *AB* and *CD* indicate positions of cross-sections below. Middle figure is a cross-section *AB*. Slumping bluff stage. The adults of *Cicindela limbalis* are distributed from *A* to *B;* the larvae, sparingly, from *E* to *F*. Other letters as in the upper figure. Lower figure is a cross-section *CD;* stage of some bluff stability and bare clay exposure. Adults of *limbalis* between *C* and *D;* larvae plentiful between *G* and *H*. Other letters as above. Reprinted from the *Journal of Morphology*.

Unless something interferes with the action of the waves the same series of events just described continues from year to year. If for some reason the action of the waves is checked, the associated processes will be checked also. At various points along the shore piers have been built out into the water at right angles to the shore for a distance of a hundred meters or more (Fig. 157). The currents in the lake are southerly in direction along the west shore. Whenever water in motion, laden with material picked up by its action against the bluff, strikes one of these piers, its velocity is decreased and a part of the material is dropped on the north side of the jetty. Materials thus deposited gradually pile up to such an extent as to protect the base of the cliff from wave-action. Thus the effect of the slumping of the springtime (which tends to reduce the angle of slope) is not fully removed from year to year.

FIG. 158.—The bluff habitats near Glencoe, Ill., showing several stages in the development of the forest on the bluff. The area to the right of a line between *a* and *b* is stable enough to support some sweet clover. Here the tiger-beetle larvae, spider, etc., are most abundant. The area between lines joining *a* and *b* and *a* and *c* is in the early shrub stage. To the left of *ac* the shrubs are denser and larger, and some trees are present. Reprinted from the *Journal of Morphology*.

2. SWEET-CLOVER ASSOCIATION

(Fig. 158) (55)

Under the condition described above, the water of rainfall, as well as the slumping, reduces the angle of slope, and the bluff becomes more and more stable. Some of the clods of turf from the top of the bank stop half way down the slope. The bluff begins to support a few xerophytic plants, such as the sweet clover, asters, etc.

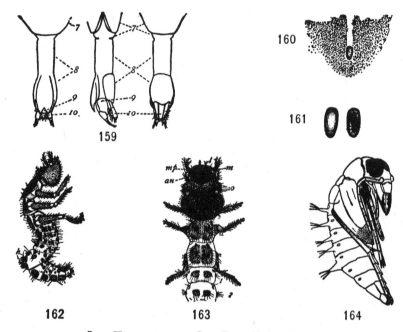

LIFE HISTORY OF THE CLAY-BANK TIGER BEETLE

(Reprinted from the *Journal of Morphology*)

FIG. 159.—From left to right—the ventral, side, and dorsal view of the ovipositor of the bluff tiger-beetle (*Cicindela limbalis*) with segments numbered; 3 times natural size.

FIG. 160.—The egg of the same in the hole in the ground made by the ovipositor; 1½ times natural size.

FIG. 161.—The egg; 3½ times natural size.

FIG. 162.—The larva, side view; *h*, hooks; 3 times natural size.

FIG. 163.—The anterior half of the larva: *an*, antennae; *mp*, maxillary palp; *m*, mandible; *o*, ocelli; 3 times natural size.

FIG. 164.—The pupa; 3 times natural size.

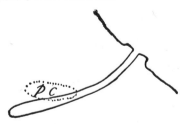

FIG. 165.—The burrow of *C. limbalis*, pupal cell; ⅓ natural size.

a) Subterranean-ground stratum.—Perhaps the most characteristic animal of the steep bluff is the bluff tiger-beetle (55, 151) (*Cicindela purpurea limbalis*) (Figs. 159–67). In the open places of this stage, the larvae, which live in curved cylindrical burrows (Figs. 165, 166), are common.

The female beetle is provided with an ovipositor (Fig. 159) adapted to making small holes in the clay in which eggs are laid (Figs. 160, 161).

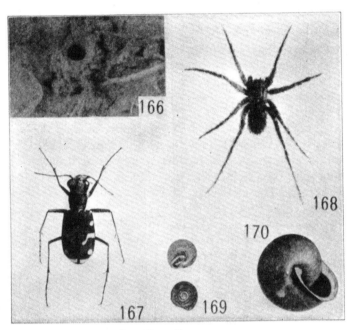

CLAY-BANK INHABITANTS

FIG. 166.—View of larval burrow of the tiger-beetle; natural size.
FIG. 167.—The adult tiger-beetle (*Cicindela limbalis*); about twice natural size.
FIG. 168.—The clay-bank spider (*Pardosa lapidicina*).
FIG. 169.—A snail of the shrub stage (*Polygyra monodon*); enlarged.
FIG. 170.—The snail (*Polygyra thyroides*); enlarged.

The larva (Figs. 162, 163) on hatching from the egg digs a burrow in the position of the ovipositor hole. The eggs, which are laid in June, hatch in two weeks and the larvae live in the spot where the eggs were laid for one year, and transform into pupae (Fig. 164) in the ground in an especially prepared cavity (Fig. 165). The adult, which is a reddish-green form (Fig. 167), appears in the autumn and lives over winter in the ground (151).

The tiger-beetle larvae are found on the bare spots and sometimes among the sweet clover (eggs are laid before the clover is full grown). They feed on any animals that crawl over the clay within reach—any that we mentioned may fall victims.

As physiographic processes go on, we find that more animals make their appearance, bristle-tails creep out of the cracks in early spring, and occasional slugs and geophilids are found hiding under clods. A large black spider (*Pardosa lapidicina*) (138) (Fig. 168) and many smaller species are present also. More rarely one of the land snails (*Pyramidula*) is present at this time of year, crawling about under the dead vegetation. The mud-dauber wasp (*Pelopoeus cementarius*) secures its mud (40) and the Carolina locust (*Dissosteira carolina*) probably breeds here.

b) Field stratum.—Under such conditions, as summer advances the sweet clover grows up, and as soon as it is of considerable size it is attacked by aphids, which form the basis for a small consocies of interdependent animals. Many coccinelids come to feed on aphids, and parts of adult coccinelids have been found in the burrows of the tiger-beetle larvae. The golden-eyed lacewing (*Chrysopa oculata*) deposits stalked (p. 291) eggs on the plant; soon its larvae—the aphis-lions—are devouring aphids, as do also the larvae of syrphus flies (164).

Crab-spiders (*Runicina aleatoria, Misumena vatia*) (138) lie in wait in the clover flowers and thus capture the nectar- and pollen-seeking flies, such as *Eristalis tenax* (Fig. 271, p. 270) and *Syrphus ribesii* Lin. (165). The common plant-bug (*Adelphocoris rapidus*) (Fig. 262, p. 266) is especially abundant in autumn. The honey-bee (*Apis mellifera*) and a bumblebee (*Bombus americanorum*) come in numbers for nectar and pollen. Grasshoppers, such as *Scudderia, Melanoplus femur-rubrum*, etc., are common, and when young may fall prey to spiders such as orb-weavers (*Epeira trivittata*). Parasitic hymenoptera (*Pimpla conquisitor* Say) are also common.

3. SHRUBS ASSOCIATION (A FOREST MARGIN SUB-FORMATION)

A little humus accumulates locally through the decay of sweet clover. The roots of plants in the soil and the undecayed trunks of the sweet clover hold this and the mineral soil in place against the action of the rain as it falls on the slope. Conditions become ripe for the germination of the seeds of other plants and for the breeding of other animals. Shrubs, such as the willow and shad-bush, appear as scattered individuals here and there, and bring with them new conditions and animal forms.

a) The subterranean-ground stratum.—In addition to *Pyramidula* mentioned above, other snails appear, especially in the more moist spots on the bank. These are *Zonitoides*, *Polygyra monodon* (Fig. 169), and *P. thyroides* (Fig. 170). Centipedes (*Geophilus*) and millipedes (*Polydesmidae*) become more numerous, while the spiders (*Pardosa lapidicina*) (Fig. 168), the tiger-beetle larvae, and other soil-inhabiting forms decrease.

b) Field stratum.—The field stratum of the shrub stage does not differ strikingly from the preceding, as it consists mainly of plants of the earlier stage scattered among the shrubs.

c) Shrub stratum.—Here we have the characteristic inhabitants of shrubs. On the young aspens and willow are the larvae of the viceroy butterfly (163). The common gall on the willow is the pine-cone gall, caused by *Cecidomyiidae* (137). Beneath the leaves of the cone we have found long slender eggs of some orthopterous insect (probably *Xiphidium ensiferum*) (40, p. 428). We have no record of the nests of birds, but many of the forest margin birds nest here (see pp. 274–75 and Table LXIV, p. 277).

4. YOUNG FOREST STAGE

(Fig. 171)

Shrubs and seedlings of trees become more and more numerous. The sweet clover and most of the animals associated with it disappear. Young trees, such as oak, hickory, hop hornbeam, etc., grow and usually give rise to a sapling forest.

a) Subterranean-ground stratum.—This stratum has all the characters of the more dense and mesophytic forest ground stratum and largely because of the springy character of the bluff which supplies much moisture. The woodchuck (*Marmota monax*) (142) sometimes digs in these banks. In the open places in which small areas of soil are covered with only a few leaves we find the larvae of the green forest tiger-beetle (*Cicindela sexguttata*) (55, 151) which lays eggs in shaded places (Figs. 172, 173). Under the leaves the snails, which were recorded in the younger stages, and sowbugs are present. We find snails and slugs (*Polygyra profunda* [Fig. 220, p. 237] and *albolabris* [Fig. 240, p. 243], *Philomycus carolinensis* [Fig. 231, p. 241]), which are commonly abundant in dense woods. The *Myriopoda* are also more numerous and belong to different species. *Fontaria corrugate* (Fig. 218, p. 237), which has the margins

THE BLUFF FOREST

FIG. 171.—An open place in the oak and hickory forest of a Tennessee mountain-side, a typical green tiger-beetle (*Cicindela sexguttata*) habitat. The individuals were seen copulating on the log in the foreground. The general aspect is very similar to that of the bluff forest. (Reprinted from the *Journal of Morphology*.)

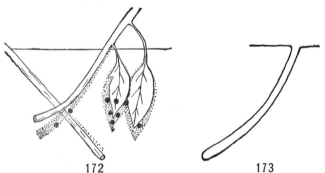

172   173

FIG. 172.—The black dots represent the distribution of the larvae of *C. sexguttata* from eggs laid in a cage. The larvae are in the exact position in which eggs are laid. The stippled area is in shadow in the middle of the day.

FIG. 173.—Diagram of a burrow of *Cicindela sexguttata*.

of the segments striped with yellow, is one of the most characteristic of moist woods, while others (*Geophilus rubens* and *Lysiopetalum lactarium*) are not uncommon. Ground beetles (*Calathus gregarius* Say) and bugs (*Reduviolus subcoleoptratus*) occur. In logs of fallen basswood we found the larvae of *Tenebrionidae* and *Cerambycidae* and of horntails, the burrowing hymenoptera, and the *Mycetophilidae* larvae (*Sciara*) (Fig. 174) (165).

c) Field stratum and shrub stratum.—The field stratum has been but little studied. We have taken a few *Scudderia* nymphs, some spiders, and bugs, but no adequate study has been carried on.

d) Tree stratum.—This has likewise been but little studied, but in these young forests, while the ground stratum is like that in the older forest, the tree stratum is poorly developed because the trees are short saplings. As time goes on, however, the forest becomes more dense. Such a forest may be seen on the bluff at Lake Bluff, Ill.

5. OTHER BARE CLAY FORESTS

Other bare clay young forests may be seen along the dumps of the drainage and Chicago-Michigan canals at Summit. Here we find practically the same stages as at Glencoe on the lake bluff. There are the steep clay bluffs with no permanent residents, the semi-stable bluffs, or weed-occupied areas. These are like the semi-stable bluffs at Glencoe but the tiger-beetle is another species and selects more nearly level places; otherwise it is very similar in habits.

FIG. 174.—One of the fungus gnats (*Sciara* sp.) the larvae of which are commonly found under the bark of trees, feeding on fungus.

The shrub stage occurs but is without the snails, since the ground-water level is lower and the moisture in the soil of the lake bluff is wanting here. This causes the development of the ground stratum to lag behind, while it is in advance in the bluff forests. Accordingly we find a sapling forest made up largely of cottonwoods. This has not been studied.

III. FOREST COMMUNITIES ON ROCK

(Station 55)

The rock exposures near Chicago are not numerous, and we have studied only those at Stony Island. There the bare rock is inhabited by incidental forms, such as the Carolina locust (*Dissosteira carolina?*) (40), with occasionally the red-legged locust (*Melanoplus femurrubrum*) and the two-lined locust (*Melanoplus bivittatus*). Under rock fragments we took the ground beetle (*Anisodactylus interpunctatus*) and the common cricket (*Gryllus pennsylvanicus*). Hancock (40) states that the smooth cockroach (*Ischnoptera inaequalis* Sauss) and the large cockroach (*I. major* Sauss) occur in such situations. We found the nest of a spider (*Agelena naevia*) attached to one of the loose rocks.

Other stages have been studied only superficially. In the cracks and crevices of rocks and rock piles, shrubs and vines grow and the young forest, field, and shrub strata have all the appearance of the *shrub stage* on clay at Glencoe. The animals are for the most part those common to thickets.

IV. Forest Communities on Sand

In chap. iii, pp. 46, 47, we discussed sand areas and their distribution. In chap. viii we noted the series of ponds and ridges with a little regarding their origin (pp. 136–40). Their general relations are indicated by Figs. 83, p. 137, and 84, p. 139. It appears that the margin of the lake may, under conditions of rapid recession, become the margin of an inland pond. Under condition of slower recession this belt may be buried and hence come to lie beneath such belts as lie farther inland. Since the sand areas about Chicago represent all the stages in the development of forests, beginning with the bare sand and ending with the beech forest, it is my purpose in the remainder of this chapter to follow the animal associations and formations of forest development. Some of the stages will be taken from till areas, but this is because these stages are more extensive than the corresponding stages on the sand deposits.

The chief stages are the wet sand of the water margin, the middle beach, the cottonwoods, the old cottonwoods and pine seedlings, the pines, the black oak, the black oak and white oak, the black oak-white oak-red oak, the red oak-white oak-hickory, the basswood-red oak-white oak-maple in moister places, and the beech and maple.

1. THE WATER MARGIN ASSOCIATION

(Stations 56, 58; Table XXXVIII)

One morning early in June, we walked along the beach of Lake Michigan for a mile and a half, for the particular purpose of studying the animals of the zone within the reach of waves. Animals were few, only stragglers of the regular residents which we have noted on p. 181.

The day was warm and a strong southeast wind was blowing. In mid-afternoon there was a small shower and the wind changed to a strong northeaster. At 4 P.M. we paid another visit to the beach. The waves were rolling moderately high and the beach was covered with a host of insects, chiefly alive, though many were dead. The beach was lined with live forms crawling away from the water. Often the live ones were still clinging to small sticks upon which they had floated ashore by the fifties. These insects represented all orders, belonging to various habitats near the lake. There were large forest margin bugs, potato-beetles, lady-beetles, horseflies, robber-flies, butterflies, water, marsh, prairie, and forest inhabitants which had been blown in the lake in the forenoon. With them were occasional fish, some with large round scars showing the work of the lampreys (166); others that had evidently died from other causes. On other occasions dead muskrats, dogs, cats, birds of all kinds have been found in these lines of drift (167). On one occasion, birds, chiefly downy woodpeckers, were so numerous that one could almost step from one to the other, had they been equally spaced over the half-mile of beach upon which they were strewn. Need-ham (168) has studied the drift and gives an account of the numerous beetles that came ashore.

In a few days after such a storm, one finds the various insects that washed ashore either lying dead, or alive under the chips, sticks, and carcasses which came with them. Flesh-flies detect the presence of the food very quickly, and often come to dead fish inside of ten or fifteen minutes (169). These flies belong to the families *Sarcophagidae* and *Muscidae*. As a result of storms which float the bodies of animals ashore from time to time, the flies always find a sufficient quantity of decaying flesh to maintain the species. The flies are in competition with a large number of scavenger beetles: e.g., a hister (*Saprinus patruelis* Lec.) which feeds on carrion (*Stereopalpus badiipennis* Lec.). Several species of rove-beetle complete a partial list of the other scavengers usually more or less abundant on the shore. The larvae of *Dermestidae* have been found under the dry remains of fish which had been worked over by the carrion-feeders.

Preying upon these and upon the insects that come ashore are the tiger-beetles (*Cicindela hirticollis* and *cuprascens*) (151, 170) which pick up the flies that they often are able to seize while alighting on the ground. They also capture the maggots of the flies when they leave the carrion, and the lady-beetles and other small insects which come ashore. Several species of the ground beetles and occasional shore bugs (*Saldidae*) are found, while the digger-wasps and robber-flies of the beach farther back come here for flies and other prey. The spotted sandpiper picks maggots from the bodies of dead fishes. Mr. I. B. Myers states that skunks visit the beach in the night and feed upon the drift.

2. MIDDLE BEACH ASSOCIATION

(Stations 57, 58, 71b) (Fig. 175)

The belt within the reach of ordinary waves is usually wet. The belt a little higher up, farther from the shore, is characterized by more permanent residents. From the often wet margin to the first cottonwoods is the middle beach (Fig. 175).

This middle beach is usually dry in summer but is reached by the waves of severe storms and often covered by snow and ice to great depths during the winter. It is the final lodging-place for the driftwood which stops temporarily farther out. This belt arises in the place of the preceding through the latter being buried by the deposition of sand. In digging into the sand here or elsewhere one usually encounters wood and other traces of organic matter.

a) *Subterranean-ground stratum.*—In the lower places where the ground is usually moist, we find the larvae of *Cicindela hirticollis* (170) which live in straight cylindrical vertical burrows about 6 in. deep. On higher ground, where there is the beginning of the incipient dunes, are the occasional larvae of the white tiger-beetle (*Cicindela lepida*) and the burrowing spider (*Geolycosa pikei*), which has a burrow similar to the tiger-beetles, but larger, and always distinguished by the presence of a tubular web at the entrance. Burrowing beneath the sand is the white carabid (*Geopinus incrassatus* Dej.) and termites or white ants. The latter

INHABITANTS OF THE MIDDLE BEACH

FIG. 175.—General view showing the line of cottonwoods and the scattered driftwood.

FIG. 176.—The larva of one of the cabbage butterflies (*Pieris protodice* Bd.); found on sea rocket; much enlarged.

FIG. 177.—Pupa of the same.

FIG. 178.—A log on the beach; favorite habitat of the termites (*Termes flavipes*).

FIG. 179.—Termites; *a*, queen; *b*, nymph of young female; *c*, worker; *d*, soldier twice natural size (after Howard and Marlatt, *Bull. 4, Div. Ent., U.S. D. Agr.*).

FIG. 180.—The older cottonwoods of the cottonwood belt.

FIG. 181.—The adult white tiger-beetle (*Cicindela lepida*); twice natural size.

FIG. 182.—The burrow of the larva of the white tiger-beetle.

INHABITANTS OF THE MIDDLE BEACH

feed on decaying wood (Fig. 178) and make their way to the under side of wood lying on the beach (Fig. 179). The bank swallow often nests in the sides of vertical sandbanks. Under the driftwood we find the scavengers and predatory species of the preceding belt. They spend their time here when the beach is not well covered with food. The sand-colored spider (*Trochosa cinerea*) (138) is a regular resident. The common toad finds shelter beneath the driftwood during the day, going forth in search of food at night. After sleeping near the beach one night we found the sand about where we had lain crossed and recrossed by the tracks of the toads and other smaller animals, such as beetles, spiders, etc. The toad finds food abundant near the shore. The white-footed mouse occasionally nests here under the largest driftwood. The spotted sandpiper and piping plover nest here occasionally.

b) Field stratum.—There are occasionally very young seedling cottonwoods. Sea rockets and some other plants grow in this belt. Occasionally we find the larvae of a cabbage butterfly (*Pieris protodice* Bdv.) (171) on the sea rocket (Figs. 176, 177). There is no shrub or tree stratum.

3. THE WHITE TIGER-BEETLE OR COTTONWOOD ASSOCIATION

(Stations 57, 58, 59; Tables L, LVI, LVII)

(Fig. 180) (115)

This begins with the line of young cottonwoods which we see in Fig. 175. The beach belt sometimes overlaps it because the large driftwood is sometimes mixed with the cottonwoods. The cottonwood belt is underlaid by the two preceding, and has succeeded them.

a) Subterranean-ground stratum.—Here the white tiger-beetles (Figs. 181, 182) reach their maximum abundance and the openings of their cylindrical burrows are numerous; the termites continue wherever there is wood for them to feed upon; the burrowing spider is commoner here than in the preceding zone (172). This is pre-eminently the zone of digger-wasps (173). Here the holes of *Microbembex monodonta* (Fig. 183) are numerous. This species is somewhat gregarious, the burrows usually being in groups. They probably store their nests with flies secured often from the beach. Another larger bembex (Figs. 184, 185) (*B. spinolae*) also stores its nest with flies. *Anoplius divisus*, the black digger, stores its nest with spiders. The velvet ant (*Mutilla ornativentris*) is present. *Dielis plumipes* appears in May and lays its eggs in the sand.

The robber-flies (*Erax*) (Fig. 186) (165) (*Promachus vertebratus*) (Fig. 187) are common; their larvae live in the sand as parasites on other species. Some bee-flies (*Exoprospa*) (Fig. 188) lay their eggs at the entrances of the burrows of *Microbembex*. The roots of the beach grasses are probably attacked by the larvae of snout-beetles (*Sphenophorus*) (Fig. 189) (174) of which several species are very common in the vicinity. The white grasshopper (*Trimerotropis maritima*) (40) and the white tiger-beetle (*Cicindela lepida*) are most characteristic. The long-horned locust (*Psinidia fenestralis*) (Fig. 189) occurs commonly.

b) *Field stratum.*—The field stratum is made up of animals that occupy the grasses, sagebrush, and a few other xerophytes. Animals

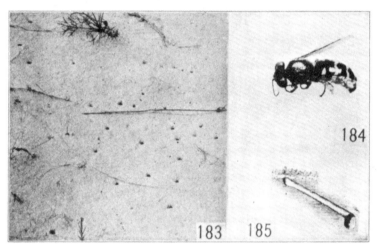

DIGGER-WASPS OF THE COTTONWOOD OR WHITE TIGER-BEETLE ASSOCIATION

FIG. 183.—Photograph of a number of the burrows of one of the digger-wasps (*Microbembex monodonta*) at Pine, Ind.

FIG. 184.—A digger-wasp (*Bembex spinolae*); about twice natural size.

FIG. 185.—A sectional drawing of a burrow of the digger-wasp (*Bembex spinolae*); reduced (after the Peckhams, Wis. Geol. and N. H. Surv.).

are few. An occasional red-legged locust (*Melanoplus femur-rubrum*) occurs here. Midges, mosquitoes, and the flies which breed on the beach rest on the leeward side of the grasses (169). Various native sparrows are common in fall, feeding on grass and weed seeds.

c) *Shrub stratum.*—On the young cottonwoods we find the crab-spider (*Philodromus alaskensis*), often with its appendages stretched out on the petiole or midrib of a leaf. The animals feeding on the cotton-wood here are few. In early spring the willow blossoms are frequented by pollen-gathering insects (*Andrenidae, Apidae,* syrphus flies, etc.). The kingbirds feed on these insects; one article of their diet, the robber-flies, is always common. A chrysomelid beetle (*Disonycha quinquevittata*) commonly feeds upon the willow. The cherry is attacked by aphids

FIG. 186.—A robber-fly (*Erax* sp.); 3 times natural size (after Williston).

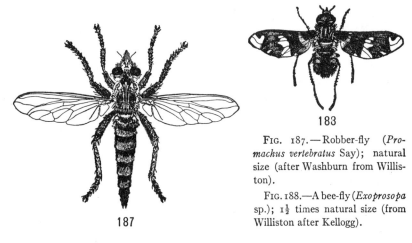

187

183

FIG. 187.—Robber-fly (*Promachus vertebratus* Say); natural size (after Washburn from Williston).

FIG. 188.—A bee-fly (*Exoprosopa* sp.); 1½ times natural size (from Williston after Kellogg).

which attract the *Coccinellidae,* and the syrphus flies. Cherries are eaten by many birds.

d) Tree stratum.—The cottonwood is attacked by many borers. The most characteristic is *Plectrodera scalator*, which is not common. There are few leaf-feeders excepting two gall aphids; the petiole gall is due to the work of *Pemphigus populicaulis*, and the terminal gall to *Pemphigus vagabundus* (137). These occur on the cottonwoods along the lake rarely, being more abundant farther inland, where they are protected from the severity of winter. The osprey nests in trees, and the tree-swallow in the dead ones.

We have noted that this association often arises through the burying of the preceding one. Deposition of sand is the chief cause of succession up to this point. When cottonwoods and grasses begin to grow and digger-wasps begin to burrow, organic matter is continually added to the soil. The grasses die down from time to time, the roots and leaves of the shrubs and other plants add humus. The myriads of digger-wasps which go elsewhere (probably commonly to the beach) for the animals with which to store their nests add a large amount of organic matter at a depth of a few inches. The grasses bind the dune sand; the conditions become favorable for other plants. At such a stage the bunch-grass and seedlings of pines appear.

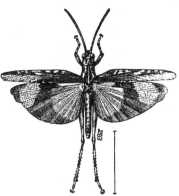

FIG. 189.—The long-horned locust (*Psinidia fenestralis*) (after Lugger).

4. TRANSITION BELT

(Station 58; Table L) (Fig. 190) (115, 170)

The stage of mixed pine seedlings, old cottonwoods, and the beginning of the bunch-grass constitutes a well-marked belt. Along the shore, from Indiana Harbor to Gary, there was formerly a ridge upon which the lakeward-facing side supported the typical community of the cottonwoods and the landward side the transitional belt. When one crosses to the landward side of such a ridge he notes a change in the animals. The white tiger-beetles and the maritime grasshopper are practically absent. Digger-wasps are abundant. The larvae of the large tiger-beetle (*Cicindela formosa generosa*) (Figs. 191–193) with their pits and crooked holes are added, but they rarely invade the dense pine areas. Another grasshopper (Fig. 194) (*Melanoplus atlanis*) and an

FIG. 190.—The cottonwood and young pine area at Buffington, Ind.

FIG. 191.—The burrow of one of the tiger-beetles resident here.

FIG. 192.—The same opened, showing the stove-pipe form of burrow opening into the side of the pit shown in Fig. 191.

FIG. 193.—The adult beetle (*Cicindela formosa generosa*).

occasional *M. angustipennis* are added (40). The burrowing spider (*Geolycosa pikei*) (Fig. 200, p. 230) continues in the open places.

5. THE CICINDELA LECONTEI OR PINE ASSOCIATION

(Stations 57, 58, 59; Tables L, LI, LVI, LVIII) (Figs. 201) (115, 170)

a) *Subterranean-ground stratum.*—Here we find the larva of the bronze tiger-beetle (*Cicindela scutellaris lecontei*) (170), with its straight, cylindrical burrow. Several digger-wasps of the earlier stage are recorded as continuing. The ant (*Lasius niger americanus*) nests beneath the sand and was seen swarming in early September. The burrowing spider continues and an occasional cicada lives deep beneath the sand. The six-lined lizard (*Cnemidophorus 6-lineatus*), the blue racer, and the pond turtle (*Chrysemys marginata*) all bury their eggs beneath the sand. There is an occasional thirteen-lined ground squirrel

FIG. 194.—The lesser migratory locust (*Melanoplus atlanis*) (after Lugger).

(*Citellus 13-lineatus*) (162), though it is never common. The surface of the ground is frequented by the adults of the tiger-beetles, digger-wasps, the six-lined lizard, and the blue racer (157). The grasshopper of the transition belt continues and two others are added, so that we have the long-horned locust, the narrow-winged locust, the lesser locust, the mottled sand-locust (*Sparagemon wyomingianum* Thom.), and sand-locust (*Ageneotettix arenosus*) (40). The ruffed grouse nests here occasionally.

b) *Field stratum.*—*Arabis lyrata* is a common herb. Shull (175) found that the larva of a cabbage butterfly feeds upon this. He watched a larva crawl on one of the bunches of bunch-grass for six hours before it began to spin the bed of silk preparatory to pupating. This was about 2 in. above the ground. Midges and mosquitoes are common and dragon- and damsel-flies are nearly always in evidence resting on the grasses and herbs and picking up the midges and mosquitoes while on the wing. Occasional *Monardas* support crab-spiders which resemble the blossoms closely (*Dictyna foliacea*). The flowers are visited by bees and flies.

c) Shrub stratum.—Here we have the young pines, the juniper, and the willows. From the evergreens we secured several spiders (*Philodromus alaskensis, Dendryphantes octavus, Theridium spirale,* and *Xysticus formosus*) (172), and with them sometimes an assassin-bug (*Diplodius luridus*). On the willows are some characteristic willow-feeders, but they appear to prefer the more mesophytic depression shrubs.

INHABITANTS OF THE PINE

FIG. 195.—The nest of the kingbird (*Tyrannus tyrannus* Linn) in a pine tree. The nest is made from the string of a fisherman's net.

FIG. 196.—The pitch mass of the pitch-moth (*Evetria comstockiana?*); twice natural size.

FIG. 197.—The larva removed from the mass.

FIG. 198.—The larva of the pine engraver beetle (*Ips grandicollis*); much enlarged.

FIG. 199.—The adult of the same, from *Pinus banksiana.*

d) Tree stratum.—The pine is attacked by many borers and few leaf-feeders. Of the borers several broad-headed grubs have been taken. The bark beetle (*Ips [Tomicus] grandicollis*) (Figs. 198, 199) (137) is common under the bark of dead and dying trees, especially on the north side, where the trees stand unprotected. The twigs are attacked by the pitch-moth (*Evetria comstockiana?*) (Figs. 196, 197) (137) which feeds on the new shoots, covering itself with a tent made of pitch and its own excreta. About the bases of the needles, or where pitch is exuding, we often find small larvae resembling *Cecidomyiidae* fly larvae, but we have found no pitch-midges, chrysomelid flea-beetles, spittle insects, or other enemy of the eastern hard pines which grow in thicker stands. More careful study of these trees at frequent intervals throughout the growing season would probably greatly increase the list of both borers and leaf-feeders.

The hairy and downy woodpeckers nest in the hollow trees. Their deserted holes are later used by the black-capped chickadee and the screech owl. Farther north the pine grossbeak and crossbill nest in the live pines. The golden-crowned kinglet and the black-throated, green, and pine warblers are abundant here during the migration period. They nest in the pines farther north, and, according to Butler (108), not infrequently at the head of Lake Michigan. Dr. Stephens photographed a kingbird's nest made from cord from a fisherman's net (Fig. 195).

The pines prepare the way for the oaks, which appear first as seedlings, usually becoming more dense with time and finally crowding out the pines.

Moving dunes and "blowouts" (depressions in the sand made by wind) are common at the head of Lake Michigan. The latter vary from a few feet square and a few inches in depth to some scores of feet n depth and diameter. Dunes, hundreds of feet high, move from place to place. On these the bare-sand conditions of the cottonwood and pine associations occur in areas generally dominated by black oak. Here continue the animals of these two belts, with the possible exception of the maritime locust. The typical black-oak forest always possesses these "blowouts," but surrounding them and under the trees we note the typical herbaceous and shrub growth, and it is with this and the oaks that we are next concerned.

6. THE ANT-LION OR BLACK-OAK ASSOCIATION
(Stations 57, 60, 61, 62; Tables L, LII, LVI, LIX)
(Fig. 202) (115, 170, 176)

Among the black oaks are open spots of relatively stable sand. These small areas may possess some of the same species as the pine areas, but other species give them individual character. In the black-oak stage proper, bare sand is limited. The bronze tiger-beetle (*Cicindela scutellaris lecontei*) (Fig. 204) which is parasitized by the larva of a bee-fly (*Spogostylum anale*) (Fig. 205) is abundant (151a.)

REPRESENTATIVES OF THE PINE AND BLACK-OAK ASSOCIATION

FIG. 200.—The burrow of a ground spider (*Geolycosa pikei*); about naturaι size.
FIG. 201.—General view in the pines. FIG. 202.—General view among the oaks.
FIG. 203.—The ant-lion and the pupa and adult into which it transforms.

FIG. 204.—The opening of the burrow of the bronze tiger-beetle (*Cicindela scutellaris lecontei*); natural size.

FIG. 205.—The bee-fly (*Spogostylum anale*); twice natural size.

a) Subterranean-ground stratum.—Several digger-wasps and parasites not found in the earlier stages occur among the more closely placed vegetation here (*Epeolus pusillus*, a parasite, *Specodes dichroa*, and *Odynerus anormis*). A megachilid or leaf-cutter makes a nicely matched thimble-shaped cell. This cell is placed at the end of a burrow about 2 in. below the surface of the sand. The burrow is about 4 in. long. The leaf-cutter is attacked by a parasitic bee (*Coeloixys rufitarsus*) which lays its eggs upon the larval cell. One sunny day we found the digger-wasp (*Ammophila procera*) (173) with a black-oak caterpillar (*Nadata*

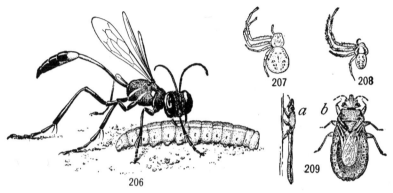

207   208

a b

209

206

REPRESENTATIVES OF THE BLACK-OAK COMMUNITY

FIG. 206.—One of the solitary wasps (*Ammophila procera*), with the oak-feeding larva (*Nadata gibbosa*), which it has carried to a point near its nest and laid upon the ground; 1½ times natural size.

FIG. 207.—Female crab spider (*Misumessus asperatus*) (after Emerton); enlarged.

FIG. 208.—Male of same.

FIGS. 209a, 209b.—The flatbug (*Neuroctenus simplex*) which lives under the bark on the dead oaks. 209a is a side view, much enlarged.

gibbosa) (Fig. 206) (137). When first observed, the larva was lying on the ground and the wasp was moving about some 6 in. away. As we approached, the *Ammophila*, apparently disturbed, seized the large caterpillar and ran into the adjoining vegetation, where it was captured.

All the forms mentioned as breeding beneath sand, feed at the surface of the soil or upon the vegetation. In open places among the black oak we find the same grasshoppers as in the earlier stages. The hog-nosed snake (40) is common; it spreads and flattens out its head when disturbed; when handled roughly it often goes into a death feint, such as the oriental snake-charmers produce in their poisonous snakes by pressure on the back of the neck.   In this state it can be handled as if dead, laid in any position, or tied into a knot.   The only movement it persists in making is that of turning its ventral side uppermost.   Ant-lions (Fig. 203) are very rarely found at the south end of Lake Michigan, except in the oak belt.   They make cylindrical conical pits in the sand (177, 179).   The most characteristic species under the bark of fallen oaks is the flatbug (Fig. 209).

   b) The field stratum.—This stratum is dominated by many flowering plants, such as *Monarda*, etc.   The addition of a host of insects and spiders not present in the earlier conditions is noticeable.   Of the grasshoppers we add six species (*Scudderia texensis, Xiphidium strictum, Chloealtis conspersa, Schistocerca rubiginosa, Oecanthus fasciatus*, and *Conocephalus ensiger*) (40).

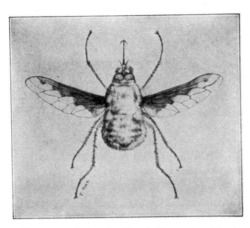

The andrenid bees (*Agapostemon splendens*) and various robber-flies are numerous.   On the *Monarda* the honey-bees, bee-flies (Fig. 210), bumblebees, and spiders (*Misumessus asperatus* [Figs. 207, 208], *Dictyna foliacea, Agriope trifasciata*, and *Epeira* sp.) are common. The blueberry is commonly one of the small herbs of the field stratum and upon it we find several characteristic galls.

FIG. 210.—A bee-fly (*Bombylius major* Linn.) (from Williston after Lugger).

   c) Shrub stratum.—This stratum is made up of the choke-cherry, young oaks, rose, etc.   The shrub which has been given most attention is the choke-cherry.   On this the lacebugs (Fig. 211) are often numerous; the puss caterpillar (*Cerura* sp.) (163) sometimes occurs.   This caterpillar has a pair of long projections at the posterior end.   When disturbed it extends and waves these projections and thus makes of itself one of the most fantastic of our caterpillars.

   Grapevines are not uncommon on the dunes and we often find a curious red petiole gall on them, which is not common elsewhere.   The large fleshy larvae of the achemon sphinx (163) are sometimes taken.

d) Tree stratum.—The black oak (137) is attacked by a large, light-green larva which has a narrow yellow stripe down its back (*Nadata gibbosa*). It is also attacked by several slug caterpillars which we have been unable to identify. The beautiful prominent larva with a saddle of red is occasionally taken. Commonly feeding on the juices of the leaves are several species of leaf-hopper (*Typhlocyba querci* var. *bifasciata*), the common grapevine leaf-hopper, and the white black-marked leaf-hopper which occurs also on the hickory. The oak tree-hopper (*Telemona querci*) (Fig. 212) is a common leaf-sucker. Squirrels are probably occasional visitors as they come to feed upon acorns. The acorns are also often attacked by weevils.

In such a set of graded forest stages as we are discussing it is possible to note many stages. The stage which we have just described passes more or less rapidly into the next, the rate of change depending upon the height above ground water and the degree to which the sand is shifted by the wind. On the parallel ridges, the next and perhaps

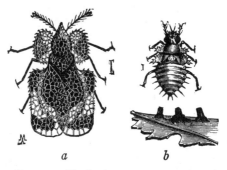

a   *b*

FIG. 211.—The lacebugs common on the oak and wild cherry in the dune region (*Corythuca arcuata*) (from Washburn after Comstock): *a*, adult; *b*, young.

most notable forest stage contains white oak and red oak and is found in places on the Tolleston, Calumet, and Glenwood beaches. The ecological age of the forest is determined by the height above ground water. Ridge 93, inside the Tolleston Beach, is low and forest has progressed as far as on the older beaches.

V. MESOPHYTIC FOREST FORMATION (115, 170)

I. HYALIODES OR BLACK OAK-RED OAK ASSOCIATION

(Station 63, also near stations 27 and 65; Tables L, LIII, LVI, LIX) (115)

This is represented at several points.

a) Subterranean-ground stratum.—In this stratum the woodchuck or groundhog is common (142). Earthworms have begun to appear. The root-borer *Prionus* (155) and several species of ants are common, while the numerous digger-wasps of the earlier stage have largely disappeared. The depressions which contain water in spring are typical forest temporary ponds. Beneath the leaves and wood are snails (*Zonitoides arboreus*), millipedes (*Polydesmus* sp.), and centipedes (*Lithobius* sp.), and in dry weather *Polygyra thyroides* and *multilineata*. Ground beetles and rove-beetles are common. One finds *Cicindela*

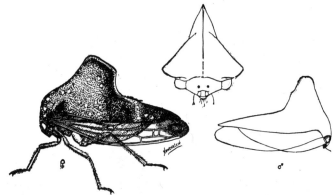

FIG. 212.—The oak tree-hopper (*Telamona querci*) (after Lugger).

sexguttata, the green tiger-beetle, here rarely; it is much commoner in later stages, however.

In the decaying logs and stumps are darkling beetles (156), numerous wireworms (*Elateridae*), and myriopods. Sometimes fungus-feeding beetles (*Diaperis hydni* and *Eustrophus tormentosus*) are present in numbers. Ants are also often abundant. Carpenter ants are common. The aphid housing ant (*Lasius umbratus* subsp. *mixtus* var. *aphidicola*) is sometimes abundant. In autumn certain galleries in the wood are crowded with woolly aphids which are the so-called "cows" which the ants house for the winter.

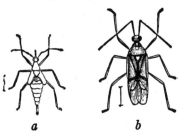

FIG. 213.—The oak plant-bug (*Hyaliodes vitripennis*) (from Washburn after Riley): *a*, young; *b*, adult.

b) Field and shrub strata.— In moist weather the snails (*Polygyra*) mentioned above are common on the herbaceous vegetation, while the tree-frogs (*Hyla versicolor* and *pickeringii*) (139) are common, and spiders are numerous.

c) Tree stratum.—The oaks (137) are affected by many of the same species as in the earlier stages. The tree-frog is sometimes found in the trees and the walking-stick (*Diapheromera femorata*) (40) is common. One of the most characteristic galls is the oak-seed gall (*Andricus seminator*), particularly abundant on white oak of this stage and not common later. Galls are very common on the white oak. The predatory capsid (*Hyaliodes vitripennis*) (Fig. 213) is usually present on the bark of the oaks, and is often in company with book-lice (*Psocus*). The squirrels, chipmunks, and birds of this association are similar to those of the next stage and will be discussed there.

Fig. 214.—General view of the white-oak red-oak hickory forest (Glencoe).

2. THE GREEN TIGER-BEETLE OR WHITE OAK-RED OAK-HICKORY ASSOCIATION

(Stations 56, 64, 65; Tables LIV, LXI) (Fig. 214)

This is the climax forest of the savanna region. The groves are largely made up of it. Though somewhat disturbed in localities where studied, it presents some variations. Areas along the north shore contain considerable basswood. The Higginbotham woods at Gaugars (Fig. 215) contain very few hickories and many maples; this type stands in closer relation to flood-plain and marsh forests than those discussed later. The woods at Suman are well invaded by beech and maple seedlings and represent the latest stages of this forest. It is thought best to treat all phases together, simply mentioning the points of difference.

a) Subterranean-ground stratum.—Earthworms, borers in the roots of trees, and cicada nymphs are numerous. The wolf, groundhog, and the red fox (*Vulpes fulvus* Des.) nest in burrows. The latter brings forth from four to nine pups in early spring.

Consocies of the under side of leaves and wood: The camel cricket

A MESOPHYTIC FOREST

FIG. 215.—General view of the Higginbotham woods near New Lenox. Woods of the flood-plain oak-hickory type.

(*Ceuthophilus*) (Fig. 216), young cockroaches, the short-winged grouse locust (*Tettigidea pennata* Morse), and the yellow-margined millipede (*Fontaria corrugate*) (Fig. 218) are most characteristic under the leaves. The large round millipede (*Spirobolus marginatus*) (Fig. 217) is common. Snails and slugs are numerous, several species (*Polygyra pennsylvanica* [Fig. 219], *P. profunda* [Fig. 220], *Zonitoides arboreus, Pyramidula alternata* [Fig. 221], *Pyramidula solitaria* [Fig. 222], *Agriolimax campestris*

Circinaria concava [Fig. 223]) are usually common and *Polygyra albolabris* is characteristic of the more mesophytic parts.

The ruffed grouse, oven-bird, and woodcock nest on the ground. The timber rattlesnake (*Crotalus durissus* Harlan) formerly occurred in rocky situations (22). The four-toed salamander (*Hemidactylium scutatum* Schl.) is found locally (22). The white-footed wood-mouse (*Peromyscus leucopus noveboracensis* Fisch.) builds a nest under fallen

INHABITANTS OF A MESOPHYTIC FOREST

FIG. 216.—The wingless wood locustid (*Ceuthophilus*); enlarged.

FIG. 217.—The common millipede (*Spirobolus marginatus*); natural size.

FIG. 218.—Another millipede (*Fontaria corrugate*); natural size.

FIGS. 219–223.—Snails from the woods. 219, *Polygyra pennsylvanica* Green; 220, *Polygyra profunda* Say; 221, *Pyramidula solitaria;* 222, *Pyramidula alternata;* 223, *Circinaria concava.*

logs and stumps (21). The gray fox (*Urocyon cinereoargenteus* Müll.) is more dependent upon heavy timber than the red fox (21). The cotton-tail (21), which belongs to forest edge, frequently winters in the woods. The bear was formerly common, nesting under fallen trees and feeding extensively on the berries.   The timber wolf had its den in similar places, though often burrowing into the ground.   In Central Illinois moles are common residents of groves near cultivated lands.   The Virginia deer (*Odocoileus virginianus* Bodd.) was formerly common and was preyed upon by the wolves and panthers.   The latter sometimes leaped upon its prey from the branches of the trees (142).

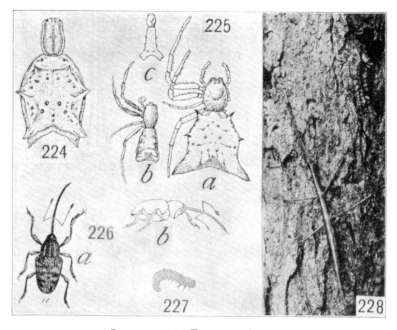

INHABITANTS OF TREES AND SHRUBS

FIG. 224.—The spiny spider (*Acrosoma gracilis*), legs wanting (after Emerton).

FIG. 225.—Another spiny spider (*Acrosoma spinea*): *a*, female; *b*, male; *c*, young (after Emerton.)

FIG. 226.—Acorn weevils: *a*, dorsal view; *b*, side view (after Riley, U.S. D. Agr.).

FIG. 227.—A red-oak sawfly larva.

FIG. 228.—A female walking-stick on the trunk of a tree, with a caterpillar (*Halisidota* sp.) on the bark above.

Consocies of logs (in wood and under bark): There is a regular succession of forms which affect any one species of the trees of the forest. The earlier forms usually attack the trees while they are standing, and accordingly belong more properly to the tree stratum.   When the bark has become loosened, however, we find practically all the small inverte-brates recorded on the ground. The small andrenid bees (*Augochlora pura*) build small cells under the bank and fill them with pollen. One egg is laid in each cell (July), and the larva feeds upon the pollen. Sowbugs (*Cylisticus convexus* and *Porcellio rathkei*) and centipedes (*Lithobius, Lysiopetalum lactarium,* and *Geophilus rubens*) are common. Numerous beetles burrow into the wood or feed on fungi under bark. Some of the chief borers are (*Cerambycidae*) *Prionus* and *Orthosoma brunneum,* and also *Passalus cornutus.* The large slug (*Philomycus carolinonsis*) is common.

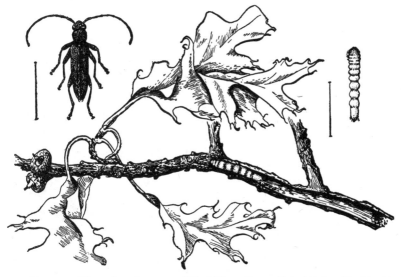

Fig. 229.—The oak twig pruner (*Elaphidion villosum* Fabr.) (after Washburn) (*17th Rept. Minn. Agr. Exp. Sta.*, p. 165, Fig. 36).

b) *Field stratum.*—After rains the slugs and snails, especially the young, crawl upon the vegetation. Several flies are common (*Sapromyza philadelphica*). A leaf-hopper (*Scaphoideus auronitens*), a damsel-bug (*Reduviolus annulatus*), the shield grasshopper (*Atlanticus pachymerus*), and a spider (*Theridium frondeum*) have all been recorded.

c) *Shrub stratum.*—Many spiders build their nests and webs in this stratum. *Epeira domicilorum* was found with a nest of leaves drawn together adjoining its web. *Epeira gigas,* the large yellow spider, builds near open places, on high shrubs. The web is a large orb, the nest in a convenient group of leaves near the upper side.

Acrosoma gracilis (Fig. 224) (138, 172) commonly stretches its web between the trunks of two small trees which stand about 4 ft. apart. The center of the orb is commonly about 6 ft. above the ground; it is nearly vertical. The spider usually hangs near the center.

THE STANDING DEAD OAK AND INHABITANTS

FIG. 230.—Showing the larva, pupa, and adult of the large wood-eating beetle (*Passalus cornutus*); about natural size.

Acrosoma spinea (Fig. 225a, b, c) (138, 172) commonly places its web in a nearly horizontal position on the upper side of leaves. The spider clings, ventral side up, on the lower side of the web. The web is usually from 1 to 3 ft. from the ground. The spider often falls to the ground when disturbed. The two *Acrosomae* are confined to mesophytic forests of the oak-hickory type. They have not been recorded north of Chicago.

A wasp (*Polistes*) builds its comb of wood pulp on the under side of the leaves. Various larvae and beetles feed upon the leaves of the undergrowth. A bug (*Acanthocephala terminalis*), a leaf-beetle (*Calligrapha scalaris*), the fork-tailed katydid (*Scudderia furcata*), the round-winged katydid (*Amblycorypha uhleri* Brun.) (40), and various other insects have been secured from shrubs, especially in slight openings. The black snake (22) (now rare) often rests on bushes in such forests. The black and yellow warblers and woodthrush nest on the shrubs.

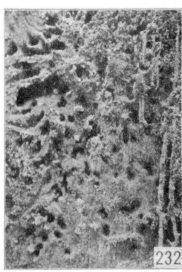

THE STANDING DEAD OAK AND INHABITANTS

FIG. 231.—The successor of *Passalus* (*Philomycus carolinensis*).

FIG. 232.—The work of a carpenter ant in the same tree.

d) *Tree stratum.*—The walking-stick (Fig. 228) (*Diapheromera femorata*) (40) is common on the tree trunks in the fall. The red oak supports the tree cricket (*Oecanthus angustipennis*), the stinkbug (*Euschistus tristigmus*), and the oak-leaf beetle (*Xanthonia 10-notata*). Felt records several insects injurious to the red oak alone. From the white oak we have taken the katydid (*Cyrtophillus perspicillatus*), the larvae of sawflies (Fig. 227) and moths (*Anisota senatoria*), and various galls. Several weevils (Fig. 226a, b) occur on acorns, and the twig-borer (*Elaphidion villosum*) (Fig. 229) in the twigs. The hickory supports many larvae, including a *Phylloxera* which forms galls on the leaves (see Fig. 277, p. 273).

The red-tailed and red-shouldered hawks, the red-headed wood-pecker, the wood-pewee, the crow, bluejay, robin, and bluebird nest in the trees. The panther and wildcat (*Lynx rufus*) were former residents.

FIG. 233.—The beech woods. Note small amount of undergrowth.

Dead standing oaks are attacked by a series of animals. As soon as the wood begins to soften, the four-legged larva of *Passalus cornutus* often appears. This is succeeded by slugs and ants (Figs. 230, 231, 232).

2. WOOD-FROG OR BEECH AND MAPLE FOREST ASSOCIATION

(Stations 70, 71, 71*a*, 71*b;* Tables LV, LXII) (Fig. 233)

The coming of this stage is indicated by the presence of seedlings of beech and maple in the oak-hickory forest, e.g., at Suman, Ind.

a) Subterranean-ground stratum.—Earthworms continue; an occasional groundhog has been seen, though they are probably much less common here than in the preceding stages. The stratum appears less closely inhabited than the preceding. Under leaves are found scattered snails, centipedes, etc. The yellow-margined millipede (*Fontaria corrugate*) is most common. There is an occasional *Centhophilus*. We have found no other *Orthoptera* in beech woods proper, though Hancock records several (40, p. 422). Animals are more abundant under logs than under leaves. Here we find the large slug (*Philomycus carolinensis*) and several species of snails which, though characteristic,

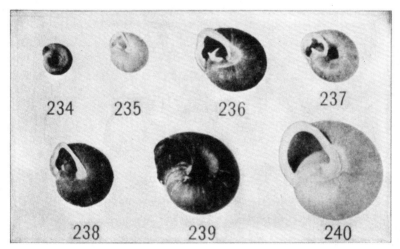

FIGS. 234–240.—Some beech woods snails: Ground stratum; 234, *Pyramidula perspectiva;* 235, *Polygyra inflecta;* 236, *Polygyra palliata;* 237, *Polygyra fraudulenta;* 238, *Polygyra oppressa;* 239, *Pyramidula solitaria,* adult; 240, *Polygyra albolabris.*

are not abundant. These snails are *Polygyra inflecta* (Fig. 235), *oppressa* (Fig. 238), *fraudulenta* (Fig. 237), *palliata* (Fig. 236), *albolabris* (Fig. 240), *Pyramidula solitaria* (Fig. 239), *alternata,* and *perspectiva* (Fig. 234), and *Zonitoides arboreus.* These species of *Polygyra* are distinguishable by the presence of characteristic "teeth" in the entrance of the shells. The large spider (*Dolomedes tenebrosus*) and millipede (*Spirobolus marginatus*) occur. Crane-fly larvae, ground beetles (*Plerostichus adoxus*), a centipede (*Geophilus rubens*), the wood-frog (*Rana sylvatica*) (Fig. 241) (139), and the red-backed salamander (*Plethodon cinereus*) (152) (Fig. 242) are common and characteristic.

Pickering's tree-frog is sometimes abundant. The oven-bird nests on the ground.

b) Field and shrub strata.—The field stratum is very poorly developed in summer, herbaceous plants being most abundant in early spring. The pawpaw supports the zebra swallowtail butterfly (*Papilio ajax* Linn.), and the spice-bush the green-clouded swallowtail (*Papilio troilus* Linn.). In the shrubbery in general we have taken snout-beetles, leaf-beetles, etc., usually as incidental occurrences, however. A lacebug (*Gargaphia tiliae*), which has been recorded on basswood, and several

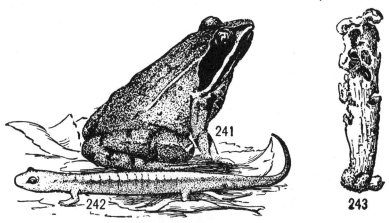

REPRESENTATIVES OF THE WOOD-FROG ASSOCIATION

FIG. 241.—The wood-frog (*Rana sylvatica*); about natural size.

FIG. 242.—The red-backed salamander (*Plethodon cinereus*); about natural size.

FIG. 243.—The remains of a fungus found growing under a pile of logs in moist woods (not beech), and the fungus-feeding beetle (*Tritoma unicolor* Say); about natural size.

species of bugs and beetles have also been taken, but all are incidental and of widely distributed species.

c) Tree stratum.—On trunks, shelf fungi are common and are usually inhabited on the under side by the tenebrionid beetle (*Boletotherus bifurcus*) (156), a curious rustic beetle. Few characteristic species have been taken from the trees. From the bark of the trunk we have taken harvestmen (*Oligolophus pictus* and *Liobunum nigropalpi*) and from the twigs woolly aphids (*Pemphigus imbricator*) (Fig. 245). There is an occasional Io larva on the leaves (Fig. 244).

The great crested flycatcher, wood-pewee, bluejay, scarlet tanager, red-eyed vireo, and woodthrush nest in the low trees and on the lower levels of the higher trees. Little is known of the mammals of the beech and maple forest. Deer, bears, wolves, foxes, hares, etc., appear to prefer forests with more undergrowth and herbaceous vegetation. Squirrels are fond of beechnuts, and are probably the chief resident mammals. The fox squirrel, gray squirrel, red squirrel, and other mammals of the preceding stages doubtless occur.

d) Consocies of the decay of a beech.—Succession: Any tree which is torn down by the wind or lightning is attacked by a series of borers,

LEAF- AND TWIG-FEEDERS

FIG. 244.—The nest of an Io caterpillar in the beech leaves; reduced.

FIG. 245.—Woolly aphids (*Pemphigus imbricator* Fitch) on the twig of the beech; reduced.

etc., each one helping to prepare the way for those that follow. To illustrate the general principles, the succession of animals in any species of tree might be presented. We have chosen the beech.

According to Felt (137), living beeches are commonly attacked by the red-horned borer (*Ptilinus ruficornis* Say) which bores into the bark and wood, and another borer (*Anthophilax attenuatus* Hald.) which lays eggs in the galleries thus formed. We have examined four stages of the decay of beech trees.

First stage: Tree freshly fallen (Fig. 246). Only forms recorded are the apple-tree engraver beetle (*Pterocyclon mali* Fitch) (Fig. 247) which makes galleries in the solid wood.

SUCCESSION IN THE BEECH LOG

FIG. 246.—The freshly fallen beech.

FIG. 247.—The first borer to enter the fallen tree (*Pterocyclon mali* Fitch); greatly enlarged (from Lugger after U.S. Dept. Agr.).

FIG. 248.—The partially decayed beech.

FIG. 249.—Closer view of the same showing the burrows of the different wood-boring larvae in the softened wood.

FIG. 250.—Shows the last stage in the decay of the beech.

Second stage (Fig. 248): Bark loosened; wood still solid or barely softened. Under the bark were the flattened *Pyrochroidae* larvae, the small snail (*Zonitoides arboreus*), a few of the four-legged larvae of the passalid (*Passalus cornutus*), many larvae of fungus-gnats (*Mycetophilidae*), and a single specimen each of the beetle (*Penthe pimelia*) and the slug (*Philomycus carolinensis*). None of these were abundant. The flattened beetle larvae were most characteristic.

Third stage (Fig. 249): The wood is thoroughly softened and the bark generally loosened. Here the animals present in the earlier stage are increased in numbers. The passalid larva is more abundant. Slugs are numerous. Snails (*Pyramidula alternata*) are found in such situations as are large enough for them to enter. Fungus-eating beetles are present (*Megalodacne heros* Say). A click-beetle larva (*Tharops ruficornis* Say) bores into the softened wood.

Fourth stage (Fig. 250): The bark fallen off; the log a mere mass of rotten wood. Such a log is only shelter for the regular inhabitants of the forest floor which we have already enumerated on the preceding pages.

VI. GENERAL DISCUSSION

A study of the tables shows several points of interest. Take first the ground stratum. Beetles which live under decaying wood are common on the beach where the decaying wood is common, but are absent through the cottonwood, pine, and black-oak stages. They appear again with the fallen leaves and moist logs of the black oak-red oak stage. Vegetation in itself is not directly important. Moist decaying wood is common, both on the beach and in the woods. Wood and moisture are evidently essential to such animals. Turning to the snails, which probably all come out into the open to feed during the night and during moist weather, we note that they do not appear until the under-log beetles put in their second appearance. In general the total number of species and of individuals increases until the oak-hickory stage is reached and falls off again in the beech and maple stage.

In general we note that as the forest passes from the bare-sand stage to the beech-maple stage, there is a great increase in the space to be inhabited by animals and the diversity of possible habitats, at least up to the oak-hickory stage.

1. CAUSES OF SUCCESSION

The causes of succession in forests are chiefly changes in physical condition with increase in denseness of vegetation, such as the increase of moisture of the atmosphere, decreased light, decreased temperature maximum in summer. The poisoning of the soil by root excretions and the modification of conditions on the ground brought about by a

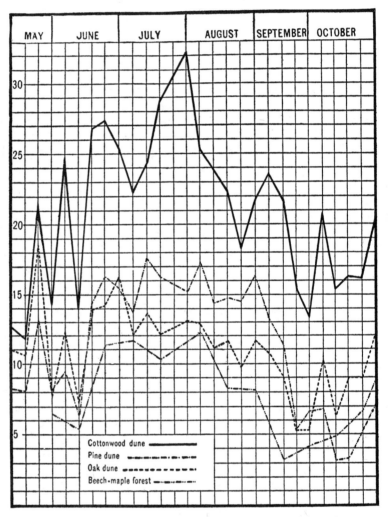

FIG. 251.—Mean daily evaporation rates (c.c. per day) in the ground stratum of four of the animal communities (after Fuller).

given set of trees are believed to prevent the germination of seeds of most of such trees, and at the same time to prepare the way for those of differently adapted species. The factors as expressed in terms of the evaporating power of the air are shown in Figs. 251, 252, and 253, which are graphic representations of the results of a season's study by Fuller (131). The graph of the cottonwood dunes is characterized by great fluctuations.

The graph for the pine dunes is decidedly lower and more regular in its contour than that of the association which it succeeds. Its four nearly equal

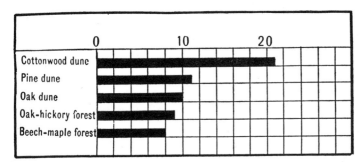

FIG. 252.—Showing the comparative evaporation rates (c.c. per day) in the ground stratum of the different animal communities from May to October (after Fuller).

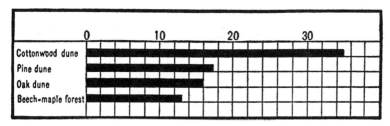

FIG. 253.—Showing the comparative evaporation rates (c.c. per day) in four of the animal communities on the basis of the maximum amount per day for any week from May to October (after Fuller).

maxima would indicate that within its limits there was, throughout the summer season, a continuous stress rather than a series of violent extremes. On the whole it shows a water demand of little more than half of that occurring in the cottonwood dunes. Its greatest divergence is plainly due to the evergreen character of its vegetation and is seen on its low range in May and the first part of June, and again in October when it falls below that of the oak dunes and is even less than that of the beech-maple forest. This would give good reasons for expecting to find within this association truly mesophytic plants [and moist forest annuals][1] whose activities are limited to the early

[1] The words in brackets are added.

spring. Evaporation in the various associations varies directly with the order of their occurrence in the succession. The differences in the rate of evaporation in the various plant associations studied are sufficient to indicate that the atmospheric conditions are most efficient factors in causing succession (Fuller, 131).

A comparison of Fuller's (131) data with the tables or lists of animals shows that the distribution and succession of animals is *clearly correlated* with the *evaporating power of the air*. Further comparison with the description of different forest stages shows that the evaporating power of the air may be taken, in this case, as an index of the materials for abode, etc.

2. CHARACTERS OF THE COMMUNITIES

It is possible to characterize the formations of the forest in physiological terms, though these cannot be of a very definite kind until the *mores* have been studied in detail, and accurate measurements made. Taking them stratum by stratum, we may note the following obvious characters:

a) Pioneer communities.—The communities of the cottonwood, pine, and black-oak stages may be designated as pioneer because of the presence of bare mineral soil.

Subterranean and ground strata: (*a*) The cottonwood community is characterized by animals which breed and spend the dark and cloudy days chiefly below the surface of the sand. They are very largely diurnal and predatory, and are exceedingly swift and wary. The burrowing spider (*Geolycosa pikei*) is one of the few nocturnal animals.

(*b*) The pine community is characterized by similar *mores*, but is to be distinguished from the preceding by the presence of many animals which prefer sand that is less shifting and which is slightly *darkened by humus* (170). Animals requiring "cover," such as the lizard, the blue racer, a few ground squirrels, etc., give character because of their absence from earlier and later communities.

(*c*) The black-oak community represents the climax of diversity of the subterranean and ground strata. The bare-sand *mores* continue in the open spaces, which we have designated as transition areas. Leaf-cutters are now present, while among the burrowers the root-borers (prionids and lucanids) work on the roots of the decaying trees. The behavior differences between this and the preceding community are differences of detail which, for the making of deductions, would require much careful study.

Field and shrub strata: The field and shrub strata of the cottonwood, pine, and oak communities are less easily characterized. The cottonwoods of the beach are far less commonly infested with aphid galls than are trees of the same species growing in less exposed situations. Furthermore we have never found any of the lepidopterous larvae such as *Basilarchia archippus* Cram. near the beach. Animals living exposed upon the trees few in number. The same general conditions obtain on and among the pines but spiders are more numerous. On the black oak the number of phytophagous animals is increased and the number of galls appears to be greater than in the later stages; the inhabitants of the herbaceous vegetation are chiefly those found in open situations such as prairies and roadsides, where the physical conditions are similar. Some animals of the same species which make up the black-oak community were taken from a roadside, and after being mixed with the inhabitants of the shrubs of the beech forest were placed in a light gradient. Soon the insects and spiders of the two communities separated sharply from each other, the beech-inhabiting species going to the darkest end while the roadside species crowded to the light.

b) Later communities.—With the coming-in of red oak, true forest with the mineral soil largely covered with humus and leaves is present and very different *mores* obtain. The diurnal diggers are practically absent. Snails, beetles, grasshoppers, spiders, and myriopods living under bark, decaying wood, and leaves, avoiding strong light and requiring moisture, are the chief types. The *mores* are typically forest in character. The differences between these and the later stages are those of detail and degree. In general with a lessening in the severity of the conditions and an increase in the denseness of vegetation, there is a proportional increase in the use of the vegetation as a place of abode.

In the field and shrub strata, we note that the animals of the cottonwood, pine, and oak stages are characteristic of open dry situations, requiring or tolerating strong light, while those animals of the red-oak, hickory, and beech stages are negatively phototactic to light of the same intensity, as shown by mixing the animals in a gradient.

The animals of the tree strata frequent a limited number of kinds of trees. Tree inhabitants are few and scattered in the cottonwood, pine, and black-oak stages while animals inclosed in galls or cases are common, if not dominant. In the red-oak, hickory, and beech stage phytophagous animals are often gregarious and numerous. Groups such as *Orthoptera*, beetles, bees, and wasps are represented more and more by species which make use of the vegetation as forest development goes on.

TABLE L. (Table XLIX follows Table L)

DISTRIBUTION OF ANIMALS BELONGING TO THE SUBTERRANEAN AND GROUND STRATA OF TWO OR MORE OF THE ANIMAL COMMUNITIES OF THE FOREST STAGES INDICATED BY NUMBERS

1, cottonwood stage; 1–2, mixed pine and cottonwood; 2, pine stage; 2–3, mixed pine and black-oak stage and open places in the black-oak forest; 3, the black-oak stage, in the later stages white oak occurs; 4, black oak-red oak stage (black oak, white oak, and red oak); 5, stages containing hickory but not beech and maple (white oak, red oak, and hickory); 6, beech and maple stage which usually contains some basswood. In the numbered columns, the star indicates that a species is present, F signifies few, C common, A abundant. Strata are indicated by letters at the heads of the columns: a, subterranean including rotten logs; b, ground; c, vegetation; in these columns F indicates feeding-place; B indicates breeding-place. Figures in column marked "Literature" refer to literature in the Bibliography.

Common Name	Scientific Name	1	1–2	2	2–3	3	4	5	6	a	b	c	Literature
Tiger-beetle	Cicindela lepida Dj.	C	F							B			151
Sand-spider	Trochosa cinerea Fabr.	C	F							B	F		138
Maritime grasshopper	Trimerotropis maritima Harr.	F	F							B	B		40
Tiger-beetle	Cicindela formosa generosa Dj.	C	C	F						B	F		151
Long-horned grasshopper	Psinidia fenestralis Serv.	C	C	C	C					B	F		40
Burrowing spider	Geolycosa pikei Marx	C	C	C	C					B	F		138
Bembex	Microbembex monodonta Say	C	C	C	C					B	F		40
Bembex	Bembex spinolae Lep.	F	F	F	F	F				B	F	F	173
Termite	Termes flavipes Koll.	C	F	*	*	F				B	F	F	178
Velvet ant	Mutilla ornativentris Oliv.	*								B			
Wasp	Dielis plumipes Dru.		?							B	F	F	177
Digger-wasp	Anoplius divisus Cress.		C	*						B	F	F	177
Sand-locust	Ageneotettix arenosus Han.		C	*						B	F		40
Ant	Lasius niger americanus Em.	*								B	F	F	54
Mottled sand-locust	Spharagemon wyomingianum Th.		C	C*						B	F		40
Migratory locust	Melanoplus atlanis Ril.	F	F*	*						B	F		40
Tiger-beetle	Cicindela scutellaris lecontei Hald.	C*	C*							B	F	F	151
Bee-fly	Spogostylum anale Say.		C*	C						B	F	F	151a
Locust	Melanoplus angustipennis Dod.		C	C	C					B	F		40
Six-lined lizard	Cnemidophorus sexlineatus Linn.		F	C	C					B	F		157

TABLE L—*Continued*

Common Name	Scientific Name	1	1-2	2	2-3	3	4	5	6	a	b	c	Literature
Andrenid	Augochlora confusa Rob			*	*					B	?	F	181
Parasitic bee	Epeolus pusillus Cress			*	*					B	?	F	181
Burrowing bee	Spechodes dicroa Sm			*	*	F				B	F	F	177
Ammophila	Ammophila procera Klg			F	C	F				B	B	F	177
Ant-lion	Cryptoleon nebulosum Oliv				*	*	C	C	C	B			156
Flat larvae	Pyrochroidae			F	?	F	C	C	F	B			91
Snails	Zonitoides arboreus Say			F	F	F	F	C	C	B			183
Centipede	Lithobius sp			F	?	F	F	C	C	B			156
Eyed elater	Alaus oculatus Linn					F	C	C	C	B			156
Borer	Orthosoma brunneum Forst					F	C	C	C	B			139
Tree-frog	Hyla pickeringii Hol					F	C	F	C	BF			139
...	Hyla versicolor Lec					F	C	C	C	BF		F	180a
Snail	Polygyra thyroides Say					C	F	*	F	B	F		
Ground beetle	Pterostichus coracinus Newm					F	*		*	B	F		
Beetle	Meracantha contracta Beau					*		C	F	B	F		182
Sowbug	Porcelio rathkei Brandt						F	*	*				
Woodchuck	Marmota monax Linn					*	*	*	*	B			91
Snail	Pyramidula alternata Say						*		C	B			
Beetle	Passalus cornutus Fabr						*		*	BF	F		
Fungus-beetle	Diaperis hydni Fab									B			
Snail	Circinaria concava Say						F	C	F	B			91
Sowbug	Cylisticus convexus De G							*	C	B	F		182
Centipede	Lysiopetalum lactarium Say							C	C	B			183
Centipede	Geophilus rubens Say							C	C	B			183
Millipede	Fontaria corrugate Wood							C	*	B			183
Millipede	Spirobolus marginatus Say							C	C	B			183
Slug	Philomycus carolinensis Bosc							C	C	B			91
Beetle	Galerita janus Fabr							*	C	B	F		156
Melandryid	Penthe pimelia Fabr								*	B	F		156
Burying beetle	Silpha surinamensis Fabr								C	B	F	F	156
Harvestman	Liobunum nigropalpi Wood								C	B		F	184
Ant	Camponotus herculeanus pennsylvanicus De G							C	*	B	F	F	54
Beetle	Ceruchus piceus Web								*	B	F	F	156
Snail	Omphalina fuliginosa Gr							?	*	B	B		91

TABLE XLIX. (Table L precedes Table XLIX)

SHOWING FOREST ANIMALS IN THE EARLY STAGES OF FOREST DEVELOPMENT OF A
CLAY BLUFF OF LAKE MICHIGAN

Subterranean and ground strata, 1, bare clay, 2, sweet clover, 3, shrubs, golden-rod, etc., 4, sapling stage, animals same as in (5) the oak-hickory forest (Station 56)

Common Name	Scientific Name	1	2	3	4	5
Tube-weaver	*Agelena naevia* Wal	*	*			
Lycosid	*Pardosa lapidicina* Em	*	*			
Carolina locust	*Dissosteira carolina* Linn	*	*			
Mud-dauber	*Pelopoeus cementarius* Dru	*	*			
Tiger-beetle larvae	*Cicindela purpurea limbalis* Klg	*	*			
Sowbugs	*Porcellio rathkei* Brandt	F	C	A	A	
Centipede	*Geophilus* sp	*	*	*	*	
Snail	*Polygyra thyroides* Say	*	*	*	*	
Snail	*Pyramidula alternata* Say	*	*	*	*	
Snail	*Polygyra monodon* Rack		*	*	*	
Tiger-beetle larvae	*Cicindela sexguttata* Fbr		*	*	*	
Snail	*Polygyra albolabris* Say		*	*	*	
Slug	*Philomycus carolinensis* Bosc		*	*	*	
Yellow-margined millipede	*Fontaria corrugate* Wood		*	*	*	
Centipede	*Lysiopetalum lactarium* Say		*	*	*	

ANIMALS RECORDED IN THE GROUND AND SUBTERRANEAN STRATA OF THE STAGES NOTED

In Tables LI–LV, in the third column B indicates breeding; F, feeding; H, hibernating, on the situation indicated in column 4. Figures in column "Literature" refer to literature cited in the special Bibliography at the end of the book. Statements made on the authority of others are in italics; those starred are by A. B. Wolcott.

TABLE LI

PINE STAGE (STATIONS 57, 58, 59)

Common Name	Scientific Name			Literature
Bee (*Andrenidae*)	*Halictus nelumbonis* Rob	B?	....	181
Larridae	*Tachytes texanus* Cres	..	....	173
Scoliidae	*Plesia interrupta* Say	..	....	...
Ceropalidae	*Anoplius marginatus* Say	..	In sand	173
Blue racer	*Coluber constrictor* Lin., Var.	B	"	157
Ground squirrel	*Citellus 13-lineatus* Mitch	B	"	21
Beetle (*Elateridae*)	*Cardiophorus cardisce* Say	..	On sand	156
Elaterid beetle	*Alaus myops* Fabr	B?	Under pine bark	...

TABLE LII

BLACK-OAK STAGE (STATIONS 57, 59, 60, 61)

Common Name	Scientific Name			Literature
Elateridae	*Lacon rectangularis* Say	B	*Under Opuntia*	*
Erotylidae	*Languria trifasciata* Say	B	"	*
Coral-winged locust	*Hippiscus tuberculatus* Beau.	B	In sand	40
Parasitic bee	*Coelioxys rufitarsus* Smith	B	Bee nest	...
Eumenidae	*Odynerus anormis* Say	..	....	...
Hog-nosed snake	*Heterodon platirhinos* Latr	B	In sand	157

TABLE LIII

BLACK OAK-RED OAK STAGE (STATION 63)

Common Name	Scientific Name			Literature
Ant	*Lasius umbratus mixtus aphidicola* Walsh	B	Log	54
Ant	*Camponotus ligniperdus noveboracensis* Fitch	B	Log	54
Ground beetles	*Pterostichus sayi* Brulle	B	Rotten log	156
Tenebrionidae	*Uloma impressa* Mels	B	Rotten log	156

(See explanation above Table LI)

TABLE LIV

RED OAK-HICKORY STAGE (STATIONS 64, 65, 69)

Common Name	Scientific Name			Literature
Green tiger-beetle....	*Cicindela sexguttata* Fabr...	B	In soil	151
White-faced hornet...	*Vespa maculata* Lin........	H	Rotten wood	179
Andrenidae.........	*Augochlora pura* Say.......	...	....	...
Scarabaeidae.......	*Geotrupes splendidus* Fabr...	B	Rotten log	156
Staphylinidae.......	*Staphylinus violaceus* Grav..	B	"	156
Elateridae..........	*Melanotus communis* Gyl...	..	"	156
Slug............. ...	*Pallifera dorsalis* Bin......	B	Log	91
Brenthid beetle......	*Eupsalis minuta* Dru	B	Solid logs	...

TABLE LV

BEECH STAGE (STATIONS 70, 71, 71*a*, 71*b*)

Common Name	Scientific Name			Literature
Frog.............	*Rana sylvatica* Le Conte......	F	Ground	139
Fly larva.........	*Pachyrhina ferruginea* Fabr...	B	Under leaves	...
Salamander.......	*Plethodon cinereus* Gr........	BF	"	...
Snail.............	*Polygyra inflecta* Say........	BF	Leaves and log	91
Snail.............	*Polygyra oppressa* Say.......	BF	"	...
Snail.............	*Polygyra fraudulenta* Pil......	BF	"	91
Snail.............	*Polygyra palliata* Say.......	BF	"	...
Snail.............	*Pyramidula solitaria* Say.....	BF	"	91
Snail.............	*Pyramidula perspectiva* Say...	?	"	...
Beetle...........	*Xylopinus saperdioides* Oliv...	B	Under bark	156, 137
Ant.............	*Aphaenogaster tennesseensis* Mayr...................	B	Rotten wood	...

TABLE LVI

DISTRIBUTION OF ANIMALS RECORDED FROM VEGETATION IN MORE THAN ONE OF THE ANIMAL COMMUNITIES OF THE FOREST STAGES INDICATED BY NUMBERS

1, the cottonwood stage; 1–2, mixed cottonwood and pine stage; 2, pine stage; 2–3, mixed pine and oak stage and open places in the oak forest; 3, black-oak stage, in its later phases white oaks occur; 4, black oak-red oak stage; 5, stages containing hickory but not beech and maple; 6, beech and maple stage.

Common Name	Scientific Name	1	1-2	2	2-3	3	4	5	6
(a) Spider (*Thomisidae*)	*Philodromus alaskensis* Key	*	*	*					
(b) Butterfly..........	*Anthocharis genutia* Fabr..		*	*	*				
(c) Spider (*Epeiridae*)..	*Epeira domicilorum* Hentz..		*	?	?	?		*	*
(d) Dusky plant-bug....	*Lygus pratensis* Lin.......			*	*	*			
(e) Phasmidae........	*Diapheromera femorata* Say					*	*	*	
(f) Spider (*Thomisidae*)	*Misumessus asperatus* Htz.			*			*	*	*	*
(g) Spider (*Dictynidae*).	*Dictyna foliacea* Hentz...					*	*	*	*
(h) Spider (*Epeiridae*)..	*Epeira gigas* Leach.......						F	C	F
(i) Spider (*Theridiidae*)	*Theridium frondeum* Hentz.						F	*	*
(j) Bug..............	*Acanthocephala terminalis* Dall.................							*	*
(k) Stinkbug.........	*Nezara hilaris* Say........							*	*
(l) Stinkbug.........	*Podisus maculiventris* Say..							*	*
(m) Fly..............	*Sapromyza philadelphica* Mac.................							*	*

The letters below at the left refer to the species opposite which they stand in Table LVI and the numbers refer to the forest stages as at the heads of the columns of Tables L and LVI. The capitals have the same meaning as in the preceding tables.

a—from cottonwoods and juniper (1, 1–2, 3) (138, 172).

b—from *Arabis lyrata* (175).

c—from pine and herbaceous vegetation (B) (4) (172).

d—herbs (174).

e—from the trunks of various trees.

f—*Monarda* (2–3), and black oak (3), maple (5) (138, 172).

g—F *Monarda* (3) (173).

h—from undergrowth (4), and beech (5) (138, 172).

i—from shrubs (4), and young beech (5) (138, 172).

j—shrubs (4) and maple trunk.

k—from red-oak trunk (4) and beech trunk (5) (*Tilia, Citrus, Gossypium* 186).

l—? (4) and beech leaves (5) (predaceous, 185).

m—herbs.

ANIMALS RECORDED FROM THE FIELD, SHRUB, AND TREE STRATA OF THE FOREST STAGES NOTED

In Tables LVII–LXII, in the third column B indicates breeding; F, feeding; H, hibernating, on the situation indicated in column 4.

TABLE LVII

COTTONWOOD STAGE (STATIONS 57, 58, 59)

Common Name	Scientific Name			Literature
Chrysomelid beetle..	*Disonycha quinquevittata* Say.	BF	Willow	156
Long-horned borer...	*Plectrodera scalator* Fab.....	BF	Cottonwood	156
Gall aphid.........	*Pemphigus populicaulis* Fitch.	BF	"	188
Gall aphid.........	*Pemphigus vagabundus* Walsh.	BF	"	188

TABLE LVIII

PINE STAGE (STATIONS 57, 58, 59)

Common Name	Scientific Name			Literature
Leaf-beetle........	*Nodonota tristis* Oliv........	F	Herbs	156, 137
....	*Bassareus lativittis* Germ....	F	"	...
Spider (*Thomisidae*)..	*Xysticus formosus* Banks....	..	Juniper	137
Spider (*Attidae*).....	*Dendryphantes octavus* Hentz.	..	"	187, 138
				173
Spider (*Theridiidae*)..	*Theridium spirale* Em.......	..	"	138, 172
Engraver beetle.....	*Ips grandicollis* Eich........	BF	Pine	137
Pitch-moth.........	*Evetria comstockiana* Fern.?..	BF	"	137

(See explanation above Table LVII)

TABLE LIX

BLACK-OAK STAGE (STATIONS 57, 59, 60, 61)

Common Name	Scientific Name			Literature
Syrphus fly........	*Milesia virginiensis* Dru.....	F	Herbs	...
Andrenid.........	*Agapostemon splendens* Lepel.	F	Primrose	177
Spider (*Thomisidae*)..	*Philodromus pernix* Black....	F	Herbs	137
Spider (*Epeiridae*)...	*Argiope trifasciata* Forsk.....	F	"	137
Sprinkled locust.....	*Chloealtis conspersa* Har.....	F	"	40
Grasshopper.......	*Schistocerca rubiginosa* Har..	F	"	40
Tree-cricket........	*Oecanthus fasciatus* Fitch....	BF	"	40
Texas grasshopper...	*Scudderia texensis* Scud......	BF	"	40
Conehead grasshopper.............	*Conocephalus ensiger* Har....	BF	"	40
Meadow grasshopper	*Xiphidium strictum* Scud....	BF	"	40
Stinkbug	*Euschistus variolarius* Pal....	BF	"	174
Flower-bug........	*Triphleps insidiosus* Say.....	F	"	185
Fork-tailed larvae...	*Cerura* sp...............	BF	Cherry	186
Fulgorid..........	*Otiocerus degeeri* Kirby.......	BF	Oak	185
Flatbug...........	*Neuroctenus simplex* Uhl.....	BF	"	185
Colydiid beetle......	*Ditoma quadriguttata* Say....	F	"	137
Prominent larva.....	*Heterocampa guttivitta* Harr...	BF	"	137
Prominent larva.....	*Nadata gibbosa* S. and A.....	BF	"	138
Tree-hopper........	*Telemona querci* Fitch (*monticola*).................	BF	"	185
Coreidae..........	*Chariesterus antennator* Fabr..	BF	"	185
Jassid............	*Typhlocyba querci* var. *bifasciata* G. and B..........	BF	"	185
Jassid............	*Phlepsius irroratus* Say......	BF	"	185

(See explanation above Table LVII)

TABLE LX

BLACK OAK-RED OAK STAGE (STATION 63)

Common Name	Scientific Name			Literature
Jumping spider......	*Maevia niger* Htz..........	F	Herbs	138
Spider............	*Gayenna celer* Htz..........	B	"	...
White-oak gall......	*Andricus seminator* Harr.....	..	White oak	188
Predaceous leaf-bug..	*Hyaliodes vitripennis* Say....	F	Tree trunks	...
Scallop-moth (larvae)	*Hydria undulata* Lin........	F	Cherry	...

TABLE LXI

RED OAK-HICKORY STAGE (STATIONS 64, 65, 59)

Common Name	Scientific Name			Literature
Rove-beetle.......	*Tachinus pallipes* Grav......	BF	Mushrooms	177
Spider (*Clubionidae*).	*Anyphaena conspersa* Key...	F	Herbs	138
Locustidae........	*Atlanticus pachymerus* Burm.	B	Grass	40
Spider (*Epeiridae*)...	*Acrosoma gracilis* Wal......	F	Shrubs	138
Spider (*Epeiridae*)...	*Acrosoma spinea* Hentz......	F	"	138
Spider (*Epeiridae*)...	*Mangora maculata* Key......	..	"	138
Jassid............	*Scaphoideus auronitens* Prov.	..	"	177
Beetle............	*Odontota nervosa* Panz.......	..	"	...
Bug (*Nabidae*)......	*Reduviolus annulatus* Reut...	B	"	...
Spider (*Linyphiidac*).	*Linyphia phrygiana* Koch....	..	"	138
Cicada............	*Cicada linnei* S. and G......	F	Young maple	177
Leaf-beetle........	*Calligrapha scalaris* Lec.....	BF	"	...
Stinkbug	*Euschistus tristigmus* Say....	BF	"	137
Arctidae..........	*Halisidota* sp..............	BF	White oak	...
Oakworm.........	*Anisota senatoria* Sm.and Abb.	BF	"	137
Tree-cricket.......	*Oecanthus angustipennis* Fitch.	BF	Red oak	40
Katydid..........	*Cyrtophyllus perspicillatus* L..	B	"	...
Leaf-beetle........	*Xanthonia 10-notata* Say.....	F	"	...
Prominent larva.....	*Symmerista albifrons* S. and A.	B	Maple	137
Prominent larva.....	*Datana angusii* G. and R	B	Hickory	137
Aphid............	*Phylloxera caryae-caulis* Fitch.	B	"	188, 137

(See explanation above Table LVII)

TABLE LXII

BEECH STAGE (STATIONS 70, 71, 71*a*, 71*b*)

Common Name	Scientific Name			Literature
Beetle............	*Boletobius cinctus* Grav.....	..	Mushrooms	177
Fungus-beetle.......	*Boletotherus bifurcus* Fabr...	..	Shelf fungus	177, 137
Cercopidae (bug)....	*Clastoptera obtusa* Say......	F	Hickory, maple, hazel	185
Leaf-hopper.......	*Gypona octolineata* Say.....	F	Hickory, maple, beech	137
Leaf-hopper.......	*Jassus olitarius* Say........	F	Maple	137
Ichneumonidae.....	*Thalessa atrata* Fabr.......	B	Larvae	137
Lacewing..........	*Chrysopa rufialbris* Burm...	B	Maple	...
Lacebug	*Gargaphia tiliae* Walsh.....	B	Beech	137
Ichneumonidae.....	*Trogus vulpinus* Cb.......	B	Larvae	177
Pentatomidae......	*Banasa calva* Say.........	F	Beech	...
Lampyrid beetle.....	*Podabrus basilaris* Say.....	..	Maple	137
Attidae............	*Wala mitrata* Hentz.......	..	"	138
Theridiidae........	*Notionella interpres* Cam....	..	"	138
Harvestman.......	*Oligolophus pictus* Wood....	..	Maple trunk	184
Syrphus fly........	*Spilomyia longicornis* Loew..	..	"	...

CHAPTER XIII

ANIMAL COMMUNITIES OF THICKETS AND FOREST MARGINS

I. Introduction

The forest margin or forest edge is a familiar natural situation. About Chicago there are groves of trees which are probably exactly as they were before settlement. The forest ends; the prairie begins. The line between the two is markedly a narrow border of shrubs and rank weeds, usually only a few feet wide. In other places the forest ends at a marsh side, lake side, or stream side, but almost always with the thicket of shrubs and rank weeds. A remarkably large number of animals belong to this forest margin. Some of these have been discussed in connection with the margins of bodies of water (chap. x), and the marsh forest (chap. x). The borders between forest and prairie remain to be discussed. These will be roughly separated into high and low forest margin, depending upon height above ground-water level. The relations of these formations to the other forest margins will be indicated in the tables.

II. Low Forest Margin Sub-Formations

(Stations 45, 49; Table LXIII) (Fig. 254)

Low forest margin is usually the border between swamp forest and low prairie. There was originally much of this in the Lake Chicago plain. One point of special study is the border of the Wolf Lake marsh forest (see p. 189).

1. SUBTERRANEAN-GROUND STRATUM

The ground is inhabited by earthworms and cicada nymphs, etc. No burrowing mammals have been recorded, but it is probable that the skunk sometimes breeds in this stratum.

The cricket (*Nemobius maculatus*) occurs under fallen leaves, sticks, etc., with an occasional snail (*Polygyra monodon*). The lubberly locust often deposits its eggs in the ground (40). Sowbugs and forest-floor forms make up most of the remaining species.

The northern yellowthroat, the song sparrow, and the common shrew sometimes nest on the ground. The skunk is sometimes a feeding resident.

2. FIELD AND SHRUB STRATA

Here two zones may be recognized. While there is no reason for separating them in the ground stratum, a rough separation is here possible.

a) Rank weeds, willow, dogwood, grape, etc.

b) Prickly ash thicket with grape and young elms.

Outside the first is a girdle of low prairie from which low prairie plants and some low prairie animals occasionally invade the forest margin.

a) *Girdle of rank weeds, dogwood, willow, etc.*—In open, grassy places the garden spiders (*Argiope aurantia* and *trifasciata*) (Fig. 255) fasten

FIG. 254.—Low forest margin at Wolf Lake, Ind. In front of *a*, low prairie area; opposite *b*, belt of rank weeds: opposite *c*, low shrubs; opposite *d*, high shrubs; opposite *e*, trees.

their webs to any firm support, such as a young shrub. Various grasshoppers occur in open situations (*Xiphidium fasciatum* and *brevipenne* belong more properly to low prairie) (Fig. 256). The long-bodied spider (*Tetragnatha laboriosa*) (138) is a common resident. On the grasses beneath the shrubs the black-sided grasshopper (*Xiphidium nigropleura*) is abundant. The snail (Fig. 257) (*Succinea ovalis*) is sometimes common.

Of the bugs which frequent the blossoms of the coarse weeds are the long-legged bug (*Neides muticus*), the buffalo tree-hopper (Fig. 259), and the candlehead (*Scolops sulcipes*) (Fig. 258). These two and especially the latter, with its curiously prolonged prothorax, are the most characteristic. The common plant-bug (*Lygus pratensis*) (Fig. 261) and an occasional dusky leaf-bug (*Adelphochoris rapidus*) (Fig. 262) are also found. The large stinkbugs (*Euschistus tristigmus* and *fissilis* Uhl.) are common. They may be predatory in the adult stage. The predatory ambush-bug (*Phymata erosa fasciata*) lies in wait for its prey in the

FIG. 255.—The garden spider (*Argiope aurantia*) on its web; about one-half natural size.

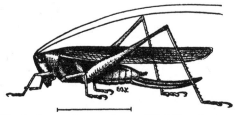

FIG. 256.—The slender meadow grasshopper (*Xiphidium fasciatum*) (after Lugger).

blossoms. A crab spider (*Mesumena vatia*) and a jumping spider (*Phidippus audax*) are common in the blossoms (40, p. 182.) Various lepidopterous larvae feed upon the rank weeds also.

On weeds and blossoms grasshoppers are numerous; we find the Nebraska conehead (*Conocephalus nebrascensis*) (see Fig. 260), the lubberly

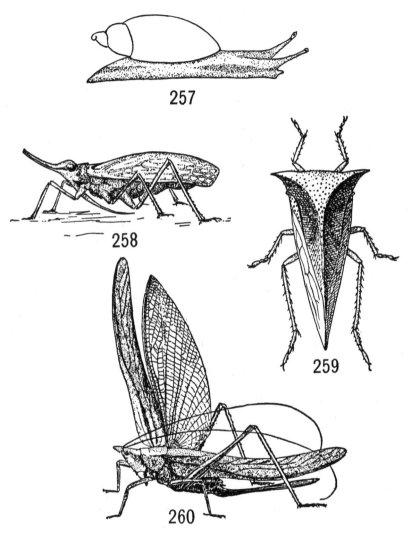

FIG. 257.—The forest-margin snail (*Succinea ovalis*); twice natural size (after Baker).

FIG. 258.—The candle-headed bug (*Scolops sulcipes*); 5 times natural size (original).

FIG. 259.—The buffalo tree-hopper (*Ceresa bubalus*); 5 times natural size (after Marlatt, U.S. Dept. Agr.).

FIG. 260.—The large cone-headed grasshopper (*Conocephalus robustus*) (after Beutenmüller [Am. Mus.] from Blatchley).

locust (*Melanoplus differentialis*), an occasional red-legged locust, and the striped shrub cricket, the short-winged brown locust (*Stenobothrus curtipennis*), the short-winged meadow grasshopper (*Xiphidium brevipenne*), and the Texas katydid (*Scudderia texensis*) (40, pp. 330, 390).

The jug-making wasp (*Eumenes fraternus*) (40, p. 207) makes its jug-like nest on the herbaceous plants. The social wasp (*Polistes*) is a frequent visitor of the flowers, and sometimes attaches its comb to the willow. The oblong leaf-winged katydid (*Amblycorypha oblongifolia*) (Fig. 263) (40, p. 391) and the fork-tailed katydid

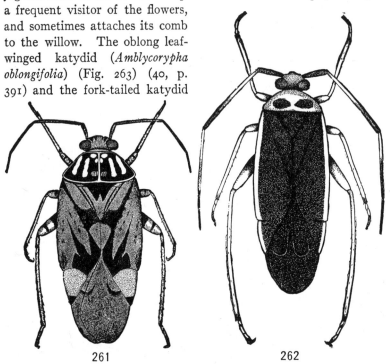

261   262

FIG. 261.—The tarnished plant-bug (*Lygus pratensis*); about one-fourth of an inch long (after Forbes).

FIG. 262.—The dusky leaf-bug (*Adelphocoris rapidus*); about one-fourth of an inch long (after Forbes).

(*Scudderia furcata*) (Fig. 264) are residents. The latter places its egg on leaves of shrubs (40). Willow leaf-feeders are numerous; several lepidopterous larvae are common. These include the brilliant larva of the smeared dagger-moth (Fig. 265), the cecropia moth, the willow sphinx, the viceroy and mourning-cloak butterflies, the maia moth (Fig. 266), the fork-tailed caterpillar (137), larva of the maia moth, and others. The small fly (*Bibio albipennis*) visits the flowers of the willow in spring (Fig. 267). Sawfly larvae are common; the large light-colored one (*Cimbex americana*) (179) has habits of special interest. The female, which is a wasp-like insect, deposits her eggs on the under sides of leaves. Blisters are formed, and a young larva lives for a time in each

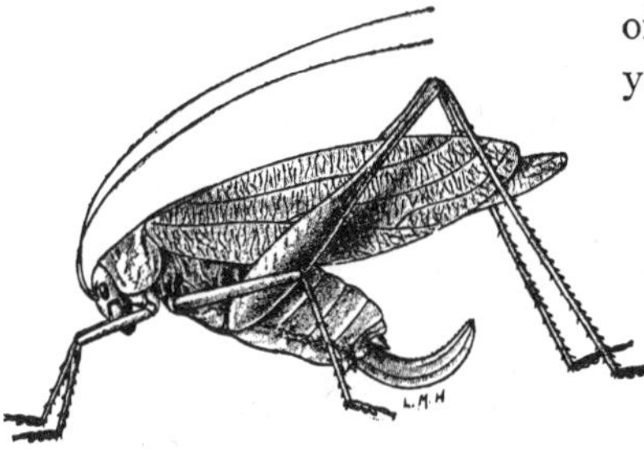

263 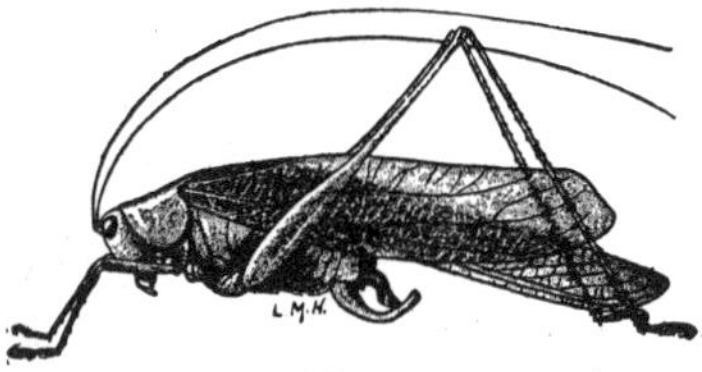   264

FIG. 263.—The oblong leaf-winged katydid (*Amblycorypha oblongifolia*); (after Forbes) natural size.

FIG. 264.—The fork-tailed katydid (*Scudderia furcata*) (after Lugger from Forbes); natural size.

of these. Later it is to be found living freely on the leaves. It usually rests with the posterior segments wrapped around a petiole or twig. Pupation takes place in a silken case. The spotted sawfly larva (*Pteronus ventralis* Say) (179) is less common.

Beetles are common on the willow. The leaves are eaten by May-beetles (189) and several leaf-feeders (*Calligrapha* and *Lina* are common). Several borers attack the twigs (*Saperda concolor*). Galls are very numerous. The trunks of small willows are commonly attacked by the larvae of the introduced snout-beetle (*Cryptorhynchus lapathi*), and the

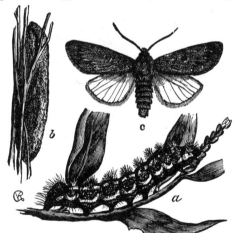

FIG. 265.—The adult and larva of the smeared dagger-moth (*Acronycta oblinita*), which feeds upon various forest-margin weeds and shrubs; natural size (after Riley).

goat-moth larva (*Prionoxystus robiniae* Peck.), which bores in the heart-wood. The sap which exudes attracts many sap-beetles (*Nitidulidae*).

The dogwood is fed upon by a few larvae. The unicorn larva (*Schizura* sp.) is occasionally found; the young of the spittle insect (*Aphrophora 4-notata*) are common. The grape and Virginia creeper are attacked by several sphinx larvae. The grapevine hog caterpillar (*Ampelophagus myron* Cram.) has been taken from the former.

Nesting in the shrubs are the goldfinch (more often in trees), the indigo bunting, the northern yellowthroat, the brown thrasher, and catbird, all of which feed in the low prairie. The song sparrow nests near the ground.

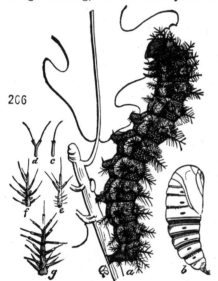

b) *The belt of prickly ash.*— This has not been so thoroughly studied. The subterranean and ground strata are similar to those of the forest adjoining (see

266

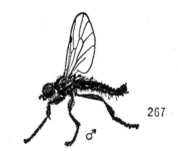

267

FIG. 266.—The larva of the maia moth (*Hemileuca maia*) which feeds on the willow; natural size (from Lugger after Riley, Div. Ent., U.S. Dept. Agr.).

FIG. 267.—*Bibio albipennis.* Early spring on the flowers of the willow. Breeds in the ground (from Williston after Washburn).

p. 269); the ground and field strata have some of the same residents. The adult *Cresphontes* butterfly (*Papilio cresphontes*) is common about the Wolf Lake forest edge and Hancock (40) has recorded the larva on prickly ash, one of its regular food plants. He also records the true tree-cricket (*Apithes agitator* Uhl.) as inhabiting prickly ash thickets.

III. HIGH FOREST MARGIN SUB-FORMATIONS

(Station 48; Table LXIV)

This surrounds the oak-hickory, black-oak, and beech forests on high ground. The witchhazel, hawthorn, sumac, and grape are the dominant shrubs; goldenrod, asters, and sunflowers are the chief herbaceous plants.

1. SUBTERRANEAN-GROUND STRATUM

Certain earthworms, cicada nymphs, and root-eating grubs belong here. This is the regular breeding-place of the skunk (*Mephitis meso-melas avia* Bang). According to Seton (143) they go in droves of six or eight, and as many as fifteen sometimes occur in a winter den. According to Seton its food consists of various insects, grasshoppers, crickets, meadow mice, snakes, and crayfishes. The short-tailed shrew in primeval conditions breeds chiefly in such tangles of bushes. It digs in moss and fallen leaves and loamy soil, and follows mouse galleries. According to Wood (21) it eats many mice. Seton (143) states it feeds on isopods, earthworms, etc. Its enemies are hawks, lynxes, and weasels.

Franklin's ground squirrel (*Citellus franklini* Sab.) burrows into the ground deeper than the ground squirrel of the prairies, but is otherwise similar in habits. It is gregarious and stores grain for winter. The chipmunk (*Tamias striatus griseus* Mear.) is a typical forest margin animal. It nests in the ground, as a rule in burrows about 6 to 10 ft. long and running diagonally down to a depth of 2 to 3 ft. (21). It stores nuts for winter. The jumping mouse (*Zapus hudsonius* Zim.) is one of the most characteristic residents; it moves by great leaps and steers its flight with its tail. The woodchuck should probably be counted here, though it belongs deeper in the forest than any of the others. The weasel is common in this situation, though it is perhaps more abundant along streams (Wood).

The ground stratum supports many of the small animals of the adjoining forest, such as centipedes, camel crickets, etc. The cottontail is one of the chief residents, as it usually breeds in such situations. The common shrew (*Sorex personatus* St. Hil.) (21) breeds on the ground, in stumps, etc. All of the mammals recorded in the preceding stratum feed here when suitable food is present. A considerable number of mammals commonly regarded as belonging to the forest are said to prefer thickets. The Virginia deer is one of these. It is probable that the elk was somewhat similar in habits.

The bobwhite and mourning dove (occasionally) breed in these situations, the former often falling a victim to the weasel (Wood). The high forest margin was probably a favorite location for the huts of the aborigines. Some of the early travelers record huts around the edges of the prairies. Such locations would supply shelter and firewood, etc., as well as sunshine.

2. FIELD AND SHRUB STRATA

Here the ground-cherry, milkweed, and thistle have a characteristic fauna. On the milkweed are the larvae of the monarch butterfly, the milkweed beetle (*Tetraopes tetraophthalmus* Forst.) (40, p. 136), and the leaf-beetle (*Doryphora clivicollis*); the latter is very characteristic. The milkweed flowers attract hosts of flies which are preyed upon by various digger-wasps; bees are numerous, gathering honey. The ground-

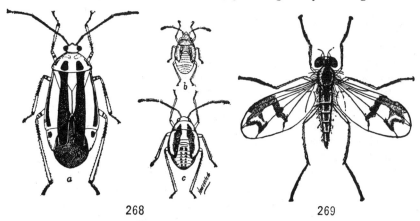

268   269

FIG. 268.—The four-lined leaf-bug (*Poecilocapsus lineatus*); *a*, adult; *b*, *c*, immature forms; 5½ times natural size (from Lugger).

FIG. 269.—A long-legged fly (*Psilopodinus sipho* Say); enlarged (from Williston after Lugger).

270   271

FIG. 270.—A large robber-fly (*Dasyllis* sp.); natural size (from Williston after Kellogg).

FIG. 271.—A syrphus fly (*Eristalis tenax*); 1½ times natural size (from Williston after Kellogg).

cherry is the food plant of the "Spanish fly" (*Epicuata*) and the Colorado potato-beetle. On the thistle we find the larvae of the cosmopolitan and painted-lady butterflies (*Pyrameis huntera* Fab. and *cardui* Lin.). One of the most characteristic bugs is the 4-lined

272

273

Fig. 272.—A leptid fly (*Coenomyia ferruginea*); enlarged (after Williston).
Fig. 273.—A large syrphus fly (*Milesia virginiensis*); enlarged (after Williston).

leaf-bug (*Poecilocapsus lineatus*) (Fig. 268). The long-legged fly (Fig. 269), the large robber-fly (Fig. 270), the common syrphus fly (*Eristalis tenax*) (Fig. 271), a leptid fly (Fig. 272), and *Milesia virginiensis* (Fig. 273) visit the flowers in numbers. The garden spider occurs; also high in the shrubs is the brilliant *Epeira gigas* found also in the forest openings. The goldenrod gall-forming fly (*Straussia longipennis*) (Fig. 274) with its beautifully marked wings is common. Professor

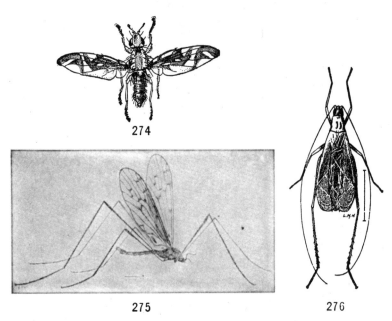

274

275   276

FIG. 274.—The goldenrod gall-fly (*Straussi longipennis*); much enlarged (from Williston after Kellogg).

FIG. 275.—One of the crane-flies (*Helobia hybrida*); enlarged (from Williston after Lugger).

FIG. 276.—The tree-cricket (*Oecanthus fasciatus*); twice natural size (after Lugger).

Williston states that the crane-fly (*Helobia hybrida*) (190) (Fig. 275) occurs. Several leaf-bugs occur; the dusky leaf-bug is common.

Several species of *Orthoptera* are characteristic. Of the tree-crickets several occur among which are *Oecanthus nivens* DeG. and *angustipennis* Fitch and *fasciatus* (Fig. 276). Two or three katydids occur; the round-winged (*Amblycorypha rotundifolia* Scud.) is most characteristic.

The grape often grows in these situations, and is especially subject to attack by the *Phylloxera* (Fig. 277) and the grapevine June beetle, the larvae of the 8-spotted forester (*Alypia octomaculata* Fabr.), and the grapevine epimens (*Psychomorpha epimensis* Drury) (163). All of these spend a part of their lives in the ground. The *Phylloxera* (Fig. 277) winters on the roots of the grape. The grape-beetle larva bores in wood. The pupae of the two moths bore into rotten wood or the ground for pupation and also to spend the winter. This may be an important cause for their presence in the forest margin. Brownie-bugs are common (Fig. 278).

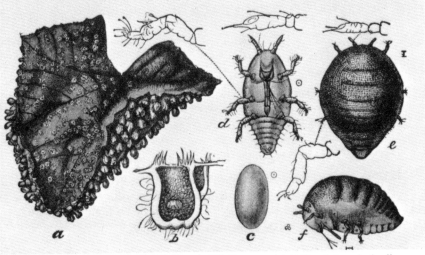

FIG. 277.—The grapevine Phylloxera (*Phylloxera vastatrix* Planch.): *a*, leaf galls; *b*, section of gall with mother louse at center with young clustered about; *c*, egg; *d*, nymph; *e*, adult female; *f*, same from side; *a*, natural size, others much enlarged (after Marlatt, Div. Ent., U.S. Dept. Agr.).

One of the most interesting forms found here is *Mantispa brunnea* (Fig. 279). This is a neuropterous insect with forelegs adapted for seizing prey. Its larva is a parasite in the egg-cases of spiders. The adult appears in July. In the autumn, after the leaves have fallen, one sees many nests of spiders on the high forest margin shrubs, so the young parasites have a good chance to secure their best food conditions here.

Hawthorns often occur, and on the trunks we find woolly plant-lice (*Schizoneura*) in great white clusters (150). The hawthorn supports many of the pests of the apple.

The birds of the high forest margin are numerous (191). The goldfinch builds a nest of thistledown, grasses, etc., on shrubs or low trees. The chipping-sparrow builds its nest of rootlets and lines it with horsehair. The Baltimore and orchard orioles nest in trees and high shrubs and feed in the open. The field sparrow sometimes builds on the rank weeds, in other cases on shrubs near the ground. The mourning dove, the indigo bunting, and the yellow warbler nest on shrubs; the latter often builds near water. The redstart builds in the forks of bushes and trees. The loggerhead shrike is common. The sparrow-hawk nests in deserted woodpecker holes near the edge of the woods and feeds in the meadow or prairie. The flicker is similar in

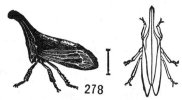

FIG. 278.—A brownie-bug (*Enchenopa binotata* Say); enlarged (after Lintner).

FIG. 279.—One of the Mantis-like neuroptera (*Mantispa brunnea*); enlarged.

habits, but uses holes of its own making. The bronzed grackle and sharp-shinned hawk nest in trees near the forest edge and feed in the prairie. The cowbird, which lays its eggs in the nests of other birds, often chooses those nests of the high forest margin.

IV. GENERAL DISCUSSION

The forest margin, as we have seen, possesses in addition to the characteristic species a considerable number of species which frequent the prairie or forest; our list includes the breeding species. The classification below shows the various types of habit in birds and mammals.

FOREST MARGIN BIRDS AND MAMMALS

(Compiled from literature cited)

H indicates high forest margin; *L*, low forest margin.

A. Breeding in the ground under the shrubs; feeding in the meadows or prairies and woods.
1. Mammals: Skunk (*H*), Chipmunk (*H*), Franklin ground squirrel (*H*), Jumping mouse (*H*). Feed chiefly in woods.
2. Birds: No birds have this habit.

B. Breeding on the ground among the shrubs and feeding in the open meadows or prairies.
   1. Mammals: Common shrew (*Sorex personatus*) (*L*), the cottontail (*H*).
   2. Birds: Bobwhite (*H*), mourning dove (*H*) sometimes, northern yellow-throat (*L*) sometimes, song sparrow (*L*) sometimes.

C. Breeding on the shrubs and feeding in the forest edge and sometimes in the open meadows or prairies.
   1. Mammals: None.
   2. Birds: (*a*) Low forest margin: song sparrow, goldfinch, indigo bunting, northern yellowthroat, brown thrasher, and catbird.
      (*b*) High forest margin: goldfinch, lark sparrow, chipping-sparrow, field sparrow, indigo bunting, yellow warbler, redstart, loggerhead shrike, mourning dove, catbird, cowbird, bronzed grackle, brown thrasher.

D. Breeding in the trees of the forest and feeding in the prairies.
   1. Mammals: raccoon.
   2. Birds: Sparrow-hawk, sharp-shinned hawk, and several other hawks, flicker, bronzed grackle, Baltimore oriole.

The list shows animals which breed in the margin of woods and often feed not only there but in the prairies. Similar relations were noted by Bates in the savannas along the middle Amazons. The advantage of the forest margin lies in the facts of: (1) shade for the nocturnal and crepuscular forms; (2) abundant space in the thickets for nests; (3) large stiff plants which accommodate the large animals: (*a*) places for the spiders to stretch their nets; (*b*) plants large enough for the roosting- and nesting-places of birds and larger insects; (4) protection from wind and from winter freezing afforded by the forest. From the standpoint of food relations many forest margin animals must be counted in with the prairie forms.

One of the most striking facts concerning the forest margin animals is (*a*) their wide distribution and (*b*) their survival under agricultural conditions. Many animals of importance as crop pests belong to forest edges rather than to the forest proper. They take possession of the roadsides when the country is cleared. Their distribution is a function of the forest margin type of habitat. While it is a characteristic feature of the forest border area, it is also to be found extending along the wooded streams into the great plains and toward the east through the forest area, as the shrubby bluff, the creek and river margin, the fired area, and the marsh margin. While local and always leading a precarious existence in unstable situations, this type of community, probably by virtue of its adaptation to such conditions, has given us a very large number of animals of very considerable economic importance. Tables LXIII and LXIV indicate the forms which we have found common to the forest margins and other situations.

TABLE LXIII

ANIMALS RECORDED FOR A MOIST LOW-GROUND FOREST MARGIN OR THICKET NEAR
WOLF LAKE (STATION 45)

The names that are starred represent animals that have been recorded from the shrubs and weeds along the margins of bogs, lakes, ponds, and streams, June 15 to August 30.

Common Name	Scientific Name
Orb-weaving spider....................	*Singa variabilis* Em.
Jumping spider......................	*Attus palustris* Peck.
*Garden spider......................	*Argiope aurantia* Lucas
*Long-bodied spider..................	*Tetragnatha laboriosa* Htz.
*Orb-weaving spider..................	*Epeira trivittata* Key
*Orb-weaving spider..................	*Epeira trifolium* Htz. (rare)
Black-sided locust...................	*Xiphidium nigropleura* Bruner
Tree-cricket........................	*Oecanthus fasciatus* Fitch
Fork-tailed katydid..................	*Scudderia furcata* Bruner
Nebraska conehead...................	*Conocephalus nebrascensis* Bruner
*Robust lubberly locust...............	*Melanoplus differentialis* Thos.
*Red-legged grasshopper..............	*Melanoplus femur-rubrum* DeG.
*Grasshopper.......................	*Melanoplus bivittatus* Say
*Oblong-winged katydid..............	*Amblycorypha oblongifolia* DeG.
Long-horned grasshopper.............	*Orchelimum indianense* Blatch.
Coreid.............................	*Protenor belfragei* Hagl.
Candlehead.........................	*Scolops sulcipes* Say
Stinkbug...........................	*Euschistus fissilis* Uhl.
*Four-lined leaf-bug.................	*Poecilocapsus lineatus* Fab.
Coreid.............................	*Corynocoris distinctus* Dal.
Solitary wasp.......................	*Odynerus tigris* Sauss
*Buffalo tree-hopper.................	*Ceresa bubalus* Fab.
Long-legged bug....................	*Neides muticus* Say
*Ambush-bug........................	*Phymata erosa fasciata* Gray
*Plant-bug.........................	*Adelphocoris rapidus* Say
*Tarnished plant-bug................	*Lygus pratensis* Linn.
Flower ground beetle................	*Callida punctata* Lec.
*Willow-beetle......................	*Lina scripta* Fab.
Willow-borer.......................	*Saperda concolor* Lec.
Elm-borer..........................	*Saperda lateralis* Fab.
Introduced beetle...................	*Cryptorhynchus lapathi* Linn.
*Goldenrod beetle...................	*Trirhabda tormentosa canadensis* Kirby
Fork-tailed larva	*Cerura* sp.
*Wasp.............................	*Polistes variatus* Cress.
*Jug-making wasp....................	*Eumenes fraternus* Say
*Sawfly............................	*Cimbex americana* Leach.
Swallowtail........................	*Papilio cresphontes* Cram.
*Maia larva........................	*Hemileuca maia* Dru.

TABLE LXIV

ANIMALS RECORDED FROM THE MEDIUM MOIST OR CLIMATIC FOREST EDGE OR
THICKET AT RIVERSIDE, ILL. (STATION 48)

Those starred have been taken from weedy and shrubby roadsides and identified
by specialists. According to the author's field identification nearly all should be
starred.

Common Name	Scientific Name	Month
Crab-spider	*Runcinia aleatoria* Htz	6 9
Jumping spider	*Maevia niger* Htz	6 9
Spider	*Pisaurina undata* Htz	6
Spider (*Dictynidae*)	*Dictyna foliacea* Htz	..
*Orb-weaving spider	*Epeira trifolium* Htz	6 9
Spider	*Atypus milberti* Walck	6
Spider	*Clubiona obesa* Htz	6
Texas grasshopper	*Scudderia texensis* S. and P.	8
Spittle insect	*Clastoptera proteus* Fitch	6
Leaf-hopper	*Diedrocephala coccinea* Forst	6
Four-lined leaf-bug	*Poecilocapsus lineatus* Fab	6
Leaf-bug	*Stiphrosoma stygica* Say	6
Leaf-bug	*Ilnacora stalii* Reut	6
Stinkbug	*Podisus maculiventris* Say	6
Long-horned beetle	*Oberea tripunctata* Sw.	6 7
Long-horned beetle	*Dectes spinosus* Say	7
Tortoise beetle	*Coptocycla bicolor* Fab	7
Tortoise beetle	*Coptocycla signifera* Herbst	7
*Old-fashioned potato-beetle	*Epicauta marginata* Fab	7
*Goldenrod blister beetle	*Epicauta pennsylvanica* DeG	7
Dock curculio	*Lixus macer* Lec	8
Leaf-beetle	*Chelymorpha argus* Herbst	...
Beetle (*Erotylidae*)	*Languria angustata* var. *trifasciata* Say	6 8
Beetle (*Erotylidae*)	*Acropterys gracilis* Newm	6
*Beetle	*Odontonus nervosa* Panz	6
*Grapevine beetle	*Pelidnota punctata* Linn	8
*Milkweed leaf-beetle	*Doryphora clivicollis* Kirby	7 8
Ground beetle	*Lebia atriventris* Say	7 8
Oak-pruning twig-borer	*Elaphidion villosum* Fab	..
Flower beetle (*Carabidae*)	*Callida punctata* Lec	..
Lantern-fly	*Megamelus marginatus* Van D.	6
Wasp	*Crabro interruptulus* D.T	5
Bee (*Halictidae*)	*Chloralictus cressoni* Rob	5
Crane-fly	*Helobia hybrida* Meig	4 5
Crane-fly	*Pachyrhina ferruginea* Fab	5
Fly	*Coenomyia ferruginea* Scop	5
Goldenrod gall fly	*Straussia longipennis* Wied	5

CHAPTER XIV

PRAIRIE ANIMAL COMMUNITIES

I. Introduction

We have noted that a part of the region about Chicago is to be classed as savanna and that the savanna is made up of trees in groves and along the streams, and of forest margin and prairie. Prairie may roughly be separated into high and low. The low prairie commonly exists in depressions in the moraine, lower places in the plain of old Lake Chicago. They are usually covered with water in the spring. The high prairie is above water and is dominated by different plants. As the depressions are filled or become better drained, high prairie plants capture the habitat.

II. Prairie Formations

We have noted that the low prairie is covered by water in spring (Figs. 280, 281). As the water dries up, which usually occurs by the middle of May, the prairie plants begin to grow and the prairie animals make their appearance. This change does not take place abruptly, but gradually. There is a succession of adult-stage animals through the summer. This is what is known as seasonal succession.

I. SEASONAL SUCCESSION

When the snow melts in March and the frost goes out of the ground, the salamander (*Amblystoma tigrinum*) comes out of the ground and soon deposits masses of eggs in the water. The young of *Eubranchipus*, *Cyclops*, and rotifers appear after a few days and often reach adult size by April 1. On April 6, 1908, Mr. Dimmit found adult *Eubranchipus*, *Cyclops*, and rotifers in the pond south of Jackson Park. The salamanders had disappeared. On April 12 three species of flatworms (*Vortex viridis, Planaria velata* Stringer, and *Dendrocoelum*) had appeared, and the first frogs were noted. On April 14 he found frogs' eggs and the red crustacean (*Diaptomus*). *Eubranchipus* was at its maximum abundance. On April 19 he found *Daphnidae*, rhabdocoel worms, and tadpoles. On May 3 but few *Eubranchipus* were found. *Diaptomus* was plentiful, perhaps at its maximum abundance. *Daphnidae* was more abundant than before. *Planaria* were near their maximum. On May 10

278

Eubranchipus serratus had disappeared and *Diaptomus* was not common. Our next record is one month later, when the grasshoppers and other prairie or land species had begun to appear. This succession is of annual occurrence. The temporary pond community is seasonally succeeded by the low prairie community. Flies which breed in water,

FIG. 280.—A prairie pond, still permanent.

FIG. 281.—A temporary prairie pond in spring. The short dead grass indicates that a crop was harvested the preceding season.

such as *Scoliocentra* (Fig. 282) and *Tetanocera* (Fig. 283), are common (also Figs. 284, 285, 286).

2. LOW PRAIRIE ASSOCIATION

a) The subterranean-ground stratum (Stations 42, 43, 44, 45; Table LXV).—Earthworms are abundant. Several of the grasshoppers deposit their eggs in the ground. The larvae of the click-beetle (*Melanotus*

282

283

Fig. 282.—A low prairie fly (*Scoliocentra helvola* Loew); enlarged.
Fig. 283.—A low prairie fly (*Tetanocera umbrarum*); enlarged.

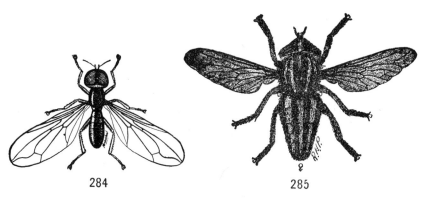

SOME LOW PRAIRIE FLIES

FIG. 284.—*Pipunculus fuscus* (after Lugger from Williston).
FIG. 285.—*Tabanus lineola* Fabr. (after Lugger from Williston).

FIG. 286.—*Spilogaster* sp. from Williston, who says it inhabits high grass.

fissilis), of the strawberry flea-beetle (*Typophorus canellus*), and the corn rootworms (*Diabrotica*) (174), and of many other insects well known in economic literature, burrow into the roots of the plants in the larval stage. Many of the grass-eating cutworms, caterpillars, and sawflies (Fig. 287) pupate beneath the surface of the ground. The salamander (*Amblystoma tigrinum*) spends ten months of each year buried in the mud of such temporary ponds. The Pennsylvania meadow-mouse (*Microtus pennsylvanicus* Or.) has been common in these situations.

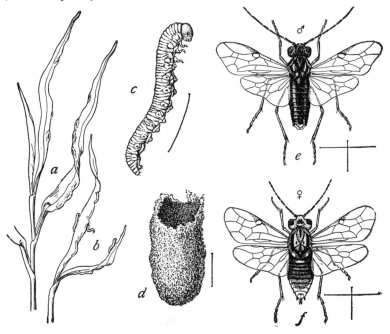

Fig. 287.—Grass sawflies: *a*, eggs; *b*, larvae (*a* and *b* natural size); *c*, larva; *d*, cocoon; *e*, adult male; *f*, adult female (*c* to *f* enlarged as indicated) (after Marlatt, *Insect Life*).

The star-nosed mole burrows beneath the sod. It is remarkable for its curiously fringed nostril. The wetness of the ground excludes other burrowing mammals.

One of the most abundant forms found here is the snail (*Succinea avara*). The ant (*Formica subpolita* var. *neogagates* Em.) is also usually common. It builds a hill and burrows below the surface of the ground also. Several snout-beetles, the adult click-beetles, and the short-winged grouse locust (*Tettigidea parvipennis* and *pennata*) are common on the ground. The 6-spotted spider (*Dolomedes sexpunctatus*) preys upon the other small animals. The common toad and the marsh tree-frog (*Chorophilus nigritus*) are common (139). The latter is particularly abundant in the autumn. Its eggs are laid in April in the temporary pools. Transformations are complete by the last of May. The prairie garter-snake (*Thamnophis radix*) was formerly common. It is known to feed upon the swamp tree-toad. The prairie water-snake (*Tropidonotus grahamii*) was formerly common in and about prairie sloughs (22).

The bobolink builds a nest here in a bunch of grass; the meadow lark and dickcissel build nests of grass and weeds, usually arched over. The bisons, residents of the high prairie, were fond of rolling in the low

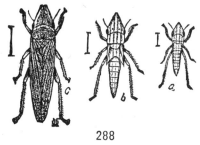

288

Fig. 288.—The large green leaf-hopper (*Draeculacephala mollipes*): *a*, young; *b*, one half-grown; *c*, adult; enlarged as indicated (after Forbes).

Fig. 289.—The six-spotted leaf hopper (*Cicadula sexnotata*); enlarged as indicated (after Forbes).

289

wet places on the prairie and covering themselves completely with mud. This must have destroyed numbers of pond animals and badly disturbed others.

b) The field stratum (Stations 42, 43, 44, 45; Table LXVI).—This is the chief stratum. While various conditions of the subterranean and ground strata, depending upon nearness to ground water, could be recognized, our studies have not been sufficiently detailed to warrant attempts at separation. A girdle of bulrushes can, however, often be distinguished.

Bulrush girdle: Two of the large green leaf-hoppers (*Draeculocephala mollipes* [Fig. 288] and *Cicadula 6-notata* [Fig. 289]) are common. The damsel-bug (*Reduviolus ferus*), which feeds upon leaf-hoppers, is sometimes taken. The slender meadow grasshopper (*Xiphidium fasci-taum*) is common, but breeds in the sedge zone. A flea-beetle (*Monachus saponatus*), the 12-spotted *Diabrotica* (*Diabrotica 12-punctata*) (156), and the salt-meadow snout-beetle (*Endalus limatulus*) (156) are the chief beetles.

The spiders (*Epeira trivittata* and *Tetragnatha laboriosa*) are common. The flies of this girdle are perhaps the most noteworthy insects. Several species of brownish or yellowish flies with conspicuously marked wings are nearly always common. They are *Sciomyzidae* (*Tetanocera plumosa* and *umbrarum*) (Fig. 283). Other characteristic flies are *Osinidae* (*Chlorops sulphurea* Leow.), midges, mosquitoes, *Dolichopodidae*, *Drosophilidae*, and *Anthomyidae*. The blue and yellow moth (*Scepsis fulvicollis*) is common.

Boneset and sedge girdle: The buffalo tree-hopper (*Ceresa bubalus*) (Fig. 259) is found here. The dusky (Fig. 261) and tarnished plant-bugs (Fig. 262) suck the juices of the mint and other plants. The ambush-bug and the damsel-bug often lie in wait in the blossoms for prey.

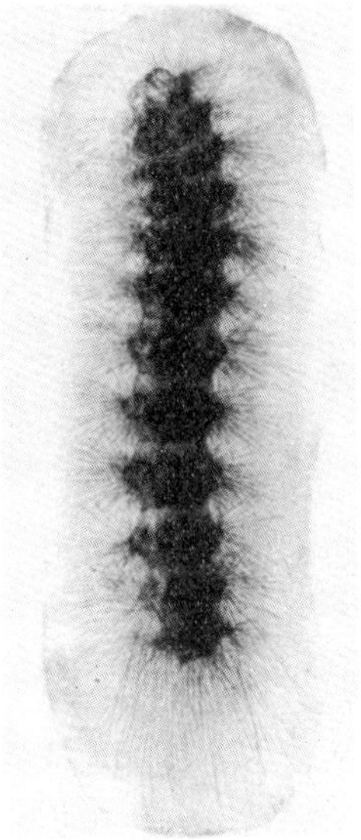

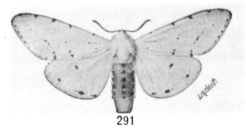

290   291

FIG. 290.—Larva of the salt-marsh caterpillar (*Estigmena acraea* Dru.); natural size (after Forbes).

FIG. 291.—Adult female of the same; natural size (after Forbes).

Aphids occur and with them are the syrphus flies, lady-beetles, and other aphid enemies (164), which are discussed more fully in connection with high prairies. The bright green beetle (*Chrysochus auratus*) feeds on the small-leafed milkweed. One of the corn "bill-bugs" (174) or snout-beetles (*Sphenophorus pertinax* Oliv.), another snout-beetle (*Cryptocephalus venustus*), common garden pests, as well as the leaf-beetle (*Typophorus canellus*) are common (174).

One of the most characteristic groups of the low prairie is that of the grass-feeding larvae. The first of these to appear in spring is the grass sawfly (Fig. 287), which is very abundant in early June. Asscoiated with this are many caterpillars (174). The greasy cutworm (*Agrotis ypsilon* Rott.) feed supon the strawberry. The army worm (*Leucania unipuncta* Haw.) feeds upon a variety of plants, and several of its near relatives occur. The larvae of the salt-marsh caterpillar (*Estigmene acraea*) (Figs. 290, 291), the yellow bear (*Diacrisia virginica* Fab.) (Fig. 292), hedgehog caterpillar (*Isia isabella* S. and A.), and *Apantesis phalterta* Harr. are common.

Of the *Orthoptera, Xiphidium fasciatum* and the 2-lined locust (*Melanoplus bivittatus*), the red-legged locust (*Melanoplus femur-rubrum*), and the short-winged brown locust (*Stenobothrus curtipennis*) (Fig. 293) are most characteristic.

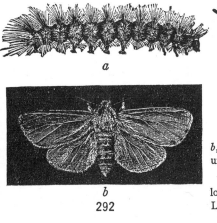

a

293

FIG. 292.—The yellow bear: *a*, larva; *b*, adult (*Diacrisia virginica* Fabr.); natural size (after Forbes).

FIG. 293.—The short-winged brown locust (*Stenobothrus curtipennis*) (after Lugger).

b

292

On the flowers are many flower-frequenting flies, viz., *Sparnopolius flavius* Wied., *Asilus* sp., *Syritta pipiens* Linn., *Coenosia spinosa* Walk., *Paragus angustifrons* Loew., *Pachryrhina ferruginea*, and *Helophilus conostoma* Will. Preying upon the various insects are the mud-dauber wasp (*Scelipron cementarius*) and the digger-wasp (*Ammophila nigricans*). Parasites, such as *Ichneumon zebratus*, *Paniscus gemminatus*, *Epeolus cressonii*, etc., occur upon the plants, and certain of them are often found engaged in depositing eggs in or on caterpillars. The onion-fly (*Tritoxa flexa*) (190) is striking because of its black body and black wings, obliquely marked with white.

Spiders, especially crab spiders, are abundant. The white *Misumena vatia* occurs on the milkweed and the flowers of the mint. *Epeira trivittata* and the long-bodied spider (*Tetragnatha laboriosa*) occur on the blossoms and stems of various plants.

HIGH PRAIRIE ASSOCIATION

(Stations 47, 48; Table LXVII) (Fig. 294)

The type of vegetation which dominates the high prairie is most noticeably characterized by the silphiums—the rosin-weed and the compass plant. The former has broad undivided leaves, the latter divided leaves which usually face east and west. This plant formation springs up throughout the temperate American forest border area on all well-drained ground. It succeeds the low prairie as the depressions occupied by the latter are filled or drained. The high prairie then succeeds the low prairie just as the bulrushes succeed the pond plants; the sedges, the bulrushes; and the boneset association, the sedges. All stages in the development of a pond into prairie may be found near Chicago. Dr. Cowles is of the opinion that shallow ponds with gently sloping sides develop into prairie, while deeper ponds with steep sides develop into forest.

a) *Subterranean-ground stratum.*—Earthworms abound. The larvae of the May-beetles and other *Scarabaeidae* are abundant, feeding on the roots of the prairie plants. The May-beetle is often parasitized by a wasp larva (*Tiphia vulgaris*) (Fig. 297, p. 289) (189). The eggs of the 2-lined locust (*Melanoplus bivittatus*) are deposited here in the ground.

The 13-lined ground squirrel (*Citellus 13-lineatus*) (21) is a slightly gregarious species, strictly diurnal, staying in during dull and cloudy days. Its burrows are from 3 to 16 in. below the surface, and often have five or six entrances into a larger cavity lined with grass. In a den studied by Thompson-Seton the nest was centrally located. Food, which includes cabbage butterflies, cutworms, grasshoppers, beetles, ants, birds (shore lark and lark bunting), and vegetation, is carried in the cheek pouches and stored. The species is non-social. A brood of about eight young are produced in April.

The prairie deer-mouse (*Peromyscus bairdii* H. and K.) (21) is still probably common. According to Thompson-Seton (143) its home range is about 100 yds. It is neither social nor gregarious. It is strictly nocturnal and active all winter, though some seeds are stored. Its food is chiefly seeds. Hawks and owls frequently prey upon it.

Of the extinct forms several are characteristic. The coyote (*Canis latrans* Say) was formerly common. According to Thompson-Seton (143), its home range is ten miles. The den is in a bank or an abandoned badger hole. The nest is a cavity 3 ft. in diameter, with an air-shaft. It is not so social as the gray wolf. Three to ten young are produced

Fig. 294.—General view of high prairie among natural groves at Riverside, Ill.

in April and are fed on disgorged food by the mother. The food consists of ground squirrels, mice, rabbits, frogs, birds, and grasshoppers.

The badger (*Taxidea taxus* Schr.), according to Thompson-Seton, digs a U-shaped burrow with two openings about 6 ft. deep. It is a very rapid burrower. It is nocturnal, but basks in the sun at the mouth of its burrow and hibernates. Its food consists of mice and ground squirrels.

The pocket gopher (*Geomys bursarius* Shaw), according to Thompson-Seton, makes a burrow 3 in. wide. It burrows with its feet and when

FIG. 295.—The nest and eggs of the prairie chicken. Photo by T. C. Stephens.

a pile of dirt has been loosened, turns about and forces it to the exterior with its head. The coyote sometimes rears its young in badger holes on the prairies.

On the ground we find ants (*Myrmica rubra scabrinodis*), one thousand of which were found by Judd (191) in the stomach of a single nighthawk. Ground beetles are common. Crickets, spiders, and weevils all frequent the ground. Most of the field stratum species hibernate on the ground under the fallen plants.

The common toad is rarely wanting near water. The garter-snake (*Thamnophis radix*) has been recorded by Ruthven (156) from such situations in Iowa. The green snake (*Liopeltis vernalis*) is the most characteristic reptile. The prairie rattlesnake or Massasauga (*Sistrurus catenatus*) was formerly common (22).

Eight nesting birds, all of which are quite familiar to everyone, occur. The bobolink nests in a bunch of grass. It feeds upon flea-beetles, weevils, ants, bees, wasps, and grasshoppers of the field stratum. The meadow lark feeds on parasitic hymenoptera, including the parasite of the May-beetle, ground beetles, crickets, grasshoppers, weevils, spiders, etc. The dickcissel is similar in habits. The grasshopper sparrow feeds on long-horned grasshoppers, flea-beetles, cutworms, and parasitic hymenoptera. The vesper sparrow feeds upon moths, flies, ants, beetles, grasshopper eggs, etc., and grain and weed seeds. The nighthawk builds no nest, flies at twilight, and feeds chiefly upon ants. The

297

296

FIG. 296—The adult of the wasp which is parasitic on the May-beetle grubs (*Tiphia vulgaris*) (after Forbes).

FIG. 297.—The larva of the same (after Forbes).

prairie chicken is the most characteristic bird. Its nest is a simple hollow in the grass (Fig. 295). The prairie horned lark builds a nest lined with thistledown and feathers. The lark bunting nests in a tuft of grass.

All of the mammals noted in the subterranean stratum should be added here, as nearly all of them feed largely in the ground and field strata.

The field-mouse (*Microtus ochrogaster* Wagner) (21) is a resident of the ground stratum. Its nest is a pile of grass fragments on the ground. The species feeds chiefly upon grasses and cultivated plants. The bison (*Bison bison* Linn.) is the most characteristic mammal. Thompson-Seton says that the bison population of North America was originally 75,000,000. This animal generally went in clans or families which are said to have had characteristics of their own. An old cow was the

usual leader of the clan. On the great plains these united
and formed the larger herds of 20,000 to 4,000,000 or more,
which have been described by travelers. The males aided
in defending the young. The cowbird is said to have followed the herds constantly.

b) Field stratum.—The lepidopterous larvae are similar
to those of the low prairie, but much less numerous. The hymenoptera are represented by *Bombus separatus,* and many of those recorded on the
low prairie. The adult of the parasite (*Tiphia
vulgaris*) of the May-beetle larva
(Figs. 296–97) occurs commonly.
Several species of aphids (Figs.
298–300) occur, especially on the

milkweeds and thistles.
These are commonly at-
tended by ants, which
stroke them and secure the honey dew from
the posterior ends of their alimentary canals.
The aphids reproduce rapidly, the young being
born in rapid succession at a very ad-
vanced state of development. They
begin sucking the juices of the plant
at once. Several small parasitic
hymenoptera (braconids) (Fig.
299) lay their eggs in the bodies
of the aphids. These finally kill
the aphids, whose bodies with

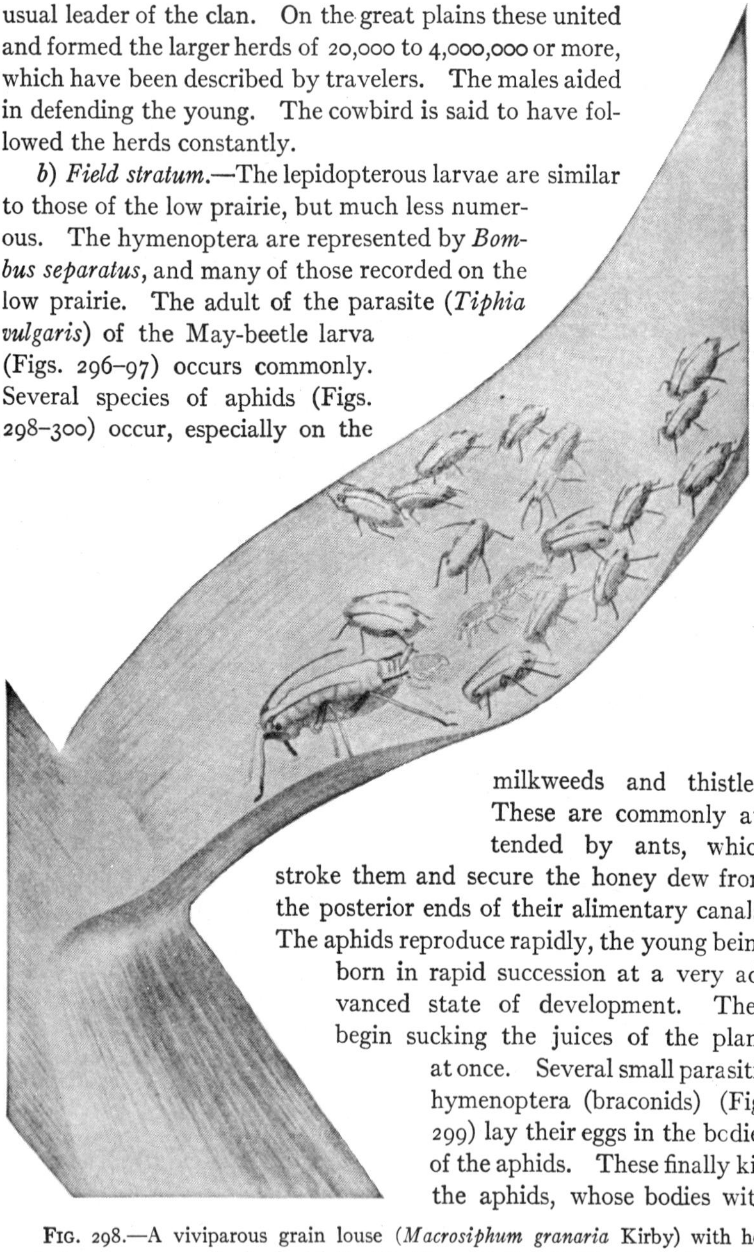

Fig. 298.—A viviparous grain louse (*Macrosiphum granaria* Kirby) with her
newly born young on a barley leaf (after Washburn, *Bull. 108, Minn. Agr. Exp. Sta.,*
Fig. 2, p. 262).

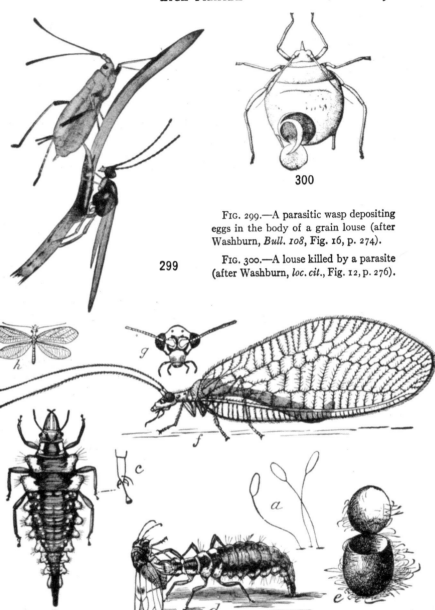

300

FIG. 299.—A parasitic wasp depositing eggs in the body of a grain louse (after Washburn, *Bull. 108*, Fig. 16, p. 274).

FIG. 300.—A louse killed by a parasite (after Washburn, *loc. cit.*, Fig. 12, p. 276).

299

FIG. 301.—The life history of the golden-eyed lacewing (*Chrysopa oculata*): *a*, eggs; *b*, the larva—"aphis-lion"; *c*, foot of the larva; *d*, the larva seizing an aphid; *e*, the pupal cocoon; *f, g, h*, the adult; *h*, natural size (after Chittenden, Div. Ent., U.S. Dept. Agr.).

small circular openings on the abdomen can often be seen sticking to the food plant (Fig. 300). The aphis-lion, which is the larva of the golden-eyed lacewing, feeds upon them (Fig. 301). The eggs of the lacewing are peculiar in that each is attached to a stalk. This is supposed to be an adaptation preventing the larvae already hatched from devouring the remaining eggs. The larva of the syrphus fly (*Mesogramma* sp.) (Fig. 302) devours the aphids in numbers. Lady-beetles, both adults and larvae (*Hippodamia parenthesis* Say, *Megilla maculata*) (Fig. 303), eat aphids.

In June the narrow leaf-bug (*Miris dolobrata*) and the dark leaf-bug (*Horcias goniphorus*) are usually very abundant; both are characteristic.

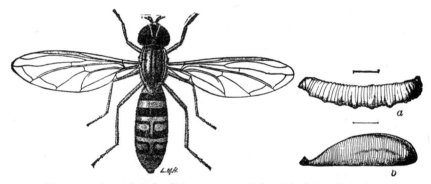

FIG. 302.—A syrphus fly (*Mesogramma polita*), adult (after Forbes): *a*, the larva which feeds on aphids; *b*, pupa; enlarged as indicated (from Forbes after Riley and Howard, Div. Ent., U.S. Dept. Agr.).

Later in the season their places are taken by several others (*Lygus pratensis* and *Adelphocoris rapidus*). The garden flea-hopper (*Halticus uhleri*) occurs on the under side of leaves. The squash-bug family is represented by *Alydus conspersus*.

The tree-hoppers are represented by the buffalo tree-hopper (*Ceresa bubalus*), and the curve-horned tree-hopper (*Campylenchia curvata*). The only lantern-fly recorded is *Amphiscepa bivittata*. Leaf-hoppers are numerous; about ten species have been taken.

The species of *Orthoptera* are mainly different from those of the low prairie. The 2-lined and short-winged brown locusts still continue. *Xiphidium strictum* (Fig. 304) takes the place of *fasciatum*. The common meadow grasshopper (*Orchelimum vulgare*) and an occasional Texas katydid (*Scudderia texensis*) are taken from the goldenrod. From the goldenrod we also take the goldenrod beetle (*Trirhabda tormentosa* var. *canadensis*) and the case-bearer (*Pachybrachys*). The lady-beetles (*Cycloneda, Hippodamia, Megilla*, etc.) are common. The clover-leaf beetle (*Languria mozardi?*) (Fig. 305) is also of common occurrence. The snout-beetles are represented by the large, elongated *Lixus* (Fig. 306), the larvae of which feed in the stalks of rank weeds.

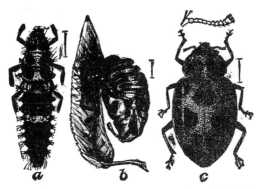

FIG. 303.—The lady-beetle (*Megilla maculata* DeG.) and its life history: *a*, larva; *b*, pupa; *c*, adult (Chittenden, U.S. Dept. Agr.); enlarged as indicated.

FIG. 304.—Meadow grasshopper (*Xiphidium strictum* Scud.); twice natural size (after Forbes).

The onion-fly occurs in connection with the prairie onion. *Eristalis tenax* is common on the flowers. Various flower-flies occur. Waiting in the flowers for such animals as may come are the ambush-bugs (*Phymata erosa fasciata*), and the crab spiders (*Misumessus asperatus* and *Runcina aleatoria*). The jumping spiders (*Phidippus podagrosus*) are also predatory (138). The orb-weavers (*Epeira trivittata, Agriope trifasciata*) build webs into which many insects fall.

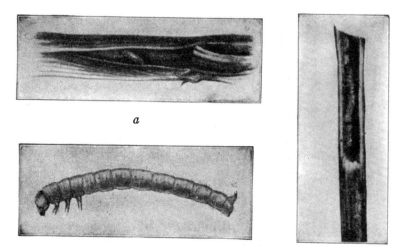

a

c

b

d   *e*

FIG. 305.—The clover-stem borer (*Languria mozardi* Lec): *a*, the egg; *b*, *c*, the larva; *d*, the pupa; *e*, the adult; much enlarged (after Folsom from Forbes).

III. General Discussion

One of the striking peculiarities of the prairie formation is the almost complete cessation of life activities of all the smaller animals in winter. In this respect the prairie animals follow the plants. In spring we find chiefly the insignificant seedling that has sprouted from bulb or seed, and the nymph that has just hatched from the egg. As the season advances the plants become adult, the majority of these reaching maturity with the animals in midsummer.

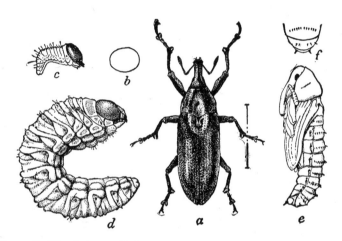

Fig. 306.—The dock curculio (*Lixus concavus* Say): *a*, adult; *b*, egg; *c, d*, newly hatched and full-grown larva; *e*, pupa; *f*, tip of pupa from above; about twice natural size (from Forbes after Chittenden, Div. Ent., U.S. Dept. Agr.).

The low prairie is of interest because of its relation to the eastern forest region. Many if not most of the low prairie forms probably originally occurred in the marshes of the eastern forest region and the river-bottom swales of the prairie and great plains. Many of them (such as place their eggs into plants) are quite independent of the ground, and therefore are most likely to survive under conditions of cultivation where mesophytic plants are favored and the cultivation of the soil does not interfere with their activities.

TABLE LXV

Low Prairie Animals Inhabiting the Ground

R = Riverside (Station 48); W = near Wolf Lake (near Station 45); J = south of Jackson Park (Stations 42, 43).

Common Name	Scientific Name	Habitat		
Crayfish	*Cambarus diogenes* Gir	R		
Crayfish	*Cambarus gracilis* Bun	R		
Ground beetle	*Chlaenius aestivus* Say	R		
Ground beetle	*Platynus affinis* Kirby	R		
Ground beetle	*Amara angustata* Say	R		
Spider	*Ozyptila conspurcata* Thor	R		
Beetle	*Diplochila laticollis* Lec	R		
Shorebug	*Salda coriacea* Uhl	R		
Salamander	*Amblystoma tigrinum* Green		J	
Frog	*Rana pipiens* Sch		J	W
Cricket-frog	*Acris gryllus* Lec		J	W
Swamp tree-frog	*Chorophilus nigritus* Lec		J	W
Toad	*Bufo lentiginosus* Shaw	R	J	W
Garter-snake	*Thamnophis radix* B. and G.?		J	

TABLE LXVI

Low Prairie and Temporary Marsh Animals Frequenting the Vegetation

B = the triangular bulrush belt about Wolf Lake (Station 45); S = the sedge belt of the same; J = sedge prairie south of Jackson Park (Stations 42, 43); R = sedge prairie near Riverside (Station 48), June 15 to August 30.

Common Name	Scientific Name	Habitat			
Spider	*Eugnatha straminea* Em	B			
Diving spider	*Dolomedes sexpunctatus* Htz	B			
Spider	*Epeira trivittata* Key	B	S	J	
Long-bodied spider	*Tetragnatha laboriosa* Htz	B	S	J	
Crab spider	*Runcinia aleatoria* Htz		S	J	
White crab spider	*Misumena vatia* Clerck				
Striped spider	*Argiope trifasciata* Forsk			J	
Garden spider	*Argiope aurantia* Luc			J	
Small orb-weaver	*Epeira trifolium* Htz	B	S	J	
Tube-weaver	*Agelena naevia* Walck			J	
Slender meadow grass-hopper	*Xiphidium fasciatum* DeG	B	S	J	R
Short-winged brown locust	*Stenobothrus curtipennis* Harr		S	J	R
Meadow grasshopper	*Orchelimum vulgare* Harr		S	J	R
Red-legged grasshopper	*Melanoplus femur-rubrum* DeG		S	J	R

TABLE LXVI—*Continued*

Common Name	Scientific Name	Habitat			
Texas grasshopper....	*Scudderia texensis* S. and P...		S		
Cockroach..........	*Blattid* sp.................		S	J	
Cricket.............	*Nemobius maculatus* Blatch...		S	J	
2-lined locust.......	*Melanoplus bivittatus* Say.....		S	J	R
Green-legged locust...	*Melanoplus viridipes* Scud....		S	J	R
Capsid bug.........	*Teratocoris discolor* Uhl......		S	J	
6-spotted leaf-hopper..	*Cicadula sexnotata* Fall.......	B	S	J	
Large leaf-hopper.....	*Draeculacephala mollipes* Say..	B	S	J	
Damsel-bug.........	*Reduviolus ferus* Linn........	B	S	J	
Leaf-hopper........	*Chlorotettix unicolor* Fitch.....	B	S	J	
Leaf-hopper........	*Helochara communis* Fitch....			J	
Leaf-hopper........	*Athysanus striolus* Fall.......			J	
Leaf-hopper........	*Chlorotettix tergata* Fitch......	B	S	J	
Tarnished plant-bug...	*Lygus pratensis* Linn.........		S	J	
Brownie bug........	*Campylenchia curvata* Fab....		S	J	
Dusky plant-bug.....	*Adelphocoris rapidus* Say.....	B	S	J	
Leaf-hopper........	*Athysanus parallelus* Van D...		S	J	
Ambush-bug........	*Phymata erosa fasciata* Gray..		S	J	
Long-horned leaf-beetle	*Donacia subtilis* Kunze.......	B			
Cornroot worm-beetle.	*Diabrotica 12-punctata* Oliv...	B	S	J	R
Marsh snout-beetle....	*Endalus limatulus* Gyll......	B			
Buprestid	*Acmaeodera pulchella* Hbst....	B	S		
Leaf-beetle.........	*Monachus saponatus* Fab.....	B	S	J	R
Click-beetle........	*Melanotus fissilis* Say........		S	J	
Green beetle........	*Chrysochus auratus* Fab......		S	J	
Chrysomelid........	*Nodonota tristis* Oliv........			J	
Chrysomelid	*Typophorus canellus aterrimus* Oliv...................		S	J	
Case-bearer.........	*Cryptocephalus venustus* Fab..		S	J	
....	*Cryptocephalus cinctipennis* Rand.................			J	
Milkweed beetle......	*Tetraopes tetraophthalmus* Forst			J	
Goldenrod beetle.....	*Trirhabda canadensis* Kirby..			J	
Snout-beetle........	*Desmoris scapalis* Lec.......			J	
Fly................	*Tetanocera umbrarum* Linn...	B		J	R
Cloudy-winged fly....	*Tetanocera plumosa* Loew....	B		J	
Long-legged fly.......	*Dolichopodidae* sp..........		S	J	
Syrphus fly.........	*Syrphus americanus* Wied....			J	
Onion-fly...........	*Tritoxa flexa* Wied.........		S	J	
Ant................	*Formica subpolita neogagates* Emery...................			J	
Syrphus fly.........	*Eristalis tenax* Linn.........		S	J	
Bee................	*Agapostemon viridulus* Fab...			J	
Bumblebee..........	*Bombus separatus* Cress......			J	
Lacewing...........	*Chrysopa albicornis* Fitch....			J	
Ant................	*Myrmica rubra scabrinodis* Nyl.			J	
Ichneumon fly.......	*Ichneumon galenus* Cress.....			J	
Moth..............	*Scepsis fulvicollis* Hbn.......	B			
Syrphus fly.........	*Mesogramma geminata* Say....		S	J	
Social wasp.........	*Polistes variatus* Cress.......		S	J	

TABLE LXVII

ANIMALS USUALLY COMMON ON COMPASS-PLANT PRAIRIE

Collections made near Riverside (Station 48) and Chicago Lawn (Station 47), June 15 to August 30.

Common Name	Scientific Name
Cricket	*Nemobius fasciatus vittatus* Harr.
Jumping spider	*Maevia niger* Htz.
Jumping spider	*Phidippus podagrosus* Htz.
Jumping spider	*Phidippus borealis* B.
Jumping spider	*Phidippus rufus* Htz.
Harvestman	*Liobunum grande* Say
Garden spider	*Argiope trifasciata* Fors.
Ant	*Formica cinerea* var. *neocinerea* Wheeler
Grasshopper	*Orphulella speciosa* Scud.
Meadow grasshopper	*Orchelimum vulgare* Harr.
Meadow grasshopper	*Xiphidium strictum* Scud.
Brown locust	*Stenobothrus curtipennis* Harr.
Conehead	*Conocephalus ensiger* Harr.
Katydid	*Scudderia texensis* S. and P.
Leaf-hopper	*Athysanus striolus* Fall.
Leaf-hopper	*Agallia 4-punctata* Prov.
Leaf-hopper	*Platymetopius acutus* Say
Leaf-bug	*Trigonotylus ruficornis* Four.
Leaf-bug	*Miris dolabrata* Linn.
Leaf-hopper	*Chlorotettix spatulata* O. and B.
Membracid	*Stictocephala lutea* Wlk.
Garden flea-hopper	*Halticus uhleri* Giar.
Stinkbug	*Euschistus variolarius* Pal. Beauv.
Leaf-bug	*Plagiognathus politus* Uhl.
Leaf-hopper	*Eutettix straminea* Osb.
Leaf-hopper	*Empoasca mali* LeB.
Negro-bug	*Thyreocoris pulicaria* Van D.
Coreid	*Alydus conspersus* Mont.
Stinkbug	*Cosmopepla carnifex* Fab.
Leaf-bug	*Garganus fusiformis* Say
Leaf-bug	*Horcias marginalis* Reut.
Beetle (Mordellid)	*Mordellistena connata* Lec.
Lady-beetle	*Cycloneda sanguinea munda* Say
Case-bearer	*Pachybrachys* sp.
Strawberry beetle	*Typophorus canellus gilvipes* Horn
Beetle	*Photinus punctulatus* Lec.
Syrphus fly	*Eristalis tenax* Linn.
Green snake	*Liopeltis vernalis* Harlan

CHAPTER XV

GENERAL DISCUSSION

I. INTRODUCTION

We have briefly presented some facts regarding the nature and environmental relations of animals, an account of the environment, and a discussion of the inhabitants of some of the type habitats of the forest and forest border regions. We noted also in preceding chapters some aspect of relations of the animals of the same and of different communities to one another, and our relations to them. We may still present (a) the relations of the different communities to one another, (b) the laws governing distribution, and (c) a discussion of the relations of ecology to broader geographic problems.

II. APPLICATION OF THE LAWS GOVERNING ANIMAL ACTIVITIES TO WORLD AND REGIONAL PROBLEMS

As was stated in the first chapter, the relative importance of different environmental factors is not definitely known, but probably in local and experimental conditions, land environments can best be measured in terms of evaporating power of the air, light, and materials for abode, aquatic environment by carbon dioxide, oxygen, and materials for abode. In explaining extensive or regional distribution, a few factors have been emphasized and these usually in the sense of barriers. Merriam (48) emphasizes temperature, Walker (128) atmospheric moisture. Heilprin (192, p. 39), like most paleontologists, emphasizes food. Nothing is, I believe, more incorrect than the idea that the same single factor governs the regional distribution of most animal species. Since the environment is a complex of many factors, every animal, while in its normal environmental complex, lives surrounded by and responds to a complex of factors in its normal activities (44, p. 193). Can a single factor control distribution?

1. REACTIONS TO SINGLE FACTORS

Considerable physiological study of organisms has been conducted with particular reference to the analysis of the organism itself, but with little reference to natural environments. Many of the factors and conditions employed in such experiments are of such a nature that the animal would rarely or never encounter them in its normal life. Other experiments are attempts to keep the environment normal, except for one factor (44, p. 180). These have demonstrated that animals are capable of responding to the action of a single stimulus.

A typical experiment to demonstrate this would consist in preparing two long receptacles in such a way that one is the normal environment of the animals in *all* respects and the other in all respects except for one factor, as, for example, temperature. The temperature conditions of the latter might be as follows: temperature at one end 10° C., at the other 35° C., with a gradient between. If then 100 animals are placed in each of the receptacles, those placed in one end of the gradient will soon show signs of stimulation and will move about until they come near the center of the pan where the temperature is 20°–25°. If, after sufficient time has elapsed for the experimental animals to take up this position, the control animals have remained equally distributed, the experiment will show that the animals have responded to temperature alone.

Certain general laws govern the reaction of animals to different intensity of the same stimulus. Take, for example, temperature. There is in most animals which have been subjected to experimentation with temperature a range of several degrees within which the activities of the animal proceed without marked stimulative features, as is suggested by the experiment outlined above. Conditions within this range of several degrees are called the optimum. As the temperature is raised or lowered from such a condition, the animal is stimulated. If the temperature is continuously raised, a point is reached at which the animal dies. The temperature condition just before death occurs is called the maximum (35). The lowering of temperature produces comparable results.

2. EXPERIMENTAL STUDIES OF HABITAT SELECTION

Animals select their habitats, and distribution is the result of this selection. To decide whether or not one factor can determine distribution, experiments, of which the following is a typical example, have been performed.

a) Methods of experimentation.—Do animals select their breeding-places? To answer this question, tiger-beetles were selected as material and adults were placed in cages containing soil of several kinds. Each kind was so arranged into steep and level parts, that about one square foot of each type was exposed. The adults placed in the cage were taken when the species was breeding (see p. 212). The soil was kept very moist up to the time the first ovipositor holes were made, because this species lays only in moist soil. After this the wetting of the soil was done very cautiously, so as not to wash the eggs from the ground in steep parts. Accordingly, the holes were not obliterated from day to day. The counts, however, are not accurate for the soil in which a large number were made, because eggs are sometimes laid very close together and adjoining holes destroyed. Some eggs are deposited in irregular cracks and crevices where they are likely to be overlooked. The greatest care was taken to discover every hole made in the soils in which larvae do not occur in nature. Soils in the different lots were arranged in different orders.

b) Results.—Table LXVIII shows the approximate number of holes made in the clay and probably the actual number made in the other soils, together with the number of larvae which appeared: 80 per cent on the steep slope, 98 per cent in clay.

The count of holes includes some in the first stages of digging, mere scratches on the ground, and others which had been excavated to the usual depth with or without eggs being laid.

TABLE LXVIII (55)

DISTRIBUTION OF OVIPOSITOR HOLES AND LARVAE OF *C. purpurea limbalis* UNDER EXPERIMENTAL CONDITIONS

S=steep; L=level.

		CLAY		CLAY, 9 PTS. HUMUS, 1 PT.		FOREST HUMUS		HUMUS, 1 PT. SAND, 9 PTS.		CLEAN SAND	
		S	L	S	L	S	L	S	L	S	L
Lot I	Holes.......	o	o	o	o	o	o	o	o	o	o
	Larvae......	9	o	o	o	o	o	o	o	o	o
Lot II	Holes.......	21	5	o	o	o	o	o	o	o	o
	Larvae......	12	1	o	o	o	o	o	o	o	o
Lot III	Holes.......	17	7	1	o	o	c	o	o	o	o
	Larvae......	24	10	1	o	o	o	o	o	o	o

c) Factors controlling habitat selection (55).—Pairs taken in coitus were placed in cages containing sand only and level clay only. No larvae appeared in either case. The experiment with the level clay has not been repeated. Females placed in cages containing rough, steep clay, deposited eggs. Eggs are also absent from dry soils, whether steep or level.

Slope, kind of soil, and soil moisture are factors governing the deficiency or absence of eggs. A deficiency or excess in any one of these respects decreases the number of eggs laid, or causes them not to be laid at all. The animals are in the condition for egg-laying for but a short period.

d) *Method of selection.*—It has been determined by opening holes that eggs are not laid in all, and in one case the first holes made by the female were empty. This would tend to show that the female beetle tries the soil before laying the eggs, but I have not been able in other cases to determine whether the first holes contained eggs or not. To determine this, it would be necessary to watch a female all of the time during several days.

3. LAW OF TOLERATION (55)

Repeated experiments with several species have shown results similar to those shown in Table LXVIII, and we have concluded that the egg-laying place of the tiger-beetles is their true habitat. The tiger-beetles which lay eggs in soil do so only when the surrounding temperature and light are both suitable, the soil moist and probably also warm. The soil must satisfy the ovipositor (egg-laying organ) tests with respect to several factors. Egg-laying, the *positive reaction*, is then probably a response to several factors. Furthermore, after the eggs are laid, the conditions favorable for egg-laying must continue for about two weeks if the eggs are to hatch and the larvae reach the surface. The success of reproduction depends upon the qualitative and quantitative completeness of the complex of conditions. This complete complex is called the *ecological optimum*. The *negative reaction*, on the other hand, appears to be different. The absence of eggs, the number of failures to lay, and therefore the number of eggs laid in any situation, can be controlled by qualitative or quantitative conditions with respect to *any one of several factors*. The presence, absence, or number of eggs laid may be governed by a single factor.

For example, all other conditions being optimum, moisture may control the presence, absence, or number of eggs laid. If the moisture be optimum, the maximum number of eggs will be laid. If it is too great few or no eggs will be laid. This factor then controls according as it is near the optimum, or near either the maximum or minimum tolerated by the species. It is, however, not necessary that but a single factor should deviate; the effect is similar or more pronounced if several vary.

The success of a species, its numbers, sometimes its size, etc., are determined largely by the degree of deviation of a single factor (or factors) from the range of optimum of the species. It is obvious that the cause of the fluctuation might be, for example, moisture due to (climatic) deficiency in rainfall, or rapid run-off, due to steep slope. The evidence for the application of the law of toleration to local distribution is good. Since the same factors are involved in the "geographic" or more extensive distribution, there is no difficulty in the application of the law to such distribution also, for, to assume that the law is not applicable is to assume that animals distinguish between the *causes which lie back of the changes in physical factors* by which they are affected. The fact that, in so far as our observation can go at present, most animals are found in similar conditions throughout their ranges is also good evidence for the application of both the laws of minimum and toleration to problems of geographic range. In fact, the law of minimum (see p. 68) is but a special case of the law of toleration. Combinations of the factors which fall under the law of minimum may be made, which make the law of toleration apply quite generally. For example, food and excretory products may be taken together as constituting a single factor. From this point of view the law of toleration applies, the food acting on the minimum side, excretory products on the maximum.

4. APPLICATION OF THE LAW OF TOLERATION TO DISTRIBUTION (55)

As has already been implied, the locality or region of optimum, or the locality or region in which the animal is most nearly in physiological equilibrium, is called the habitat (ecological optimum) when it refers to ecological or local distribution, and the *center of distribution* when it refers to extensive areas. The so-called centers of distribution are often only areas in which conditions are optimum for a considerable number of species. The distribution and number of individuals of any species may be graphically represented as below:

Minimum Limit of Toleration		Range of Optimum		Maximum Limit of Toleration	
	←««	Habitat or center of distribution	»»→		
Absent	Decreasing	Greatest abundance	Decreasing	Absent

On account of the nature and distribution of climatic and vegetational conditions, it follows that as we pass in one direction from a center, one factor may fluctuate beyond the range of toleration of a species under consideration; but as we pass in another direction the fluctuating factor is very likely to be *different*.

a) *Governing the limit of local and geographic range.*—The geographic or local range of any species is limited by the fluctuation of a single factor (or factors) beyond the limit tolerated by that species. In non-migratory species the limitations are with reference to the activity which takes place within the narrowest limits (usually breeding). In migratory species this activity limits the range during only a part of the life history.

b) *Governing the distribution area and habitat area* (55).—The distribution area of a species is the distribution of the complete environmental complex in which it can live, as determined (1) by the activity which takes place within the narrowest limits and the animal's power of migration, and (2) by barriers in which some factor of the complex fluctuates beyond the limits of toleration of the species in all periods of its life history.

If these statements are borne out by further investigation it follows that every *study of animal behavior* which is *related to measured physical factors or to natural environments is directly related to problems of distribution.*

III. AGREEMENT BETWEEN PLANTS AND ANIMALS

In recent years the ecology of plants has received much attention and the subject has made great progress. In animal ecology but little progress has been made, and students (and teachers) have been inclined to expect relations and conditions in animals parallel with those in plants. Little progress has been made, largely because workers have not recognized the important phenomena in animals as compared with plants.

1. ECOLOGICAL AGREEMENT OF INDIVIDUALS

Organisms may be divided on the basis of their ability to move about, into *sessile* or *fixed*, and *motile* forms. All organisms are of course capable of movement of some sort, even though it be only mechanical movement dependent upon turgor. There are also all degrees of ability to move from place to place. Some motile plants and animals move about only very slowly, and the division of organisms into sessile and motile is a somewhat artificial classification, as many forms are difficult to place in either group. Some are sessile at one period of their lives and motile at another. Comparable difficulty arises, however, in the separation of plants from animals.

The animals with which we, as inland people, are most familiar, are the highly motile forms, and the plants with which we are most familiar are sessile forms. We are all also somewhat familiar with numerous marine animals, such as polyps, sea plumes, etc., which are sessile, like plants. Sessile animals are probably all aquatic. Logically, ecology cannot be divided into plant and animal ecology, but it may be divided into the ecology of sessile and motile organisms.

An appreciation of the likenesses and differences of sessile and motile organisms is an important thing in ecology. The plant and the animal groups contain both sessile and motile types together with types intermediate between the two and thus taken as a whole plants and animals *are in agreement in the matter of response.* However, since the vast majority of animals with which we deal are motile, their activities are evident because of their ability to move about. On the other hand the majority of plants are sessile, and sessile individuals usually can change the position of the whole or its parts only by growth. Changes in the relation and character of parts are the results of the application of stimuli to sessile plants. Movement is the chief result of the application of stimuli to animals. Animal ecology has very much in common with plant ecology. Diatoms, flatworms, and many other marine animals and plants meet the same conditions in the same or similar ways (72, p. 121; 53*a*, p. 156; 53*b*, p. 155). Sessile animals, such as reef-forming corals, show growth form differences (193, 194, 195) under different conditions, just as sessile plants do. Comparable plants and animals show comparable responses. The physiological life history aspect of plant ecology (52) is parallel with the same phenomenon in animals, but the activities of motile animals correspond roughly to the growth form phenomena in sessile plants (55, p. 593).

All the way through the study of ecology we look for behavior or activity difference in motile organisms (chiefly animals), when considering the species of two different habitats, while, when making a comparison of the sessile organisms (chiefly plants) of two habitats, we look for differences in form and structure. To be sure an occasional sessile plant can move some of its parts and likewise some motile animals change color, size, or form with differing conditions during development, but these are of secondary rather than primary importance and we must look mainly to *form changes as "plant response" and behavior, or activity changes as "animal response."*

2. AGREEMENT OF COMMUNITIES

Are physical conditions sometimes similar when vegetation and landscape aspect are very different? That they are is clearly suggested when we compare the forest and the shrub-covered bluff where forest animals occur. Plants grow from seeds only under a very limited range of conditions. However, if trees are given a few years' growth under favorable conditions they will be successful under a great range of conditions. The great age to which trees often live and the slowness with which they grow make it possible for conditions to change while the trees still live on with changes only in leaf structure. It is to be expected that the distribution of animals is correlated with the occurrence of seedlings or of quick-growing plants or at least with leaf structure types rather than strictly with species of trees. These facts suggest that there are two types of cases in which physical conditions and forest conditions are not in accord. In the first case atmospheric conditions become favorable for forest animals before any woody plants have been able to grow; in the second, woody plants remain after conditions have become unfavorable for forest animals; both are due to lagging behind of vegetation; both are very local and of minor significance.

The reasons for the wide distribution of some animals in the forest stages which we have considered are no doubt various. For example *Zonitoides arboreus* (Table L, p. 252) is rare in the early stages and is confined to the lower and moister localities. If *Epeira domicilorum* is a species of stable physiological makeup we can offer no explanation for its peculiar distribution (Table LVI, p. 257). A species may have its critical period in the early spring when the leaves are off the trees and the condition of the atmosphere similar in all stages (see Fig. 251, p. 248) or may live at higher levels in the denser and older stages, and thus be surrounded by similar atmospheric conditions, but we are not warranted in assuming either of these causes here.

Another striking feature of the distribution of many beetles, bugs, spiders, and *Orthoptera* is the fact that they are found in open woods, edges of woods, on the vegetation of marshes, and over the water of small ponds in which vegetation is growing. In this way many species are found to occur in what at first appear to be very unlike situations. *Lygus pratensis, Triphleps insidiosus,* and *Euschistus variolarius,* which occur on the vegetation of the margins of swamps, of the black-oak forest dunes, and on prairies and agricultural lands, may serve as examples. Shull has pointed out similar facts as one of the difficulties in the way of ecological classification of *Orthoptera* and *Thysanoptera.* Such species as the bugs mentioned above are said to occur "everywhere," although they are rarely found in moist woods or in any situation in which they are not fully exposed to the sun and may always live in similar conditions.

Some investigators have questioned the importance of vegetation to animals and we note here that the distributions of plant and animal species are not always correlated. If one refers to *species* of *plants* and *species* of *animals* then the vegetation very often is not correlated with the distribution of the animals. If on the other hand one means that the plants are controllers of physical conditions, then vegetation can be said to be of very great importance.

Before discussing the problem of agreement between plant and animal communities, it is necessary to state what is meant by agreement. According to present developments of the science of ecology *plant and animal communities may be said to be in full agreement when the growth form of each stratum of the plant community is correlated with the conditions selected by the animals of that stratum.* Questions of agreement are primarily questions for experimental solution. Two types of disagreement are to be expected. We may illustrate the first by a bog or marsh community. Considering plants rooted in the soil we note that water is secured from the soil by the roots and is lost through the leaves and twigs. Accordingly since bog soil is unfavorable, due to the presence of toxins or to other causes, plants growing in it do not secure water easily even when the quantity of soil water is great. *Such plants have xerophytic structures (which tend to check the loss of water) developed far beyond the requirements of the atmospheric conditions surrounding their vegetative parts.* It is improbable that the animals inhabiting a bog-vegetation field stratum would *select* atmospheric conditions such as produce equally xerophytic structures under *favorable soil* conditions. We may therefore expect disagreement. The smaller plants such as fungi, algae, etc., are related to the strata of soil and atmosphere exactly as the smaller animals and as *much disagreement* is to be expected between such plants and the rooted vegetation as between the rooted vegetation and animals. It must also be noted that the xerophytic structures of the plants of *unfavorable* soils may have important influence upon ectophytic plants and animals and in part counteract the effect of favorable atmospheric conditions.

The second type of disagreement is represented by cases in which the vegetation lags behind. We have already noted that on the clay bluff (pp. 209–17) conditions become favorable for inconspicuous plants and forest animals as soon as the growth of the pioneer vegetation gives shade to the soil. In other cases woody vegetation remains in situations where the conditions have become unfavorable for it and the less conspicuous plants and some of the animals have disappeared. We may expect lack of accord within and between plant and animal communities under such conditions. In these cases, however, conditions are only *temporarily out of adjustment*, due to rapid physiographic changes, and we note from the data presented that plant and animal communities are usually in agreement. The exceptions are often apparent only and due to the emphasis of *species* instead of *mores* and *growth form*. From this viewpoint and with such exceptions as are noted, plant and animal communities are probably in agreement the world over.

IV. RELATIONS OF COMMUNITIES

1. SUCCESSION—CAUSES

Succession is no doubt one of the most important and widespread of the phenomena discovered by the ecologists up to the present time (120, 197). Simply stated, it means that on a given fixed area organisms succeed one another, because of changes in conditions. These changes make impossible the continued existence of the forms present at any given time; with the death or migration of such forms, others adapted to the changed conditions occupy the area, whenever such adapted forms are available. The changes referred to result from physical or biological causes, or combinations of the two. It is probable that the causes of the changes are frequently complex combinations of various factors.

We have among the physical causes changes in climate and changes in topography. All degradation of land is a cause of succession. Such geological processes are well understood and treated in textbooks on geology and physiography.

The biological causes of succession lie chiefly in the fact that organisms frequently so affect their environments that neither they themselves nor their offspring can continue to live at the point where they are now living. Every organism adds certain poisonous substances to its surroundings, and takes away certain substances needed by itself. It frequently thus so changes conditions that its offspring cannot live and grow to maturity in the same locality as the parents. However, by these same processes it prepares the way for other organisms which can live and grow in the conditions thus produced.

Obviously, those organisms whose decaying bodies and excretory materials are not removed or distributed by their wanderings will modify their environments most. Organisms which remain in one place do nothing which tends to remove the results of their own existence, and frequently modify their environments in manners detrimental to themselves.[1] On the land, plants are the dominant sessile forms, and often profoundly modify the conditions in which they live, so that they cannot succeed themselves. When will the process of succession stop? Obviously, it must cease when there are no available species to take the places of those which have destroyed their own habitats. There are species which are immune to their own products and the products of the species which are associated with them. Obviously, when a condition in which these species can live is reached, and they come to occupy the place which is thus made ready for them, the formation which they constitute can, so far as the plants are concerned, *last indefinitely*. This is theoretically true of all climax or geographic formations, and has been established for the beech and maple forest of eastern America.

2. MOTILE AND SESSILE ORGANISMS IN SUCCESSION

Motile Organisms	Fixed Organisms
a) Motile organisms affect their own environments by the destruction of materials of abode and food supply and the pollution of their habitats by waste products (196, 114, and citations).	*a)* Sessile organisms modify their own environments largely through growth of their own bodies, cutting off light, interfering with circulation in surrounding medium and accumulation of waste products (195, 120).
b) The changes under (*a*) make the continued existence of the group in question impossible and prepare the way for other differently adapted (succession) forms.	*b)* The same as for motile organisms (197).
c) Succession is a succession of breeding-places.	*c)* Breeding and living places are not contrasted as young stages usually thrive only where adults can live.

Succession can take place only where forms adapted to the changed conditions are available.

3. CONVERGENCE

The work of running water, for example, is in a measure convergent. When a new body of land is uplifted, streams begin to work their way into the new land mass and cut deep valleys. The formation of numerous tributaries (92 and citations) isolates portions of the upland in the

---

[1] In the sea (195) sessile forms are chiefly animals and animals are probably the chief cause of succession there. Coral polyps cannot build upward indefinitely, as they soon reach the surface and can no longer exist. By reaching the surface they prepare the way for other forms.

SAND RIDGE	CLAY BLUFF
Cottonwood	Aspen
Gray pine	Cottonwood
Black oak	Hop-Hornbeam
White oak	White oak
Red oak	Red oak
Hickory	Hickory

BEECH AND SUGAR MAPLE

Tulip	Hickory
Basswood	Red oak
White elm and White ash	Bur oak
Swamp white oak	Basswood
Buttonbush	Hawthorn
Cattail and Bulrush	Slippery elm and White elm
Water-lily and Water Mill-foil	River maple
Chara	Black willow

POND	FLOOD-PLAIN

DIAGRAM 8.—Showing the convergence of four types of habitat, to the beech and maple forest. Read from the extremities toward center. (Prepared with the assistance of Dr. Cowles and from his writings.)

form of hills. These hills are broken up into smaller hills by the smaller tributaries, and the resulting hills into still smaller ones, until the upland is all removed and the country reduced to a generally level condition known as a peneplain. The process of peneplanation then tends to fill all low lakes and ponds and drain all high ones. It works over all the materials of the upland and lays them down as alluvial deposits, which process tends to make the surface materials of a uniform nature. Associated with this, and more or less independent of it, the process of plant succession makes the conditions converging (Diagram 8) to a still greater degree (13).

The principle of convergence, while not generally established, is believed to be of wide application. It has been suggested for the tropical forest of the Philippines by Whitford (198), for the coniferous forest regions of North America by Adams and by Gleason, and for the arid Southwest by Ruthven. Theoretically at least, in all the varied types of land habitats of any large area, communities are tending toward some one type which is primarily adjusted to the climate of the region when its topography approaches base level. Such a *climatic type* of community rapidly displaces the communities of all the varied kinds of soil of a newly uplifted area which is only a few hundred feet above the sea. In these situations the climatic communities dominate sterile soil by process of successional development extending over a few score or hundreds of years.

V. GENERAL RELATION OF COMMUNITIES OF THE SAME
CLIMATE (13)

In each climatic realm of the world there are relations between communities of two sorts, (a) physiological relations, best defined as physiological similarities, and (b) successional or evolutionary relations. Diagram 9 shows both types of relations for the temperate American forest border area. Single-pointed arrows show the directions of succession, double-pointed arrows show similarities of conditions and the occurrence of several or many of the same species in considerable numbers in communities between which such arrows extend. Broken lines indicate less definite relations than the solid lines. Starting with the aquatic communities, we note that spring-fed and intermittent stream communities converge with physiographic aging to small, permanent, swift-stream communities, and permanent swift-stream communities are succeeded by base-level stream communities. The characteristic communities of small permanent streams and base-level streams are indicated above. Taking up another line, we note that the large-lake communities are succeeded by the small-lake communities. Rocky-shore communities of the large-lake areas have features in common with those of the rocky rapids of the stream. The sand, gravel, and vegetation communities of the base-level stream and the small lake have many things in common, while the silt and humus bottom communities are distinguishing features of the two. Communities of ponds originating

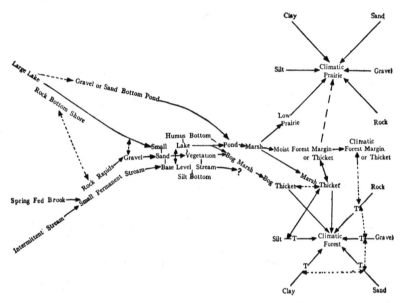

DIAGRAM 9.—Showing some relations of the chief animal communities of the forest-border region of Central North America. The word community or communities is to be understood as following all the words appearing in the diagram. For full description see text.

by very rapid physiographic changes pass through a series of stages comparable to those found in the different parts of the small lake. The lake communities pass to the pond community stage or give rise to a floating-bog marsh community which is displaced by a floating-bog thicket community. Cowles states that this takes place in deep lakes, while the shallow ones become ponds which give rise to marshes with firm substrata. Such a marsh community may be displaced wholly by a low prairie community, in part by a thicket forest margin community, or wholly by a thicket community which will be succeeded by a forest community. In the savanna or prairie climate the marsh margin thicket may become a climatic thicket or forest margin. In the savanna or prairie climate the communities of all the various soils and the low prairie community may converge to the prairie climate community, or to the forest community as is shown below for the forest climate. In the forest climate and locally in the savanna climate the communities of all the various soils pass through a thicket community stage (T), related to a climatic forest. The thicket communities of all the dry soils are related to the forest margin thicket community of the savanna climate.

I. CORRESPONDENCE OF COMMUNITIES OF DIFFERENT PARTS OF THE WORLD (55)

The botanists have abundant evidence for the correspondence of the formations of similar climates (58*a*). The vegetation of different parts of the world which have similar climates is similar and the plants though usually belonging to different taxonomic groups are similar in growth, form, and appearance. Correspondence and similarity of vegetation is not limited to the climatic or extensive formations, but applies also to strictly local situations wherever the *physical conditions are similar*. On the animal side we have less trustworthy evidence of similarity or correspondence. If the physiological similarity occurs in the *same* community, due, as has just been stated, to selection of habitat and modification of behavior, we conclude that it occurs in *all* communities occupying similar conditions and that *similar situations* in different parts of the world have *physiologically similar communities*, and identical situations approximately identical communities.

The direct evidences for the correspondence of formations in different parts of the world are as follows: (*a*) the existence of identical or closely corresponding species has long been known to naturalists (3, 199, 192); (*b*) similarity of physiological life histories of many species is well known, as, for example, corresponding species in the United States and Europe or Japan, and a general concentration of breeding in the rainy season in all arid climates, etc.; (*c*) certain animals in similar environments in different parts of the world appear from the accounts of naturalists to behave alike with reference to the physical condition of different parts of the day, year, and different weather. For example, it appears that there is a close physiological and ecological similarity between certain antelopes of the savannas of Africa and certain savanna kangaroos of Australia (200). In other words certain kangaroos are ecologically and physiologically similar to some antelopes. As has already been stated, the zoölogist is usually unduly impressed with specificities such as mode of movement of limbs, body, etc. Now if my reader pictures an *African antelope running gracefully from a pack of Cape hunting dogs* (102, pp. 119–23), *and an old-man-kangaroo leaping from a pack of dingoes* (202, pp. 41, 243), *noting mainly the specific peculiarities of the movement of limbs and body of the pursued in each case,* he will be dwelling upon *specificities of little ecological significance and missing the point of view of the ecologist altogether.* These specificities of behavior are matters of little ecological significance; it matters not if one animal progresses by sommersaults so long as the two are in *agreement* in the matter of *reactions to physical factors* as indicated by the *manner of spending the day* (200), *avoidance of forests, swamps, cold mountain tops, etc.,* entirely *available to them,* and in the mode of meeting enemies as indicated by the reaction to the approaching hunter or enemy.

a) Distribution of land communities represented in Central North America.—The following climatic formations are represented at Chicago and distributed as given below:

Temperate Deciduous Forest Formations: Forest with broad, thin leaves which are shed in autumn; near Chicago, oak, hickory, beech, and maple (58*a*).
　　　Distribution: Eastern North America, north to the Great Lakes; Chili, north to 35°; Europe, north of the Alps, and south of 60°; Japan and the vicinity of Okhotsk (58*a*).

Temperate Savanna Formations: Grasslands with scattered trees, or trees in groves surrounded by thickets, and with dense forests along larger streams. Near Chicago, the grassland is prairie and the trees chiefly oak and hickory.
　　　Distribution: A narrow belt in North America surrounding the great plains on the east, north, and west; Uruguay, South Australia, South Africa, and Eastern Siberia.

Formations of Forests with Narrow Thick Leaves: Coniferous forest. Dense evergreen forests with little undergrowth. Lies just to the north of Chicago and was represented locally in the parts of Michigan shown on Map I (frontispiece).
　　　Distribution: North America north of the Great Lakes and Columbia River extending southward into the mountains; Eurasia north of 60°, extending southward into the mountains.

The localities which are in agreement are indicated by distribution of the different types of formation. It will be noted that the deciduous forest animal formation with which we have dealt is found in *several parts of the world, this animal community being essentially duplicated* in these differently located areas. This correspondence is probably much more striking physiologically than in the matters of interrelation of species because in some formations certain groups, as, for example, antelopes in African steppes, are especially numerous, while in a corresponding situation in South America they are very few.

As has already been suggested, correspondence is not limited to the gross characters of extensive formations, but is equally true of the more local communities. In matters of correspondence of species there are often striking correspondences within the groups of formation indicated above. For example, there is a striking correspondence in behavior between the meerkats of the steppes of East Africa (3) and the prairie dogs of our own steppe, both being grasslands but differing in climate. Considering a local formation, as that of the sandy beaches of the sea and very large lakes, we note that along the New England coast and around the shores of Lake Michigan the moist, sandy beaches are inhabited by the larvae of the beach tiger-beetle (*Cicindela hirticollis*) (Fig. 134, p. 179). Along the Gulf Coast at Galveston, Texas, we find the larvae of *C. saulcyi* inhabiting almost identical situations, holes of about the same depth, etc., while Dr. Horn (203) describes a different larva in like situations and with like habits on the coast of India.

Still, with all that has been said, matters of agreement of different animal communities in different parts of the world are largely theoretical, and while apparently logically well grounded, the general statement must be treated with due caution and subjected to experimental test as soon as possible. Such testing will involve careful experimental study of the communities of two like environments under rigidly controlled and carefully measured conditions.

VI. Relations of Ecology to Other Biological Subjects

The environmental processes which we are discussing are those in which organisms have existed since their origin on earth. The stresses and strains to which organisms have been subjected have been in the same direction for long periods. Now that we have learned much concerning organic response to environment, such as physiological response, behavior response, and structural response, we note at once that processes of adjustment and equilibration of living substance may bear important relations, on the one hand to environmental processes, and on the other to the physiological aspect of biological phenomena.

Ecological matters are then worthy of the attention of the student of morphology, heredity, and evolution.

What is the significance in the fact that the white tiger-beetle (*Cicindela lepida*) belongs to the first association in the development of a forest community on sand, which we may say corresponds to a family, and to the subterranean ground stratum (corresponding to genus) and to the white tiger-beetle *mores*? Furthermore, that *Cicindela lecontei* and the green tiger-beetle (*Cicindela sexguttata*) belong respectively to different and older situations or associations? We note that the habitats in which the species occur are characterized by distinctly different soils, moisture, amounts of shade and light. We note, furthermore, that these animals are possessed of unusual powers of flight and are able to *select* conditions suited to their physiological constitution. Their *mores* characters are definite characters, which can be measured in terms of reactions to measured complexes of physical and other environmental factors. They are as clearly defined as any morphological taxonomic characters and can be measured with the accuracy of any physical phenomena.

Doubtless to the student of genetics or evolution, the question of the origin of such characters and their fixation in heredity is a leading question. At this point we know little or nothing. Since nearly all species have definite habitat preferences and since many varieties differ slightly from the related species form in the matter of habitat preference, it is probable that origin of a slight change in habitat preference, meaning a slight change in reaction to physical factors, a change in ecological optimum, is usually an early correlative of the origin of new races. Still the so-called taxonomic characters may remain apparently unchanged, while marked changes in habitat preference and in reaction to physical factors are being brought about in plastic animals (56). On the other hand, the segregation in the pure lines and races accomplished in experimental breeding often appears to take place without any regard to environment (204). These two facts, accepted as they stand, are in full accord and we might conclude that there are no relations between primary ecological characters and taxonomic characters. Such, however, can hardly be strictly true, but we cannot see what the real relations may be. If our point of view is correct the *ecological* characters of a race experimentally segregated, or experimentally produced, must in practice consist *primarily* of *reaction to physical factors* or *combinations of physical factors* or to entire environmental complexes; secondly of a definite rate of metabolism, time of appearance or the like;

thirdly of specificity of behavior, and fourthly of structural characters modifying behavior. Relatively fixed taxonomic integumentary characters have no bearing on ecological matters, not even according to the broadest definitions of the subject. The characters which are not related to the environment and which are of no ecological value are the ones quite generally used in breeding work, specificity of behavior standing second, and plastic structure third, *primary ecological matters usually receiving no adequate attention or only such attention as comes incidentally with the handling of the material.* The results consist of noted differences in reaction to light of doubtful intensity and quality, or similar inaccurately measured temperature differences, etc. The testing of primary ecological characters can be easily conducted and will answer the question before us.

With all of its imperfections and uncertainties, the ideas of phylogeny which are presented in our phylogenetic system of taxonomy are an important asset in zoölogical thinking from the point of view of structure and development. The classification which ecologists are striving to build up will serve a purpose in behavior, physiology, and ecology, analogous in this respect to that served by the phylogenetic classification in morphological thought, but should be flexible rather than rigid and true to fact rather than to schemes. Figuratively speaking, an ecological classification cuts taxonomy vertically, showing many structural adaptations as matters of stratum or over-adaptations (205) or lack of adjustment to conditions (206, 206a). It also cuts it again horizontally, showing ecological similarity in organisms structurally and phylogenetically diverse. It therefore provides a new and different means of organization of data.

In this work we have sharply separated evolution and structure, on the one hand, from physiology and behavior, on the other. Space, clearness, and the condition of the subjects have forbidden that we attempt to unite them here. While it may be expedient to continue in this manner until our knowledge of physiology and behavior is commensurate with that of the other subjects, the following of such a course indefinitely, with respect to either morphological or physiological aspects of biology, cannot, if it be general, bring about the best development or unification of biological science. Indeed, its present lack of unity is traceable to such a course followed until recently by zoölogists generally.

If our understanding of the data of physiological cytology be correct, we may expect to find so-called structures of some sort within or among the cells concerned in function, which stand for or are correlated with each physiological state and physiological condition to which we have referred. Our methods may not, at present, be sufficiently delicate to detect such structure, or the processes which lie back of it, but we may, it is believed, confidently expect the necessary methods for the detection of such structures and processes, and especially their correlation with and relation to the more permanent and more easily recognizable morphological conditions.

We classify the responses and changes in animals as evolution, modification by the environment, behavior, and physiological response. Are not all these, after all, but different expressions of the same or similar processes? Future investigations must answer this question, and it is around this question that the future of much that is known as biology hinges.

VII. RELATIONS OF ECOLOGY TO GEOGRAPHY

Ecology is primarily the study of the *mores* of animals and animal communities. It is fundamentally a branch of physiology—the physiology of the relations of animals to their environments. While we may study in the field and in the laboratory, both types of study are commonly conducted with reference to natural environments. Natural environments are used as the basis for study, because when natural environments are destroyed, animals which can live in the new conditions select some one of several possibilities which approach the normal habitat. Habits appear particularly variable under these conditions. Little can be gained from the study of the relations of animals to man-made environments, except in cases where the species has long been living under such conditions and has become fully adjusted to them.

Ecology being a subject or branch of physiology, and including all of the sociological side of animal life, its relations to human geography are particularly intimate. Indeed, geographers have been disappointed with the data which zoölogy has furnished them, as these data are almost exclusively data concerning the taxonomy and morphology of animals. The parallelism between the geographic phenomena in animals and the "relation of culture to environment" lies not in the color and structural adaptations of animals, but in the behavior-characters of animals which enable them to live under a given set of conditions, and the behavior which those conditions produce (207, 208, 209).

While attempting to make comparisons between human society and man on the one hand, and plants and animals on the other, geographers, sociologists, and psychologists—in so far as I have been able to read their writings along this line—have compared structure in plants and animals with what is obviously not structure in man, namely, his culture and mental makeup. Waxwieler (210) compares human society with the whole animal kingdom, as constituting another society. McGee (211) takes a similar position. In discussing the relation of culture to environment he says:

When the law of biotic development is extended to mankind, it appears to fail; for the men of the desert and shore land, mountain and plain, arctic and tropic, are ceaselessly occupied in strife against environmental conditions which transform their subhuman associates; yet men remain essentially unchanged, some taller, some stouter, some swifter of foot, some longer of life than others, yet all essentially *Homo sapiens* in every characteristic.

More careful examination indicates that the failure of the law when extended to man is apparent only. The desert nomads retain certain common physical characteristics, but develop arts of obtaining water and food and these arts are adjusted to the local environment. . . . .

He continues with the citation of other cases. Such adjustment of arts (212) is comparable to the adjustment of animals with regard to food, nest-building, materials used in nest-building, and other features of ecology and behavior. Finally, animal ecology offers the material and methods with which many ideas of geography may be experimentally verified (213, 214).

APPENDIX

METHODS OF STUDY

Methods used in the study of environment, while not new, involve the methods of several sciences. To determine the gross features, the methods of dynamic and historic geology and physiography, or of plant ecology, must be applied. For further analysis the methods of meteorology and special methods for measuring the environment physically and chemically must be employed, where other sciences have given us no data and method (see Clements). These consist of methods of studying the rate of evaporation, water content of the soil, and the application of meteorological methods to climatic details. The special chemical methods, aside from chemical methods of the study of the soil, consist of detection of the presence of excretory products in the soil or water. The best discussion of special methods is given in the references (35a, 43, 69, 74, 76, 77, 117, 118, 121, 124, 125, 129, 130, 131).

METHODS OF STUDYING ANIMALS IN THE FIELD AND LABORATORY

a) Observation.—One important thing in ecological study is simply to sit quietly and watch animals, and record what they do. This requires much time, and the best observers often sit for hours before making the desired observations, but the reward is always adequate. Some good ecological knowledge has thus been acquired. One difficulty is encountered in this work. When the observer is watching one animal whose actions are not of especial interest at that moment another animal often suddenly appears and does something which seems of importance or which is of especial interest. The observer's attention is diverted from its original object of observation. "Which shall I continue to watch?" is often asked by the student. No definite rules can be laid down. In general it is probably better to follow the original object. The answer depends entirely upon the relative ease with which the two animals before the worker can be observed. The beginner cannot answer this question and only experience can decide which should be followed.

b) Experimentation.—Investigation in ecology requires, in preparation, long training in both the biological and physical sciences. Persons not possessing such training cannot hope to make important contributions to the science. Ecology is a field often requiring very complicated experimental methods. Animal behavior and some aspects of physiology are fundamental in ecology. We can sketch out here only such methods as are modifications of the usual method of these branches of biological science in such a way as to be intelligible to those somewhat familiar with such laboratory methods.

(a) Experiments in the field are of prime importance in ecological work. Here smaller animals can be secured in numbers and subjected to experimental conditions before their physiological state has been modified by bad treatment. Any student competent to undertake ecological investigation will find no difficulty in devising apparatus which can be carried into the field and which will enable him to do work of a high degree of scientific accuracy. Each experiment should be accompanied by a control. That is, the same number of animals should be put under the same conditions as in the experiment, except for the one factor which is to be varied. For example, in an experiment designed to determine the reaction of animals to light, the control should be either equally lighted or entirely dark (more easily accomplished), and the experiment which is exactly the same except that the light ranges from darkness to bright sunlight.

The apparatus which we have just begun to develop for this purpose is still in need of much perfecting. Thus far it consists of granite-iron and galvanized-iron containers about 13 in. long, 3 in. deep, and 4 in. wide. These are provided with galvanized-iron covers, somewhat larger, and a little deeper. One of these is provided at one end, with an adjustable slide which may be used to open a slit to admit light when desired. In connection with this slit a mirror is provided with which the sunlight may be projected into the pan as nearly vertically as possible. The rays are allowed to pass through a water screen to cut out the heat. For work with temperature the same receptacles have been used and temperature differences secured by placing one end of the experimental tank in contact with hot soil and the other with cold soil. Land animals are confined in tubes 11 in. long by 1¾ in. in diameter with round bottom and close-fitting cap, shaped like the bottom. Reactions to gravitation have been tested with the use of wire cylinders for land animals, and glass cylinders lined with screen for aquatic animals. Black covers are used to exclude light in various ways as a check. For the study of reactions to current two long galvanized boxes (24×5×4 in.) have been used, one having screen ends and the other tight ends. They are placed in the stream side by side, one serving as an experiment, the other as a control. Large tin pans have been used in connection with the long boxes; the water in the experiment being stirred so as to produce a circular current, while the control is left undisturbed. The study of reactions to contact has been carried on by the use of pans described in connection with light and temperature and with the use of mica chips, leaves, etc.

In all experiments the containers are divided into several divisions and the number of animals noted in each division counted at each reading. About ten readings are taken, the number being determined by the number of animals used, which is determined by the number that can be observed before they can move any considerable distance. This is a function of the speed of movement, which also determines the frequency of reading. Readings should be taken at such intervals as to enable the animals to completely adjust their positions with reference to the conditions in the interim.

The most effective method of study is that of mixing animals of different habitats; this removes the necessity of accurate measurement for rough comparison. The degree of accuracy of such experiments is determined almost entirely by the ingenuity and care exercised by the experimentor. Accuracy of measurement can be acquired, but in the case of some factors, such as light, with some difficulty. Such accuracy should, however, be the constant aim of the worker.

While a high degree of accuracy may be attained in the field in the case of some factors and reactions, it is, in other cases, necessary to perform experiments in the laboratory also. As a rule all experiments should be performed in both field and laboratory.

(*b*) To determine the most important activities: The first step in field observation is the continuous watching of animals throughout a number of life cycles. Experimentation is almost always necessary also. It is only under unusually favorable conditions that the relative importance of the various periods of the life history of an animal can be ascertained without experimentation. On the other hand, experimentation must be correlated with field observation. Simple experimentation on the behavior of animals in the laboratory does not illuminate this matter to any appreciable extent.

To determine the habitat preference of animals, they should be placed in cages, in which they find several different sets of natural conditions, and the selection made by the animal noted.

METHODS OF TAKING A CENSUS

Species are of importance because each usually has a physiological makeup and habitat preference differing from other species. To make a census of the animals present in a given habitat it is necessary to visit the place at various times of day and night and at various times of the year, to overturn and open all loose objects. It is necessary therefore to collect animals which have been observed in nature in such a manner that the correct names can be applied later. It is customary to assign numbers to the animals. The method commonly used is as follows:

Loose sheets of ruled paper are filled in with the locality, date, weather, etc., carbon copies usually being made as a matter of safety and convenience. Next, an animal, say a spider, is observed as fully as time permits, the observations are recorded, and the specimen, if small, is placed in a 4-drachm homeopathic vial containing alcohol. The notes are written in abbreviated form on a slip, and the same number assigned to the notes and to the slip which is put in the bottle. Animals too large to put into bottles are prepared in the same way by tying a tag to each. In due time the bottle is sent to a specialist who assigns the name, which is recorded in a blank space on the note sheet. A new sheet is filled out for each different habitat, and later all the sheets relating to one kind of a situation can be brought together.

Nearly all animals can be sufficiently well preserved to permit identification by specialists, in the following manner:

a) Vertebrates, in 10 per cent formalin, the abdomen opened to permit the fluid to enter.

b) Crustaceans, most insects, spiders, worms, and lower forms by dropping into 80 per cent alcohol.

c) Insect larvae and pupae must be subjected to high temperature, 80° C., or they will turn black. Vials or bottles containing them with corks removed should be set in a pan of hot water for 20 minutes immediately after returning from the field.

d) Flies must be killed by poison fumes, pinned in the field, and the pins set in suitable boxes.

e) Moths and butterflies must be killed by fumes and pinned; the partial spreading of one pair of wings will suffice and save much time.

BIBLIOGRAPHY

[References are numbered in the order of first citation in the text, beginning with Chapter I.]

CHAPTER I

1. Ritter, W. E. The Marine Biological Station of San Diego. Circulated by the Station, La Jolla, Cal. 1910.
2. Brehm, A. E. From North Pole to Equator. London. (Tr. by Thomson.) 1896.
3. Roosevelt, T. African Game Trails. Scribners. 1909–10.
4. Haddon, A. C. The Saving of Vanishing Data. Pop. Sci. Mon., LXII, 222–29, 1903.
5. Webb, Sidney. The Diminution and Disappearance of the Southeast Fauna and Flora. Southeast Nat., VIII, 48. 1903.
5a. Forbes, S. A. The Native Animal Resources of the State. A Symposium on Conservation. Tr. Ill. Ac. Sci., Vol. V. 1912.
6. Shelford, V. E. Ecological Succession. III, A Reconnaissance of Its Causes in Ponds with Particular Reference to Fish. Biological Bull., XXII, No. 1, and citations. 1911.
7. Ruthven, A. G. Variation and Genetic Relationships of the Garter Snakes. U.S. Nat. Mus., Bull. 61. 1908.
8. Beal, F. E. L. The Relations between Birds and Insects. Yearbook of Dept. of Agri. for 1908, pp. 343–50.
9. Surface, H. A. Serpents of Pennsylvania. Penn. State Dept. Agri., Div. Zoöl., IV. Nos. 4 and 5. 1906.
9a. Surface, H. A. Lizards of Pennsylvania. Ibid., V, No. 8. 1908.
10. Kirkland, A. H. The Usefulness of the American Toad. U.S. D. of Agri., Farmers' Bull. 196. 1905.
11. Forbes, S. A. The Food Relations of the Carabidae and Coccinellidae. Bull. Ill. State Lab. Nat. Hist., Bull. No. 6, pp. 33–64. 1883.
12. Warming, E. Ecology of Plants. An Introduction to the Study of Plant Communities, and citations. Oxford. (Tr. by Percy Groom.) 1909.
13. Shelford, V. E. Ecological Succession. V, Aspects of Physiological Classification. Biol. Bull., Vol. XXIII, 331–70. 1912.
14. Kirkland, Joseph. Chicago Massacre of 1812. Chicago. 1893.
15. Blanchard, Rufus. Discovery and Conquests of the Northwest. Chicago. 1898–1900.
16. Report of the Commissioner of Indian Affairs. 1890.
17. Sparks, Jared. Life of Robert Cavalier de LaSalle. Boston. 1848.

18. Parkman, F. LaSalle and the Discovery of the Great West. Boston. 1872.

19. Mason, E. G. Early Chicago and Illinois. Chicago Hist. Soc. Coll. No. IV. Chicago. 1890.

20. Ellsworth, H. L. Illinois in 1837. Philadelphia. 1837.

20a. Reynolds, John. My Own Times. Chicago. 1855.

21. Wood, F. E. A Study of the Mammals of Champaign County, Ill. Bull. Ill. State Lab. Nat. Hist., VIII, Art. V, 501–613. 1910.

22. Kennicott, R. Catalogue of Animals Observed in Cook County, Ill. Trans. Ill. Agri. Soc. I, 557–95. 1855.

23. Jones, A. Illinois and the West. Boston. 1838.

24. Marsh, M. C. The Effects of Some Industrial Wastes on Fishes. U.S. Geol. Surv., Water Supply and Irrigation Paper No. 912 (The Potomac River Basin), pp. 337–48. 1907.

25. Nichols, W. R. Water Supply. New York. 1894.

26. Forbes, S. A. Some Interactions of Organisms. Bull. Ill. State Lab. Nat. Hist., I, 3–18. 1880.

27. Park, W. H. Pathogenic Microorganisms Including Bacteria and Protozoa. Philadelphia. 1905.

27a. McFarland, Jos. The Relation of Insects to the Spread of Disease. Medicine, January, 1902.

28. Hortag, M., and others. Cambridge Natural History. Vol. I. London. 1906.

29. Braun, Max. Animal Parasites of Man. New York. 1906.

30. Darwin, Charles. Vegetable Mould and Earthworms. London. 1892.

31. Kunz, G. F. On Pearls and the Utilization and Application of the Shells in Which They Are Found. Bull. U.S.F.C., 1893, pp. 439–57. 1897.

32. Stevenson, C. H. Aquatic Products in Arts and Industries. Rep. U.S.F.C., p. 177. 1902.

33. Kingsley, J. S. (ed.). Riverside Natural History. II. 1884.

34. Sharp, D. Cambridge Natural History. Vol. VI. London. 1899.

Chapter II

35. Verworn, Max. General Physiology. London. (Tr. by F. S. Lee.) 1899.

35a. Adams, Charles C. Guide to the Study of Animal Ecology. New York. 1913.

36. Parker, T. J., and Haswell, W. A. A Text-book of Zoölogy. Vol. II. London. 1897.

37. Child, C. M. Studies on Regulation, II. Jour. Exp. Zoöl., I, 95–133. 1904.

37a. Child, C. M. Regulatory Processes in Organisms. Jour. Morph., XXII, 171–222. 1911.

38. Osborn, H. F. From the Greeks to Darwin. New York. 1894.

39. Cope, E. D. The Primary Factors of Organic Evolution. Chicago. 1896.

40. Hancock, J. L. Nature Sketches in Temperate America. Chicago. 1911.

41. Eigenmann, C. Adaptation. Fifty Years of Darwinism: Modern Aspects of Evolution. New York. 1909.

42. Standfuss, M. On the Causes of Variation and Aberration in the Imago State of Butterflies. (Tr. by F. A. Dixey.) Entomologist, XXVIII, 69–76, 102–14, 142–50. 1895.

43. Hill, L., Moore, B., Macleod, J. J. R., Pembrey, M. S., and Beddard, A. P. Recent Advances in Physiology and Biochemistry. London. 1908.

44. Jennings, H. S. Behavior of the Lower Organisms. Macmillan. Bibliography. 1906.

45. Mast, S. O. Light and the Behavior of Organisms. New York. 1911.

46. Riddle, O. The Genesis of Fault Bars in Feathers. Biol. Bull., XIV, 328–70. 1908.

47. Johnstone, James. Conditions of Life in the Sea. Cambridge. 1908.

48. Merriam, C. H. Results of a Biological Survey of the San Francisco Mountain Region and the Desert of the Little Colorado, Arizona. U.S. Dept. of Agri., N.A. Fauna, No. 3. 1890.

49. Herrick, F. H. The Home Life of Wild Birds. A New Method of the Study of the Photography of Birds. New York. 1902.

50. Reighard, Jacob. Methods of Studying the Habits of Fishes, with an Account of the Breeding Habits of the Horned Dace. Bull. Bur. Fish, XXVIII, 1112–36. 1910.

51. Semper, K. Animal Life. New York. 1881.

52. Ganong, W. F. Organization of the Ecological Investigation of the Physiological Life Histories of Plants. Bot. Gaz., XLIII, 341–44. 1907.

53. Allee, W. C. An Experimental Analysis of the Relation between Physiological States and Rheotaxis in Isopoda. Jour. Exp. Zoöl., XIII, 269–344. 1912.

53a. Bohn, G. Naissance de l'intelligence. Paris (Bibliothèque de philosophie scientifique). 1910.

53b. Holmes, S. J. Evolution of Animal Intelligence. New York. 1911.

53c. Abbot, C. C. Notes on Fresh-Water Fishes of New Jersey. Am. Nat., IV, 99–117. 1870.

54. Wheeler, W. M. Ants, Their Structure, Development, and Behavior. New York. 1910.

55. Shelford, V. E. Physiological Animal Geography. Jour. of Morph. (Whitman Volume), XXII, 551–618, and citations. 1911.

56. Allee, W. C. Seasonal Succession in Old Forest Ponds. Trans. Ill. Ac. Sci., IV, 126–31. 1911.

57. Salisbury, R. D. Physiography. Henry Holt, New York. 1907.
58. Cowles, H. C. The Plant Societies of Chicago and Vicinity. Bull. II. Geog. Soc. Chicago. 1901. Also Bot. Gaz., XXXI, 73–108, 145–82.
58a. Schimper, A. F. W. Plant Geography upon a Physiological Basis. Oxford. (Tr. by W. R. Fisher.) 1903.

Chapter III

59. Leverett, F. Illinois Glacial Lobe. U.S. Geol. Surv. Monograph, 38. 1899. Maps: p. 420, Chicago and vicinity; p. 340, Southwestern Michigan and Northern Indiana; p. 284, Fox River region.
60. Salisbury, R. D., and Alden, W. C. The Geography of Chicago and Environs. Bull. I. Geog. Soc. Chicago. 1899.
61. Alden, W. C. Chicago Folio; No. 81, Geological Atlas of the United States, U.S. Geol. Surv., Maps. 1901.
62. Atwood, W. W., and Goldthwait, J. W. The Physiography of the Evanston-Waukegan Region. Bull. 7. Ill. Geol. Surv. 1908.
63. Goldthwait, J. W. Abandoned Shorelines of Eastern Wisconsin. Wisconsin Geol. and Nat. Hist. Surv., Bull. No. 17. Sc. Ser. 5. 1907.
64. Goldthwait, J. W. Physical Features of the DesPlaines Valley. Ill. Geol. Surv., Bull. 11. 1909.
65. Lane, A. C. Surface Geology of Michigan. Rep. Geol. Surv. Mich. pp. 98–143. 1907.
66. Chamberlin, T. C., and Salisbury, R. D. Geology, III. Henry Holt, New York. 1907.
67. Adams, C. C. Postglacial Dispersal of North American Biota. Biol. Bull., IX, 53–71. 1905.
68. The Climatology of the United States. U.S. Dept. Agri., Weather Bur., Bull. Q. 1906.
69. Transeau, E. N. The Relation of Plant Societies to Evaporation. Bot. Gaz., XLV, 217–31. 1908.
70. ———. Forest Centers of Eastern America. Am. Nat., XXXIX, No. 468, pp. 875–88. 1905.

Chapter IV

71. Marsh, M. C. Notes on the Dissolved Content of Water in Its Effect on Fishes. Bull. U.S.F.C., 1908. International Fish Congress. 1910.
72. Loeb, J. Dynamics of Living Matter. New York.
73. Shelford, V. E., and Allee, W. C. The Reaction of Fishes to Gradients of Dissolved Atmospheric Gases. Jour. Exp. Zoöl., XIV, 207–66. 1913.
74. Birge, E. A., and Juday, C. The Inland Lakes of Wisconsin; the Dissolved Gases of the Water and Their Biological Significance. Wis. Geol. and Nat. Hist. Surv., Bull. No. 22., Sc. Ser. 7. 1911.

75. Ward, H. B. Biological Examination of Lake Michigan in the Traverse Bay Region. Bull. Mich. Fish. Comm., No. 6. With appendices by Jennings, Walker, Woodworth, and Kofoid. 1897.

76. Forel, F. A. Le Leman, monographie limnologique. In three volumes. Lausanne. 1892–1904.

77. Kofoid, C. A. The Plankton of the Illinois River. Part I. Quantitative Investigations and General Results. Bull. Ill. St. Lab. Nat. Hist., VI, 95–629. 1903.

78. Marsh, C. D. The Plankton of Lake Winnebago and Green Lake. Wis. Geol. and Nat. Hist. Surv., Bull. No. 12, Sc. Ser. 3. 1903.

79. Forbes, S. A., and Richardson, R. E. The Fishes of Illinois. Nat. Hist. Surv. of Ill., Vol. III. (State Lab. Nat. Hist.) 1908.

CHAPTER V

80. Smith, S. I. Sketch of the Invertebrate Fauna of Lake Superior. Rept. U.S.F.C., 1872–73, pp. 690–707. 1874.

81. Milner, J. W. The Fisheries of the Great Lakes. Rep. U.S.F.C., 1872–73, pp. 1–75. 1874.

82. Stimpson, W. Notes on the Deep Water Fauna of Lake Michigan. Am. Nat., IV, 403. 1870.

82a. Hoy, P. R. (On Dredging in Lake Michigan.) Trans. Wis. Acad., I, 100. 1872.

83. Adams, Charles C., and others. An Ecological Survey of Isle Royale, Lake Superior, 1908.

　　1. Adams, Charles C. General, pp. 1–57; Birds, pp. 121–54; Beetles, pp. 157–203; Mammals, pp. 389–96.

　　2. Gleason, H. A. Ecological Relation of the Invertebrate Fauna, p. 57.

　　3. Wolcott, A. B. Supplementary List of Beetles, p. 204.

　　4. Holt, W. P. Notes on Vegetation, p. 217.

　　5. Walker, Bryant. List of Mollusca, p. 281.

　　6. Morse, A. P. Report on Orthoptera, p. 299.

　　7. Needham, J. G. Neuropteroid Insects, p. 305.

　　8. Hine, J. S. Diptera, p. 308.

　　9. Titus, E. S. Hymenoptera, p. 317.

　　10. Wheeler, W. M. Ants, p. 325.

　　11. Ruthven, A. G. Cold-blooded Vertebrates, p. 329.

　　12. Peet, Max, M. Birds, pp. 97 and 337.

84. Meek, S. E., and Hildebrand, S. F. Synoptic List of the Fishes Known to Occur within Fifty Miles of Chicago. Field Mus. Nat. Hist., Pub. 142. Zoöl. Ser. 7, No. 9. 1910.

85. Forbes, S. A. The Lake as a Microcosm. Bull. Peoria Sci. Assoc., 1887.

86. Snow, Julia W. The Plankton Algae of Lake Erie. Bull. U.S.F.C., XXII, 371. 1902.
87. Jennings, H. S. The Rotatoria of the United States. Bull. U.S.F.C., XX, 67–104. 1900.
88. ———. On the Protozoa of Lake Erie. *Ibid.*, XIX, 105. 1899.
89. Forbes, S. A. Some Entomostraca of Lake Michigan and Adjacent Waters. Am. Nat., XVI, 640. 1882.
90. Clark, F. N. A Plan for Promoting the White Fish Production of the Great Lakes. Bull. Bur. Fish., XXVIII, 637–43. 1910.
91. Baker, F. C. The Mollusca of the Chicago Area. Bull. III, Chicago Acad. Sci. In 2 parts. 1898–1902.
91a. Moore, J. P. Classification of the Leeches of Minnesota. Geol. and Nat. Hist. Surv., Zoöl. Series No. 5, pp. 67–128. 1912.

CHAPTER VI

92. Shelford, V. E. Ecological Succession. I, Stream Fishes and the Method of Physiographic Analysis. Biol. Bull., XXI, 9–25. 1911.
93. Smith, B. G. The Spawning Habits of Chrosomus erythrogaster. Biol. Bull., XV, 9–18. 1908.
94. Lyon, E. P. On Rheotropism. I, Rheotropism in Fishes. Am. Jour. Phys., XII, 149–61. 1904.
95. Needham, J. G., and others. Aquatic Insects in the Adirondacks. N.Y. State Mus., Bull. 47. 1901.
96. ——— Aquatic Insects in New York. N.Y. State Mus., Bull. Entomology, 18. 1903.
97. Reeves, C. D. The Breeding Habits of the Rainbow Darter. Biol. Bull., XIV, 35. 1907.
98. Needham, J. G., and others. May Flies and Midges. N.Y. State Mus., Bull. Entomology, 23. 1905.
99. Lefevre, George, and Curtis, W. C. Reproduction and Parasitism in the Unionidae. Jour. Exp. Zoöl., IX, 79–115. 1910.
99a. Isely, F. B. Preliminary Note on the Ecology and Juvenile Life of the Unionidae. Biol. Bull., XX, 77–80. 1911.
99b. Kingsly, J. S., Editor. Riverside Natural History, Vol. V. Mammals. 1888.
99c. Sherman, J. D. Some Habits of the Dytiscidae. Jour. N.Y. Ent. Soc., XXI, 43–54.
100. Baker, F. C. The Ecology of Skokie Marsh with Particular Reference to Mullusca. Bull. Ill. State Lab. Nat. Hist., VIII, 441–97. 1910.
101. Ortmann, A. E. The Crawfishes of the State of Pennsylvania. Mem. Carn. Mus. Pittsburgh, II, 343–523. 1907.
101a. Pearse, A. S. The Crawfishes of Michigan. Mich. Geol. and Biol. Surv. Pub. Biol. Series No. 1, pp. 9–22. 1910.

102. Weckel, A. L. Freshwater Amphipods of North America. Proc. U.S. Nat. Mus., 1907, pp. 25–58.
103. Adams, Charles C. Baseleveling and Its Faunal Significance. Am. Nat., XXXV, 839–52. 1901.

CHAPTER VII

104. Juday, Chauncey. Diurnal Movement of Plankton Crustacea. Tr. Wis. Acad. Sci. Arts and Letters, XIV, 524–68. 1904.
105. Hankinson, T. L. Walnut Lake. Biol. Surv. of Michigan (Lansing). Rep. Geol. Surv., 1907, pp. 157–271.
106. Gill, T. Parental Care among Freshwater Fishes. Smithsonian Report for 1905, pp. 403–531. 1907.
107. Newman, H. H. The Habits of Certain Tortoises. Jour. Comp. Neur., XVI, 126. 1906.
108. Butler, A. W. Birds of Indiana. Rep. Ind. Dept. Geol. and Nat. Resources, XX, p. 515. 1897.
109. McGillivray, A. D. Aquatic Chrysomelidae. N.Y. State Mus. Bull., LXVIII, pp. 288–312. 1903.
110. Juday, C., and Wagner, George. Dissolved Oxygen as a Factor in the Distribution of Fishes. Wis. Acad. Sci. Arts and Letters, XVI, Part I. 1908.
111. Juday, C. Some Aquatic Invertebrates That Live under Anaerobic Conditions. *Ibid.* 1908.

CHAPTER VIII

112. Shelford, V. E. Ecological Succession. II, Pond Fishes. Biol. Bull., XXI, 127–51. 1911.
113. Titcomb, J. W. Aquatic Plants in Pond Culture. Rep. U.S. Bur. Fish. 1908.
114. Colton, H. S. Some Effects of Environment on Growth of Lymnaea Columella Say. Proc. Ac. Nat. Sci., Philadelphia, pp. 410–48. 1908.
114a. Dachnowski, A. The Toxic Properties of Bog Water and Bog Soil. Bot. Gaz., XLVI, 130. 1908.

CHAPTER IX

115. Shelford, V. E. Ecological Succession. IV, Vegetation and the Control of Land Animal Communities. Biol. Bull., XXIII, No. 2, pp. 5–99. 1912.
116. Van Hise, C. R. A Treatise on Metamorphism. U.S.G.S. Monograph, XLVII. 1904.
117. Briggs, L. J., and McLane, J. W. The Moisture Equivalents of Soils. Bull. 45, Bureau of Soils, U.S. Dept. Agri. 1907.

118. Briggs, L. J., and Shantz, H. L. The Wilting Coefficients for Different Plants and Their Indirect Determination. Bull. 230, Bureau of Plant Industry, U.S. Dept. Agri. 1912.

119. Fuller, G. D. Soil Moisture in the Cottonwood Dune Association of Lake Michigan. Bot. Gaz., LIII, 512–14. 1912.

119a. McNutt, W., and Fuller, G. D. The Range of Evaporation and Soil Moisture in the Oak-Hickory Forest Association of Illinois. Trans. Ill. Acad. Sci. 1912.

120. Cowles, H. C. The Causes of Vegetational Cycles. Bot. Gaz., XLI, 161–83. Also Ann. Ass. Am. Geog., Vol. I. 1911.

121. Schreiner, O., and Reed, H. S. Some Factors Influencing Soil Fertility. U.S. Dept. Agri., Bull. Bur. Soils, 40. 1907.

122. Transeau, E. N. The Bogs and Bog Flora of the Huron River Valley. Bot. Gaz., XL, 351–428. 1906.

123. Congdon, E. D. Recent Studies upon the Locomotor Responses of Animals to White Light. Jour. Comp. Neur. and Psych., XVIII, 309–28. 1908.

124. Zon, R., and Graves, H. S. Light in Relation to Tree Growth. U.S. Dept. Agri., Forest Service, Bull. 92. 1911.

125. Hann, J. Hand Book of Climatology. Part I. (Tr. by R. de C. Ward.) New York. 1903.

126. Cohnheim, O. Physiologie des Alpinismus II. Ergebnisse der Physiologie, Bd. 12. 1912.

127. Huntington, E. The Effect of Barometric Variation upon Mental Activity. Preliminary Program 8th Ann. Meeting of the Ass. Am. Geographers. 1911. See Annals of the same society.

128. Walker, A. C. Atmospheric Moisture as a Factor in Distribution. S.E. Nat., VIII, 43–47. 1903.

129. Yapp, R. H. Stratification of the Vegetation of a Marsh and Its Relation to Evaporation and Temperature. Ann. Bot., XXIII, 275–319. 1909.

130. Livingston, B. E. The Relation of Desert Plants to Soil Moisture and to Evaporation. Carnegie Inst. of Wash., Publication 50. 1906.

130a. Evaporation and Plant Habitats. Plant World, XI, 1–10. 1908.

130b. A Rain-Correcting Atmometer for Ecological Instrumentation. Plant World, XIII, 79–82. 1910.

130c. Operation of the Porous Cup Atmometer. Plant World, XIII, 111–19. 1910.

131. Fuller, G. D. Evaporation and Plant Succession. Bot. Gaz., LVII, 195–208. 1911.

131a. —— Evaporation and Stratification of Vegetation. Bot. Gaz., LIV, 424–26. 1912.

131b. Fuller, G. D., and others. The Stratification of Atmospheric Humidity in the Forest. Trans. Ill. Acad. Sci., VI.

132. Greeley, A. W. On the Analogy between the Effect of Loss of Water and Lowering of Temperature. Am. Jour. Phys., VI, No. 2. 1901.

133. Bachmetjew, P. Ueber die Temperature der Insekten nach Beobachtung in Bulgarien. Zeit. f. wiss. Zool., LXVI, 521–604. 1899.

134. Shelford, Victor E. Reactions of Certain Animals to Gradients of Evaporating Power of Air. A Study in Experimental Ecology. With a Method of Establishing Evaporation Gradients by V. E. Shelford and E. O. Deere. Biol. Bull. July, 1913.

135. Shimek, B. The Prairies. Bull. Lab. Nat. Hist. State Univ. Iowa, April, 1911, pp. 169–240.

136. Sherff, E. E. The Vegetation of Skokie Marsh. Bot. Gaz., LIV, pp. 415–35. 1912.

137. Felt, E. P. Insects Affecting Park and Woodland Trees. N.Y. State Mus., Mem. VIII, 2 vols. 1906.

CHAPTER X

138. Emerton, J. H. Common Spiders. Boston. 1902.

139. Dickerson, M. C. The Frog Book. New York. 1907.

140. Hine, J. S. Habits and Life Histories of Some Flies of the Family Tabanidae. U.S.D.A., Div. Ent., T.S., 12. 1906.

141. Woodruff, F. M. Birds of the Chicago Area. Bull. VI. N.H. Surv., Chicago Academy of Sciences. 1907.

142. Merriam, C. H. Mammals of the Adirondack Region. New York. 1886.

143. Seton, E. Thompson-. Life Histories of Northern Animals. New York. 1909.

144. Pearl, R. The Movements and Reaction of Fresh Water Planarians. Qr. Jour. Micro. Sci., XLVI, 509–714. 1903.

145. Smith, J. B. Mosquitos Occurring in the State and Their Habits, Life History, etc. Rep. N.J. Exp. Sta., 1904.

146. Marsh, C. D. A Revision of the N.A. Species of Cyclops. Trans. Wis. Ac. Sci. Arts and Letters, XVI, 1067–1134. 1910.

146a. ———. A Revision of the N. A. Species of Diaptomus. *Ibid.*, XV, 380–516. 1907.

147. Sharpe, R. W. A Further Report on the Ostracoda of the United States National Museum. Proc. U.S. Nat. Mus., XXXV, 399–430.

148. Holmes, S. J. Description of a New Species of Branchipus from Wisconsin, with Observations on Its Reactions to Light. Wis. Ac. Sci. Arts and Letters, XVI, 1252–55. 1910.

149. Wolcott, R. H. A Review of the Genera of the Water Mites. Trans. Am. Micro. Soc., pp. 161–243. 1905.

150. Lugger, Otto. Bugs Injurious to Cultivated Plants. Bull. 69, Minn. Agri. Exp. Sta. 1900.

151. Shelford, V. E. Life Histories and Larval Habits of the Tiger Beetles (Cicindelidae). Linn. Soc. Jour. Zoöl., XXX, 157–84. 1909.

151a. Shelford, Victor E.   The Life-History of a Bee-Fly (*Spogostylum anale* Say) Parasite of the Larva of a Tiger Beetle (*Cicindela scutellaris* Say var. *Lecontei* Hald.).   Ann. Ent. Soc. of Am., VI, 213–25.   1913.

152. Ruthven, A. G., Thompson, C., and Thompson, H.   The Herpetology of Michigan.   Mich. Geol. and Biol. Survey Pub. 10, Biological Series 3.   1912.

153. Reed, C. A.   Bird Guide; Birds East of the Rockies.   Worcester, Mass.   1908.

CHAPTER XI

154. Packard, A. S.   Insects Injurious to Forest and Shade Trees.   U.S. Ent. Com. Bull. 7.   1881.

155. Lugger, O.   Beetles Injurious to Fruit-producing Plants.   Bull. 66, Minn. Agri. Exp. Sta.   1899.

156. Blatchley, W. S.   On the Coleoptera Known to Occur in Indiana.   Bull. I, Ind. Dept. Geol. and Nat. Res.   1910.

157. Ditmars, R. L.   The Reptile Book.   New York.   1904.

157a. Fowler, H. W.   The Amphibians and Reptiles of New Jersey.   Report N.J. Museum, 1906.

158. Jones, F. M.   Pitcher-Plant Insects.   Ent. News, XV, 14.   1904.

159. Banks, N.   Catalogue of Nearctic Spiders.   U.S. Nat. Mus. Bull. 72. 1910.

160. Hopkins, A. D.   Report on Investigations to Determine the Cause of Unhealthy Conditions of Spruce and Pine from 1880–93.   Bull. 56, W.Va. Agri. Exp. Station.   1899.

161. ———.   Insect Enemies of Forest Reproduction.   Year Book of U.S. Dept. Agri., pp. 249–56.   1905.

162. Stone, W., and Cram, W. E.   American Animals.   New York.   1905.

163. Lugger, Otto.   Lepidoptera of Minnesota.   4th Ann. Rep. State Exp. Sta. Minn.   1899.

CHAPTER XII

164. Folsom, J. W.   Insect Pests of Alfalfa and Clover.   25th Ann. Rep. State Ent. Ill., pp. 41–123; also Exp. Sta. Bull. No. 134.   1909.

165. Williston, S. W.   Manual of North American Diptera.   New Haven. 1908.

166. Surface, H. A.   Lampreys of Central New York.   Bull. U.S.F.C. 1897, pp. 209–15.   1898.

167. Snow, Laetitia M.   The Microcosm of the Drift Line.   Am. Nat., XXXVI, 855–64.   1902.

168. Needham, J. G.   The Beetle Drift on Lake Michigan.   Canadian Entomol., p. 294.   1904.

169. Herms, W. B.   An Ecological and Experimental Study of the Sarcophagidae with Relation to Lake Beach Débris.   Jour. Exp. Zoöl., IV, 45.   1907.

170. Shelford, V. E. Preliminary Note on the Distribution of the Tiger Beetles and Its Relation to Plant Succession. Biol. Bull., XIV, pp. 9–14. 1907.

171. Scudder, S. H. Butterflies of Eastern United States and Canada. 3 vols. Cambridge , Mass. 1889.

172. Banks, Nathan. Spiders of Indiana. 31st Rep. Ind. Dept. Geol. and Nat. Res., pp. 715–49. 1906.

173. Peckham, G. W., and E. G. Instincts and Habits of Solitary Wasps. Wis. Geol. and Nat. Hist. Surv., Bull. No. 2, Scientific Series 1. 1898.

174. Forbes, S. A. Insect Injuries to Indian Corn. 23d Rep. Ill. State Entomologist. 1905.

175. Shull, C. H. The Life History and Habits of *Anthocharis olympia*. Edw. Ent. News, XIX, March, 1907.

176. Hart, C. A., and Gleason, H. A. Biology of the Sand Areas of Illinois. Bull. Ill. State Lab., VII, Art. VII, pp. 137–272. 1907.

177. Smith, J. B. Insects of New Jersey. (27th Rep. of State Board of Agric.) 2d ed., Rept. N.J. State Museum, 1909.

178. Marlatt, C. L. The White Ant. U.S. Dept. of Agri., Div. of Entomology. Circular 50, 2d series. 1902.

179. Howard, L. O. Insect Book. New York. 1902.

180. Ruthven, Alex. Amphibians and Reptiles. A Biological Survey of the Sand Dune Region on the South Shore of Saginaw Bay. Mich. Geol. and Biol. Surv., Publication 4, Biol. Ser. 2, p. 257. 1911.

180a. Baker, H. B. Mollusca: Biological Survey of the Sand Dune Region on the South Shore of Saginaw Bay. *Ibid.*, Ruthven, p. 121. 1911.

181. Robertson, C. Flowers and Insects. Bot. Gaz., XXVIII, 27–45. 1899.

182. Richardson, H. Monograph on the Isopods of North America. Bull. 54, U.S. Nat. Mus. 1905.

183. Bollman, C. H. The Myriapoda of North America. U.S. Nat. Mus., Bull. 46. 1893.

184. Weed, C. M. A Descriptive Catalogue of the Phalanginae of Ill. Bull. Ill. State Lab., III, 79–87. 1887.

185. Wirtner, P. M. Preliminary List of the Hemiptera of Western Pennsylvania. Ann. Carnegie Mus., III, 133–228. 1904.

186. Kirkaldy, G. W. Catalogue of the Hemiptera, Heteroptera. Vol. I, Cimicidae. Berlin. 1909.

187. Peckham, G. W., and E. G. Sense of Sight in Spiders with Some Observations on Color Sense. Wis. Ac. of Sci., X, 231–61. 1895.

188. Beutenmüller, Wm. Insect Galls within 50 Miles of New York. Guide

CHAPTERS XIII TO XV

Leaflet, No. 16, Am. Mus. Nat. Hist. 1904.

189. Forbes, S. A. 24th Report Ill. State Entomologist. 1906.

190. Washburn, F. L. Diptera of Minnesota. 10th Ann. Rep. State Ent., Minn. 1905.

191. Judd, Sylvester D. Birds of a Maryland Farm. U.S. D. Agr., Biol. Surv., Bull. 17. 1902.
192. Heilprin, A. The Distribution of Animals. Appleton. 1887.
193. Cowles, H. C. A Textbook of Botany, Part III, "Ecology." New York. 1911.
194. Hickson, S. J. On the Species of the Genus *Millepora*. P.Z.S. London, pp. 246, 257. 1898.
195. Wood-Jones, F. Coral and Atolls. London. 1910.
196. Woodruff, L. L. Observations on the Origin and Sequence of the Protozoan Fauna of Hay Infusions. Jour. Exp. Zoöl., XII, 205–64. 1912.
197. Clements, F. E. Research Methods in Ecology. Lincoln, Neb. 1905.
198. Whitford, H. N. The Vegetation of the Lamoa Forest Reserve. Philippine Jour. of Sci., I, No. 4, pp. 373–428; No. 6, pp. 437–682. 1906.
199. Beddard, F. E. Zoögeography. Cambridge. 1895.
200. Lydekker, A. Natural History, II and III. Mammals. Bears no date.
201. Selous, F. C. African Nature Notes and Reminiscences. London. 1908.
202. Ward, T. Rambles of an Australian Naturalist. 1907.
203. Horn, W. Entomologische Reise Brief aus Ceylon. Deut. Ent. Zeit., 228–30. 1899.
204. Cockerel, T. D. A. Aspects of Modern Biology. Popular Science Monthly, December, pp. 540–48. 1908.
205. Coulter, J. M. The Theory of Natural Selection from the Standpoint of Botany. Fifty Years of Darwinism, 56–71. 1908.
206. Wallace, A. R. Malay Archipeligo. London. 1869.
206a. Hudson, W. H. The Naturalist in La Plata. (Ed. of 1903, Dent, London.) 1892.
207. Craig, Wallace. North Dakota Life; Plant, Animal and Human. Am. Bull. Geog. Soc., XL, 321–415. Bibliography. 1908.
208. ———. The Voices of Pigeons Regarded as a Means of Social Control. Am. Jour. Sociol., XIV, 86–100. 1908.
209. Tarde, Gabriel. Inter-Psychology. Internat. Quar., VII, 59–84. 1903.
210. Waxweiler, E. Equisse d'une sociologie. Inst. Solvay de Soc. Notes et Mem., Fasc. 2, 306, Bruxelles. 1906.
211. McGee, W. J. The Relation of Institution to Environment. Smithson. Rep., 1895, pp. 701–11. 1896.
212. Mason, O. T. Influence of Environment upon Human Industries or Arts. Smithson. Rep., 1895, pp. 639–65. 1896.
213. Tower, W. S. Scientific Geography; the Relation of Its Content to Its Subdivisions. Bull. Am. Geog. Soc., XLII, 801. 1910.
214. Goode, J. Paul. Human Response to the Physical Environment. Jour. Geog., pp. 333–43. 1894.

BIBLIOGRAPHICAL APPENDIX
(March 4, 1937)

CLASSIFICATION OF COMMUNITIES

Shelford, V. E. 1932. Basic Principles of the Classification of Communities and Habitats and the Use of Terms. Ecology, 13:105–20.

The classification of communities used in this book was based on physical conditions and physiological response. Since 1913 this viewpoint has been abandoned because no experiments have been conducted on more important species of most communities to support a physiological view. The present-day nomenclature is based on the important organisms. The equivalents are as follows:

Animal Communities	Present-Day
Extensive formation	Biotic formation or Biome
Formation	Associes
Association except three noted below	Associes
Subformation	Associes
Stratum	Layer Socies or Society
Consocies	Assembly (provisional)
Mores	Mores or life-habit

The following biomes are recognized:

1. Deciduous forest biome; two associations (climax) are present—beech-maple-(wood frog) and oak-hickory-(green tiger-beetle).

2. The grassland or prairie biome; only one association represented.

3. Large lake biome; the deeper communities of Lake Michigan are sometimes regarded as a biome in a permanent or climax condition. In the case of these climaxes the term "society" is applied to subordinate groups of animals. For the nomenclature applied to organisms of various degrees of importance, etc., see the next citation.

Shelford, V. E., and Olson, S. 1935. Sere, climax, and influent animals with special reference to the transcontinental coniferous forest. Ecology 16:375–402.

CHAPTER I

1. HUMAN ECOLOGY

Murchison, C. 1935. A handbook of social psychology. 1095 pp. Worcester.

Read, C. 1920. The origin of man and of his superstitions. 350 pp. Cambridge.

2. SECONDARY COMMUNITIES

Van Deventer, W. C. 1936. Bird and mammal communities of pastures and field borders in northern Illinois. Bull. Ecol. Soc. Amer. 17:28.

3. GENERAL

Bird, R. D. 1930. Biotic communities of the aspen parkland of central Canada. Ecology 11:358–442.

Downing, Elliott R. 1922. A naturalist in the Great Lakes region. University of Chicago Press. 328 pp. Chicago.

Ewing, H. E. 1909. A system and biological study of the *Acarina* of Illinois. Univ. Ill. Studies 3 (6):359–472, with 8 pls.

Forbes, Stephen A., and Gross, A. O. 1922. The numbers and local distribution in summer of Illinois land birds. Bull. Ill. Nat. Hist. Surv. 4:187–218.

Frison, T. H., and Miller, R. B. 1926. *See* Shelford (1926).

Hebard, Morgan. 1934. The Dermaptera and Orthoptera of Illinois. Ill. State Nat. Hist. Surv. Bull. 20:125–279; chapter by H. H. Ross: "Biology and habits of the orders, and ecological factors affecting Orthoptera," pp. 125–35.

Sanborn, Colin C. 1922. Chicago winter birds. Field Mus. Nat. Hist., Zoöl. Lflt. No. 2. 12 pp.

———. 1925. The mammals of the Chicago area. *Ibid.* No. 8. 21 pp.

Shelford, V. E. 1915. The original habitat and distribution of our native insect pests. Jour. Econ. Ent. 8:171–74.

——— 1926. Naturalists' Guide to the Americas. Baltimore, by various authors; "Illinois," p. 469.

Van Cleave, H. J. 1927. A study of the characters for the identification of the snakes of Illinois. Trans. Ill. State Acad. Sci. 20:133–36 (41 spp. enumerated).

CHAPTER IV

AQUATIC CONDITIONS

Shelford, V. E. 1918. Conditions of existence. Ward and Whipple, Freshwater Biology, chap. ii, 1111 pp. New York.

CHAPTER V

LARGE LAKES

Adamstone, F. B. 1924. The distribution and economic importance of the bottom fauna of Lake Nipigon with an appendix on the bottom fauna of Lake Ontario. Univ. Toronto Studies. "Biol. Series" 25:35–100.

Clements, F. E., and Shelford, V. E. 1937. Bio-ecology. In Press.

Eddy, S. 1927. Plankton of Lake Michigan. Bull. Ill. Nat. Hist. Surv. 17:203–32.

———. 1932. The plankton of the Sangamon River in the summer of 1929. Ill. St. Nat. Hist. Surv. 19:469–86.

———. 1934. A study of freshwater plankton communities. Ill. Biol. Mono. 12:1–93.

Eggleton, F. E. 1935. The deep water bottom fauna of Lake Michigan. Mich. Acad. Arts Sci. and Let. 21:599–612.

Hubbs, C. L. 1926. Check list of the fishes of the Great Lakes and tributary water. Univ. Mich. Mus. Zoöl. Misc. Pub. 15:1–77.

CHAPTERS VI AND VII

STREAMS AND PONDS

Baker, F. C. 1928. Freshwater Mollusca of Wisconsin. Wis. Geol. and Nat. Hist. Surv. Bull. in two parts. 70:1–494 and 1–482.

Cahn, A. R. 1929. The effect of carp on a small lake. The carp as a dominant. Ecology 10:271–74.

Frison, T. 1929. Fall and winter stoneflies or Plecoptera of Illinois. Ill. Sta. Nat. Hist. Surv. Bull. 18:345–409.

———— 1935. The stoneflies, or Plecoptera, of Illinois. *Ibid.* 20:281–471.

Peterson, W. 1926. Seasonal succession in a chara-cattail pond. Ecology 7:371–78.

Richardson, R. E. 1928. The bottom fauna of the middle Illinois River, 1913–1925: Its distribution, abundance, valuation, and index value in the study of stream pollution. Ill. Sta. Nat. Hist. Surv. Bull. 17:391–472.

Shelford, V. E. 1914. An experimental study of the behavior agreement among animals. Biol. Bull. 26:294–315.

Shelford, V. E. and Eddy, S. 1929. Methods for the study of stream communities. Ecology 10:382–91.

Smith, F. 1921. Distribution of freshwater sponges in the United States. Bull. Ill. Sta. Nat. Hist. Surv. 14:11–31.

Thompson, D. H., and Hunt, F. D. 1930. The fishes of Champaign County. Ill. Sta. Nat. Hist. Surv. 19:5–101.

Van Cleave, H. J. 1927. The fairy shrimps of Illinois. Trans. Ill. State Acad. Sci. 20:130–32.

Van Cleave, H. J., and Markus, H. C. 1929. Studies on the life history of the blunt-nosed minnow. Amer. Nat. 53:530–39.

Welch, P. S. 1935. Limnology. New York.

CHAPTERS XII AND XIII

Beal, Geoffrey. 1935. Study of Arthropod population by the method of sweeping. Ecology 16:216–25.

Blake, I. H. 1931. A comparison of the animal communities of coniferous and deciduous forest. Ill. Biol. Mon. 10:371–520.

Goellner, E. J. 1931. A new species of termite, *Reticulitermes arenincola*, from the sand dunes of Indiana and Michigan, along the shores of Lake Michigan. Proc. Ent. Soc. Washington 33(9):227–34.

Holmquist, A. M. 1926. Studies in arthropod hibernation. I. Ecological survey of hibernating species from forest environments of the Chicago region. Ann. Ent. Soc. Amer. 19:395–428.

Hubbell, Theodore H. 1929. The distribution of the beach grasshoppers. *Trimerotropis huroniana* and *Trimerotropis maritima interior* in the Great Lakes region (Orthoptera, Acrididae). Jour. N.Y. Ent. Soc. 37:31–39.

Park, Orlando. 1929. Taxonomic studies in Coleoptera, with notes upon certain species of beetles in the Chicago area. I. Jour. N.Y. Ent. Soc. 37:429–36.

———. 1929. Ecological observations upon the myrmecocoles of *Formica ulkei* Emery, especially *Leptinus testacus* Mueller. Psyche 36:195–215.

———. 1929. *Reticulitermes tibialis*, Banks, in the Chicago area. Proc. Ent. Soc. Wash. 31:121–26. From upper beach, Indiana dunes; behavior, life hist., etc.

———. 1930. Studies in the ecology of forest Coleoptera (I). Ann. Ent. Soc. Amer. 23:57–80.

———. 1931a. Studies in the ecology of forest Coleoptera. II. Species associated with fungi in the Chicago area. Ecology 12:188–207.

———. 1931b. The measurement of daylight in the Chicago area and its ecological significance. Ecol. Mono. 1:189–230.

Pearson, J. F. W. 1933. Studies on the ecological relations of bees in the Chicago region. Ecol. Mono. 3:373–441.

Sanders, N. J., and Shelford, V. E. 1922. A quantitative and seasonal study of a pine dune animal community. Ecology 3:306–20.

Smith, Vera G. 1928. Animal communities of a deciduous forest succession. Ecology 9:479–500. (6 different habitats selected in a strip mine area.)

Smith-Davidson, Vera G. 1930. The tree layer society of the maple-redoak climax forest. Ecology 11:601–6.

———. 1932. Effect of seasonal variability upon animal species in total populations in a deciduous forest succession. Ecol. Mono. 2:305–33.

Strokecker, H. F. 1937. A survey of soil temperatures in the Chicago area. Ecology 18:162–68.

———. 1937. An ecological study of some Orthoptera in the Chicago area. *Ibid.* No. 2 (in press).

Talbot, Mary. 1934. Distribution of ant species in the Chicago region with special reference to ecological factors and physiological toleration. Ecology 15:416–39.

Weese, A. O. 1925. Animal ecology of an Illinois elm-maple forest. Ill. Biol. Mono. 9:345–438. (University Woods, 5 mi. from Urbana.)

CHAPTERS XIV AND XV

Adams, Charles C. 1915. An outline of the relations of animals to their inland environments. Bull. Ill. Sta. Lab. Nat. Hist. 11:3–32.

Adams, Charles C. 1915. An ecological study of prairie and forest inverte-brates. Bull. Ill. Sta. Lab. Nat. Hist. 11:33–280. *See* Hankinson (1915).

Eifrig, C. W. G. 1913. Notes on some of the rarer birds of the prairie part of the Chicago area. The Auk 30:236–40.

———. 1919. The birds of the sand dunes of northwestern Indiana. Proc. Ind. Acad. Sci. (1918), 289–303.

———. 1919. Notes on birds of the Chicago area and its immediate vicinity. The Auk 36:513–24.

Flint, W. P. 1934. The automobile and prairie wild life. Ill. State Nat. Hist. Surv., Biol. notes No. 3. 8 pp., mimeographed. (Destruction of animals on highway east of Urbana.) Univ. of Ill. Library.

Hankinson, T. L. 1915. The vertebrate life of certain prairie and forest regions near Charleston, Illinois. Bull. Ill. Sta. Lab. Nat. Hist. 11:281–303. *See* Adams (1915).

Hendrickson, G. O. 1930. Studies on the insect fauna of Iowa prairies. Iowa State Coll. Jour. of Sci. 4:49–180.

——— 1930. Notes on vertebrates of Iowa prairies. Proc. Iowa Acad. Sci. 37:398–99.

Shackleford, Martha W. 1929. Animal communities of an Illinois prairie. Ecology 10:126–54.

PRAIRIE AND FOREST EDGE

Carpenter, J. R. 1935. Fluctuation of biotic communities. I. Prairie and forest ecotone of Central Illinois. Ecology 16:203–12.

METHODS

Forbes, Stephen A. 1928. Concerning certain ecological methods of the Illinois Natural History Survey. Trans. Ill. Sta. Acad. Sci. 21:19–25 (1929). Earlier version in brief in Science 66:405–6. 1927.

Shelford, V. E. 1930. Ways and means of improving the quality of investiga-tion and publication in animal ecology. Ecology 11:235–237.

———. 1929. Laboratory and Field Ecology. 608 pp. Baltimore.

INDEXES

INDEX OF AUTHORS AND COLLABORATORS

Page numbers followed by figures in parentheses are the pages of the Bibliography, the parenthetical figures being the title numbers; the numbers following the parentheses are the pages on which the articles are cited by number. Page numbers occurring with no parenthesis in connection are those on which the authors and collaborators are referred to independently of the Bibliography.

INDEX OF SUBJECTS

HISTORY OF ECOLOGY
An Arno Press Collection

Abbe, Cleveland. **A First Report on the Relations Between Climates and Crops.** 1905

Adams, Charles C. **Guide to the Study of Animal Ecology.** 1913

American Plant Ecology, 1897-1917. 1977

Browne, Charles A[lbert]. **A Source Book of Agricultural Chemistry.** 1944

Buffon, [Georges-Louis Leclerc]. **Selections from Natural History, General and Particular, 1780-1785.** Two volumes. 1977

Chapman, Royal N. **Animal Ecology.** 1931

Clements, Frederic E[dward], John E. Weaver and Herbert C. Hanson. **Plant Competition.** 1929

Clements, Frederic Edward. **Research Methods in Ecology.** 1905

Conard, Henry S. **The Background of Plant Ecology.** 1951

Derham, W[illiam]. **Physico-Theology.** 1716

Drude, Oscar. **Handbuch der Pflanzengeographie.** 1890

Early Marine Ecology. 1977

Ecological Investigations of Stephen Alfred Forbes. 1977

Ecological Phytogeography in the Nineteenth Century. 1977

Ecological Studies on Insect Parasitism. 1977

Espinas, Alfred [Victor]. **Des Sociétés Animales.** 1878

Fernow, B[ernhard] E., M. W. Harrington, Cleveland Abbe and George E. Curtis. **Forest Influences.** 1893

Forbes, Edw[ard] and Robert Godwin-Austen. **The Natural History of the European Seas.** 1859

Forbush, Edward H[owe] and Charles H. Fernald. **The Gypsy Moth.** 1896

Forel, F[rançois] A[lphonse]. **La Faune Profonde Des Lacs Suisses.** 1884

Forel, F[rançois] A[lphonse]. **Handbuch der Seenkunde.** 1901

Henfrey, Arthur. **The Vegetation of Europe, Its Conditions and Causes.** 1852

Herrick, Francis Hobart. **Natural History of the American Lobster.** 1911

History of American Ecology. 1977

Howard, L[eland] O[ssian] and W[illiam] F. Fiske. **The Importation into the United States of the Parasites of the Gipsy Moth and the Brown-Tail Moth.** 1911

Humboldt, Al[exander von] and A[imé] Bonpland. **Essai sur la Géographie des Plantes.** 1807

Johnstone, James. **Conditions of Life in the Sea.** 1908

Judd, Sylvester D. **Birds of a Maryland Farm.** 1902

Kofoid, C[harles] A. **The Plankton of the Illinois River, 1894-1899.** 1903

Leeuwenhoek, Antony van. **The Select Works of Antony van Leeuwenhoek.** 1798-99/1807

Limnology in Wisconsin. 1977

Linnaeus, Carl. **Miscellaneous Tracts Relating to Natural History, Husbandry and Physick.** 1762

Linnaeus, Carl. **Select Dissertations from the Amoenitates Academicae.** 1781

Meyen, F[ranz] J[ulius] F. **Outlines of the Geography of Plants.** 1846

Mills, Harlow B. **A Century of Biological Research.** 1958

Müller, Hermann. **The Fertilisation of Flowers.** 1883

Murray, John. **Selections from *Report on the Scientific Results of the Voyage of H.M.S. Challenger During the Years 1872-76.*** 1895

Murray, John and Laurence Pullar. **Bathymetrical Survey of the Scottish Fresh-Water Lochs.** Volume one. 1910

Packard, A[lpheus] S. **The Cave Fauna of North America.** 1888

Pearl, Raymond. **The Biology of Population Growth.** 1925

Phytopathological Classics of the Eighteenth Century. 1977

Phytopathological Classics of the Nineteenth Century. 1977

Pound, Roscoe and Frederic E. Clements. **The Phytogeography of Nebraska.** 1900

Raunkiaer, Christen. **The Life Forms of Plants and Statistical Plant Geography.** 1934

Ray, John. **The Wisdom of God Manifested in the Works of the Creation.** 1717

Réaumur, René Antoine Ferchault de. **The Natural History of Ants.** 1926

Semper, Karl. **Animal Life As Affected by the Natural Conditions of Existence.** 1881

Shelford, Victor E. **Animal Communities in Temperate America.** 1937

Warming Eug[enius]. **Oecology of Plants.** 1909

Watson, Hewett Cottrell. **Selections from *Cybele Britannica.*** 1847/1859

Whetzel, Herbert Hice. **An Outline of the History of Phytopathology.** 1918

Whittaker, Robert H. **Classification of Natural Communities.** 1962

Date Due

			UML 735